# State of Knowledge of Endocrine Disruptors and Pharmaceuticals in Drinking Water

Awwa Research Foundation

## About the Awwa Research Foundation

The Awwa Research Foundation (AwwaRF) is a member-supported, international, nonprofit organization that sponsors research to enable water utilities, public health agencies, and other professionals to provide safe and affordable drinking water to consumers.

The Foundation's mission is to advance the science of water to improve the quality of life. To achieve this mission, the Foundation sponsors studies on all aspects of drinking water, including supply and resources, treatment, monitoring and analysis, distribution, management, and health effects. Funding for research is provided primarily by subscription payments from approximately 1,000 utilities, consulting firms, and manufacturers in North America and abroad. Additional funding comes from collaborative partnerships with other national and international organizations, allowing for resources to be leveraged, expertise to be shared, and broad-based knowledge to be developed and disseminated. Government funding serves as a third source of research dollars.

From its headquarters in Denver, Colorado, the Foundation's staff directs and supports the efforts of more than 800 volunteers who serve on the board of trustees and various committees. These volunteers represent many facets of the water industry, and contribute their expertise to select and monitor research studies that benefit the entire drinking water community.

The results of research are disseminated through a number of channels, including reports, the Web site, conferences, and periodicals.

For subscribers, the Foundation serves as a cooperative program in which water suppliers unite to pool their resources. By applying Foundation research findings, these water suppliers can save substantial costs and stay on the leading edge of drinking water science and technology. Since its inception, AwwaRF has supplied the water community with more than $300 million in applied research.

More information about the Foundation and how to become a subscriber is available on the Web at **www.awwarf.org.**

# State of Knowledge of Endocrine Disruptors and Pharmaceuticals in Drinking Water

Prepared by:
**Shane A. Snyder** and **Brett J. Vanderford**
Southern Nevada Water Authority
Henderson, NV 89015

**Jörg Drewes** and **Eric Dickenson**
Colorado School of Mines, Environmental Science and Engineering Division
Golden, CO 80401

and

**Erin M. Snyder**[*†], **Gretchen M. Bruce**[*], and **Richard C. Pleus**[*]
[*] Intertox, Inc., Seattle, WA 98121
[†] Black & Veatch Corp., Kansas City, MO 64114

Sponsored by:
**Awwa Research Foundation**
6666 West Quincy Avenue, Denver, CO 80235-3098

Published by:

Awwa Research Foundation

Distributed by:

American Water Works Association

IWA Publishing

**DISCLAIMER**

This study was funded by the Awwa Research Foundation (AwwaRF). AwwaRF assumes no responsibility for the content of the research study reported in this publication or for the opinions or statements of fact expressed in the report. The mention of trade names for commercial products does not represent or imply the approval or endorsement of AwwaRF. This report is presented solely for informational purposes.



ISBN 978-1-84339-241-5

# TABLE OF CONTENTS

# TABLES

# FIGURES

## FOREWORD

The Awwa Research Foundation is a nonprofit corporation that is dedicated to the implementation of a research effort to help utilities respond to regulatory requirements and traditional high-priority concerns of the industry. The research agenda is developed through a process of consultation with subscribers and drinking water professionals. Under the umbrella of a Strategic Research Plan, the Research Advisory Council prioritizes the suggested projects based upon current and future needs, applicability, and past work; the Collaborative Research, Research Application, and Tailored Collaborations programs; and various joint research efforts with organizations such as the U.S. Environmental Protection Agency, the U.S. Bureau of Reclamation, and the Association of California Water Agencies.

This publication is a result of one of these sponsored studies, and it is hoped that its findings will be applied in communities throughout the world. The following report serves not only as a means of communicating the results of the water industry's centralized research program but also as a tool to enlist further support of the nonmember utilities and individuals.

Projects are managed closely from their inception to the final report by the foundation's staff and a large cadre of volunteers who willingly contribute their time and expertise. The foundation serves a planning and management function and awards contracts to other institutions such as water utilities, universities, and engineering firms. The funding for this research effort comes primarily from the Subscription Program, through which water utilities subscribe to the research program and make an annual payment proportionate to the volume of water they deliver and consultants and manufacturers subscribe based upon their annual billings. The program offers a cost-effective and fair method for funding research in the public interest.

A broad spectrum of water supply issues is addressed by the foundation's research agenda: resources, treatment and operations, distribution and storage, water quality and analysis, toxicology, economics, and management. The ultimate purpose of the coordinated effort is to assist water suppliers to provide the highest possible quality of water economically and reliably.

The true benefits are realized when the results are implemented at the utility level. The foundation's trustees are pleased to offer this publication as a contribution toward that end.

David E. Rager
Chair, Board of Trustees
Awwa Research Foundation

Robert C. Renner, P.E.
Executive Director
Awwa Research Foundation

# ACKNOWLEDGMENTS

The project team would like to acknowledge Frankie Lewis, Linda Parker, David Rexing, Rebecca Trenholm, Brett Vanderford, Ron Zegers, and Janie Zeigler-Holady from the Southern Nevada Water Authority (SNWA) and Christiane Hoppe from the Colorado School of Mines for their assistance in this project. Hongxia (Dawn) Lei was a post-doctoral researcher at SNWA during the initial phase of this project and provided guidance and insight for the project. The team would also like to thank Robert Hulsey and Bruce Long from Black & Veatch Corporation for in-kind support and expert review for this project. The project team is indebted to Jane Smith, Joyce Donohue, and Octavia Conerly of the U.S. Environmental Protection Agency and to William (Willie) Owens of Procter & Gamble Company for discussions related to drinking water criteria and bioassays for endocrine disrupting chemicals. The team would also like to thank the Project Advisory Committee: Edward Bouwer from Johns Hopkins University; Yeomin Yoon from $CH_2M$ Hill and Jerry Ongerth from East Bay MUD for their review and suggestions for this report. The team greatly appreciated the support and leadership of Djanette Khiari from AwwaRF; her tenacity and spirited leadership made this project possible.

# EXECUTIVE SUMMARY

## INTRODUCTION

Over the past decade, a wealth of information regarding levels of endocrine disrupting chemicals (EDCs) and pharmaceuticals and personal care products (PPCPs) has been published. The earliest reports of steroids and pharmaceuticals in water were published more than 30 years ago; however, only in the last decade has the topic garnered public and scientific interest. While a multitude of reports have been published, most are quite focused on specific aspects (i.e., detection, occurrence, treatment) of this emerging topic of concern. This utility guide provides a comprehensive, yet concise, review of the available information related to the analysis, occurrence, treatment, and possible human health and environmental effects of trace endocrine disrupting chemicals (EDCs) and pharmaceuticals and personal care products (PPCPs).

## PROJECT APPROACH

This project was developed to provide the water industry with a current status of the science that aggregates the relevant information available on EDCs and PPCPs into a comprehensive, yet succinct, format. This body of work is expected to continue and develop as awareness increases that 'clean' water is a valuable and limited resource. It was not feasible, nor particularly useful, to review every chemical that could be considered a suspected EDC or a PPCP. Therefore, a subset of key chemicals was chosen that are representative of the subclasses and types of EDCs and PPCPs that are of greatest scientific and public interest. These indicator chemicals were used through the report for consistency. While these indicators are the focus in the current report, the general information and approach presented here is applicable to essentially all EDCs and PPCPs

Peer reviewed published manuscripts, reports, and books were reviewed for pertinent information about the specific indicator chemicals as well as fundamental principles related to analytical methods, treatment, and risk assessment. The resulting utility guide summarizes key studies broadly applicable to water utilities and provides references where additional information can be found. The report consists of the following chapters:

1. Introduction and Background
2. Analytical Methods
3. Biologically-Based Assays
4. Occurrence
5. Water Treatment
6. Human Health
7. Communication

## CHAPTER 1: INTRODUCTION AND HISTORY

The earliest documentation regarding the presence of estrogens in the environment originated from a study at Harvard University in 1965 (Stumm-Zollinger and Fair 1965). The first report on pharmaceuticals in the environment was published by the U.S. Environmental Protection Agency (EPA) in 1975 (Garrison, Pope et al. 1975). These early reports help provide important perspective to this issue and are valuable to public outreach professionals. This introduction discusses the environmental impacts that have been reported, such as "feminized fish" that led to increased public awareness and concern over the past decade. This chapter also discusses potential health impacts and regulatory actions related to EDCs/PPCPs. Sources of EDCs and PPCPs in the environment also are presented in Chapter 1 along with basic information about treatment and prevention options.

## CHAPTER 2: ANALYTICAL METHODS

Several excellent references have been published on various aspects of analytical methods for EDCs and PPCPs (Ternes, Hirsch et al. 1998; Snyder, Keith et al. 1999; Laganá, Bacaloni et al. 2000; Ahrer, Scherwenk et al. 2001; Ternes, Bonerz et al. 2001; Ternes 2001; Bruchet, Prompsy et al. 2002; Snyder, Vanderford et al. 2003; Zwiener and Frimmel 2004; Zwiener and Frimmel 2004). However, the majority of these publications focus only on a few compounds or analytical techniques. Chapter 2 extracts key information from these and other reports and builds a concise guide to the most feasible techniques for the vast number of structures that comprise EDCs/PPCPs. The chapter begins with a discussion of the structural complexity of these contaminants and demonstrates that EDC/PPCP compounds include a wide range of physical/chemical properties. This chapter continues with a discussion of the primary analytical tools for concentration, separation, and detection of chemical contaminants in water. The emphasis in this chapter is on practical techniques that can be implemented by utilities and/or contract laboratories. The types of chemicals that are amenable to analysis by each technique are described. This chapter is meant to serve as a reference that will aid in choosing an analytical method as new compounds are continually discovered in the environment. This chapter also provides critical information related to quality assurance and quality control measures. Some of the obstacles and solutions for monitoring these trace organic contaminants are discussed. For instance, the advantages and disadvantages of isotope-dilution methodology are presented. Proper sample handling and preservation issues for maintaining sample integrity also are included. This chapter is valuable to water professionals who want practical and concise information on the analytical issues surrounding EDCs and PPCPs.

## CHAPTER 3: BIOLOGICALLY-BASED ASSAYS

Several biologically-based assays have been employed for the identification and characterization of various classes of contaminants in water. This chapter focuses mainly on EDCs, but also will discuss some of the latest assay-based technologies for pharmaceuticals, such as DNA microarrays. The discussion is limited to those biologically-based assays that are most useful for detection and characterization of EDCs/PPCPs of potential concern for humans exposed via drinking water (i.e., detailed discussions of assays intended for protection of aquatic organisms are not provided). The chapter begins with explanation of basic terms, such as *in vitro* and *in vivo* assays. The uses and limitations of various *in vitro* and *in vivo* assays currently available for detection of EDCs/PPCPs in drinking water are discussed, and some techniques under development are explored. Specific emphasis is placed on the battery of assays promulgated by the U.S. EPA for the Endocrine Disruption Screening Program (EDSP). *In vitro* assays of interest in this review include receptor binding assays, enzyme immunoassays, radioimmunoassays, and hormone-responsive mammalian and yeast-based cellular assays. *In vivo* assays evaluated include acute and chronic rodent assays, non-mammalian (*e.g.*, fish or amphibian) bioassays, and DNA microarray technologies. The use of bioassays in conjunction with analytical chemistry techniques in bioassay-directed fractionation and identification or toxicity identification and evaluation schemes also are discussed.

## CHAPTER 4: OCCURRENCE

The sources of EDCs and PPCPs in the environment are diverse but are for the most part related to human activities. The most important sources for release of these compounds into surface and groundwater are discharges from municipal wastewater treatment plants (WWTPs), industrial manufacturing processes, leaky sewers, combined sewage overflows (CSOs), onsite wastewater systems (OWS), and confined animal feeding operations (CAFOs) (Drewes and Shore 2001; Kim and Carlson 2006). As a consequence, the occurrence level of EDCs and PPCPs in drinking water sources primarily depends upon the degree of wastewater impact in the watershed. Additional factors affecting occurrence are usage pattern for different compounds as well as prescription practices for pharmaceuticals, which can vary with region; per-capita water consumption, which can result in different levels of dilution; and substitution and phase-out programs for specific chemicals. These variable factors explain why the occurrence pattern of currently detected compounds is not static.

In order to compile information on the occurrence of trace organic compounds of interest in secondary or tertiary treated wastewater effluents, more than 1,000 references reporting occurrence in studies across the globe were screened. This comprehensive review considered only articles that were peer-reviewed and reported both analytical methods employed and detailed experimental conditions. Based on the findings of this survey, pharmaceutical residues, antibiotics, steroid hormones, and fragrances are the most commonly reported trace organic compounds currently observed to occur in secondary and tertiary treated municipal effluents as well as in surface water and groundwater receiving these wastewater discharges. Most studies considered in this report show occurrence of certain compounds in source water receiving various degrees of wastewater discharge, which likely explains the rather wide range of concentrations reported for individual compounds. Although the exact degree of impact is mostly unknown, the reported concentrations usually increase significantly with closer proximity to the point of discharge. The highest concentrations are usually observed in recycled water used for planned indirect potable reuse applications. Dilution with non-impacted source water and photolysis and biodegradation processes can frequently reduce concentrations close to or below the detection limits. In general, median concentrations reported for the compounds of interest in receiving streams are significantly lower than those observed in treated wastewater effluents. Due to differences in prescription practices and per-capita water consumption, certain pharmaceutical residues are present in lower concentrations in source waters in the U.S. as compared to Europe. Steroid hormones usually occur in source waters in Europe and North America at similar concentrations, frequently close or below the detection limits employed in the studies.

## CHAPTER 5: WATER TREATMENT

This chapter reviews the efficiency of individual unit operations within drinking water treatment trains with regard to their ability to remove EDCs and PPCPs from impacted source water. The chapter is structured by key operations employed in conventional and advanced drinking water treatment applications, such as coagulation/flocculation, disinfection processes, granular and powdered activated carbon, advanced oxidation processes, low-pressure and high-pressure membranes, and riverbank filtration. Removing PPCPs and EDCs from water obtained from impaired sources requires a sequence of diverse treatment processes that can tackle the wide range of physicochemical properties of PPCPs and EDCs. No single treatment process alone can provide an absolute barrier to PPCPs and EDCs in drinking water.

Conventional drinking water treatment commonly employed for surface water, consisting of coagulation/flocculation followed by sedimentation and filtration, is not capable of removing EDCs and PPCPs occurring at the parts-per-trillion level. Removal of EDCs and PPCPs, however, can be expected during drinking water disinfection using chlorine, chlorine dioxide, and ozone. Steroid hormones are particularly amenable to desctruction during chlorine disinfection. No removal is achieved by UV radiation applied at doses typically used for disinfection purposes. Stronger oxidants, such as ozone or advanced oxidation processes (AOP), can remove a wide range of PPCPs and EDC but are limited in their ability to remove chlorinated flame retardants (e.g, TCEP, TCPP) or the insect repellent N,N-diethyl-m-toluamide (DEET). Activated carbon can remove a wide range of hydrophobic PPCPs and EDCs including steroid hormones but is less effective for polar compounds, such as x-ray contrast agents. Low-pressure membranes (microfiltation (MF) and ultrafiltration (UF)) are ineffective in removing PPCPs and EDCs from water. High-pressure membranes (nanofiltration (NF) or reverse osmosis (RO)), however, can remove a wide range of PPCPs and EDCs. Problematic for high-pressure membranes are low-molecular weight organics such as N-nitrosamines or certain pharmaceuticals (i.e., acetaminophen, phenacetine). Natural processes, such as riverbank filtration (RBF) or soil-aquifer treatment (SAT), can be employed either as an additional treatment step for wastewater treatment prior to discharge to the environment or as a pre-treatment to subsequent drinking water treatment. RBF and SAT both provide a significant barrier for a majority of PPCPs and EDCs, with the exception of recalcitrant compounds that are not amendable to a biological attack.

## CHAPTER 6: HUMAN HEALTH

Human health risk assessment for chemicals requires data describing both exposure and toxicity potential. Limited research has been conducted on the potential for long-term exposure to EDCs and PPCPs at trace concentrations found in drinking water to impact human health. Toxicity testing for chemicals commonly involves doses that are much greater than those attained through environmental exposures, including drinking water. The effect database typically is greater for pharmaceuticals than for EDCs or personal care products. However, testing for pharmaceuticals usually is conducted at relatively high therapeutic doses, and potential effects in non-target populations (e.g., pregnant women, infants, and children) receiving unintentional exposure often are not known. Screening-level risk assessments conducted to date have not indicated that the trace concentrations of pharmaceuticals detected in drinking water pose a risk to consumers, and likewise, there is no convincing evidence that EDCs at levels occurring in drinking water have caused adverse effects in humans (Snyder, Pleus, and Snyder 2005). Reliable data describing occurrence of EDCs/PPCPs in drinking water are sparse, and human exposure to these chemicals through other routes (*e.g.*, diet, environmental exposure, medical treatment) must be considered. Most likely, drinking water will only contribute a small amount to the the total EDC/PPCP exposure experienced by most people. Risk assessment is complicated by the fact that people are exposed to mixtures of contaminants in drinking water and in other environmental matrices (air, ingested food) and occupational settings, and constituents of these mixtures may interact with or overwhelm the potential effects of

EDCs/PPCPs in drinking water. Chapter 6 reviews available evidence for potential human health effects due to trace concentrations of EDCs and pharmaceuticals in drinking water. Data gaps (such as lack of occurrence and toxicity data) that prevent or limit human health risk assessment are identified.

## CHAPTER 7: COMMUNICATION

Regulatory agencies in the United States have not yet established risk-based guidelines for acceptable levels of most PPCPs and EDCs (other than perchlorate) in water. As a result, drinking water utilities are left to respond to public concerns about the increasing detection of EDCs and PPCPs in waste waters and drinking waters on their own. This chapter provides information to assist utility managers and public relations staff in responding to these concerns, particularly as they relate to human health risks and questions about the future direction of treatment and research. The chapter outlines the current regulatory status of this issue with regard to protecting human health, describes how the potential health risk of these chemicals at detected levels compares to other types of common risks, and outlines some considerations for establishing a strategy to address public concerns. It is hoped that this information will facilitate positive interactions with the concerned public and help guide water treatment professionals in establishing research and treatment priorities.

## SUMMARY

This utility guide provides the water industry with a synopsis of the state of the science on EDCs and PPCPs in drinking water. For utilities, this guide presents information regarding: methods to analyze trace contaminants in water, occurrence of indicator EDC/PPCPs, EDC/PPCP removal efficiency through conventional and advanced treatment processes, applicable bioassays for drinking water, human health risk assessment, and key speaking points for communication. While this guide provides comprehensive information, it is important to note that the state of knowledge on this topic is rapidly evolving. To remain current, water industry professionals should request frequent updates from regulators and the scientific community.

## REFERENCES


Adams, C., Y. Wang, et al. (2002). "Removal of antibiotics from surface and distilled water in conventional water treatment processes." Journal of Environmental Engineering **128**(3): 253-260.

Ahrer, W., E. Scherwenk, et al. (2001). "Determination of drug residues in water by the combination of liquid chromatography or capillary electrophoresis with electrospray mass spectrometry." Journal of Chromatography A **910**: 69-78.

Bruchet, A., C. Prompsy, et al. (2002). "A broad spectrum analytical scheme for the screening of endocrine disruptors (EDs), pharmaceuticals and personal care products in wastewaters and natural waters." Water Science and Technology **46**(3): 97-104.

Buser, H. R., T. Poiger, et al. (1999). "Occurrence and environmental behavior of the chiral pharmaceutical drug ibuprofen in surface waters and in wastewater." Environmental Science & Technology **33**(15): 2529-2535.

Drewes, J. E. & Shore, L. S. (2001). Concerns about pharmaceuticals in water reuse, groundwater recharge, and animal waste. In: Ch. Daughton and T. L. Jones-Lepp (Eds.) American Chemical Society Symposium Series 791 "Pharmaceuticals and personal care products in the environment" No. 791, Washington, D.C., 206-228.

Drewes, J. E., P. Fox, et al. (2001). "Occurrence of iodinated X-ray contrast media in domestic effluents and their fate during indirect potable reuse." Journal of Environmental Science and Health Part a- Toxic/Hazardous Substances & Environmental Engineering **36**(9): 1633-1645.

Drewes, J. E., T. Heberer, et al. (2002). "Fate of pharmaceuticals during indirect potable reuse." Water Science and Technology **46**(3): 73-80.

Garrison, A. W., J. D. Pope, et al. (1975). GC/MS analysis of organic compounds in domestic wastewaters. Chem. Congr. North Am. Cont., Ann Arbor Science.

Heberer, T., K. Reddersen, et al. (2002). "From municipal sewage to drinking water: fate and removal of pharmaceutical residues in the aquatic environment in urban areas." Water Science and Technology **46**(3): 81-88.

Laganá, A., A. Bacaloni, et al. (2000). "Trace analysis of estrogenic chemicals in sewage effluent using liquid chromatography combined with tandem mass spectrometry." Rapid Commun. Mass Spectrom. **14**: 401-407.

Metcalfe, C. D., B. Koenig, et al. (2000). Drugs in sewage treatment plant effluents in Canada. ACS National Meeting, American Chemcial Society.

Reemtsma, T. and Jekel, M. (eds.) 2006. Organic pollutants in the water cycle. Wiley-VCH, Weinheim.

Sacher, F., B. Haist-Gulde, et al. (2000). Behavior of selected pharmaceuticals during drinking water treatment. 219th ACS National Meeting, San Francisco, CA, American Chemical Society.

Snyder, S. A., T. L. Keith, et al. (1999). "Analytical methods for detection of selected estrogenic compounds in aqueous mixtures." Environmental Science & Technology **33**(16): 2814-2820.

Snyder, S. A., J. Leising, et al. (2004). "Biological attenuation of EDCs and PPCPs: Implications for water reuse." Ground Water Monitoring & Remediation **24**(2): 108-118.

Snyder, S. A., B. J. Vanderford, et al. (2003). "Analytical methods used to measure endocrine disrupting compounds in water." Practice Periodical of Hazardous,Toxic, and Radioactive Waste Management **7**(4): 224-234.

Snyder, S. A., P. Westerhoff, et al. (2003). "Pharmaceuticals, personal care products, and endocrine disruptors in water: implications for the water industry." Environmental Engineering Science **20**(5): 449-469.

Snyder, E.M., R.C. Pleus, and S.A. Snyder. 2005. Pharmaceuticals and EDCs in the US water industry - an update. *Journal of the American Water Works Association*, 97(11):32-36.

Stumm-Zollinger, E. and G. M. Fair (1965). "Biodegradation of steroid hormones." Journal of the Water Pollution Control Federation **37**: 1506-1510.

Tabak, H. H., R. N. Bloomhuff, et al. (1981). "Steroid hormones as water pollutants II. Studies on the persistence and stability of natural urinary and synthetic ovulation-inhibiting hormones in untreated and treated wastewaters." Dev. Ind. Microbiol. **22**: 497-519.

Ternes, T., M. Bonerz, et al. (2001). "Determination of neutral pharmaceuticals in wastewater and rivers by liquid chromatography-electrospray tandem mass spectrometry." Journal of Chromatography a **938**(1-2): 175-185.

Ternes, T. A. (2001). "Analytical methods for the determination of pharmaceuticals in aqueous environmental samples." Trends in Analytical Chemistry **20**(8): 419-434.

Ternes, T. A., R. Hirsch, et al. (1998). "Methods for the determination of neutral drugs as well as betablockers and $\beta_2$-sympathomimetics in aqueous matrices using GC/MS and LC/MS/MS." Fresenius' Journal of Analytical Chemistry **362**: 329-340.

Ternes, T. A., P. Kreckel, et al. (1999). "Behaviour and occurrence of estrogens in municipal sewage treatment plants - II. Aerobic batch experiments with activated sludge." The Science of the Total Environment **225**: 91-99.

Ternes, T. A., M. Meisenheimer, et al. (2002). "Removal of pharmaceuticals during drinking water treatment." Environmental Science & Technology **36**(17): 3855-3863.

Ternes, T. A., J. Stüber, et al. (2003). "Ozonation: a tool for removal of pharmaceuticals, contrast media and musk fragrances from wastewater?" Water Research **37**(8): 1976-1982.

Ternes, T. A., M. Stumpf, et al. (1999). "Behavior and occurence of estrogens in municipal sewage treatment plants - I. Investigations in Germany, Canada and Brazil." The Science of the Total Environment **225**: 81-90.

Ternes, T. and Joss, A. (eds.) (2006). Human pharmaceuticals, hormones and fragrances. IWA Publishing, London.

Zwiener, C. and F. H. Frimmel (2000). "Oxidative treatment of pharmaceuticals in water." Water Research **34**(6): 1881-1885.

Zwiener, C. and F. H. Frimmel (2004). "LC-MS analysis in the aquatic environment and in water treatment - a critical review. Part I: Instrumentation and general aspects of analysis and detection." Analytical and Bioanalytical Chemistry **378**(4): 851-861.

Zwiener, C. and F. H. Frimmel (2004). "LC-MS analysis in the aquatic environment and in water treatment technology - a critical review. Part II: Applications for emerging contaminants and related pollutants, microorganisms and humic acids." Analytical and Bioanalytical Chemistry **378**(4): 862-874.

# CHAPTER 1
# INTRODUCTION

## BACKGROUND AND HISTORY

Recent advances in analytical chemistry methods have enabled the detection of exceedingly minute concentrations of contaminants in water. The detection of 0.000000001 g of a particular substance dissolved into a liter of water (ng/L) is becoming routine. Certain chemicals have the ability to mimic or block the action of natural hormones in animals. Collectively, these compounds have been termed endocrine disrupting chemicals (EDCs). Trace concentrations of certain EDCs (specifically natural and synthetic hormones) in wastewater treatment plant effluents (WWTPs) have been linked to reproductive impacts in fish; however, no evidence of human impact at the concentrations found in drinking water have been established. More recent studies have documented a plethora of pharmaceuticals and personal care products (PPCPs) occurring ubiquitously in WWTP outfalls at ng/L concentrations. Less commonly, EDCs and PPCPs have been reported in finished drinking water. The occurrence of these chemicals in water has prompted concern among the public, drinking water utilities, and regulatory agencies.

While EDCs and PPCPs represent large classes of chemicals with a diversity of chemical structure and behavior, this report will focus on a subset of representative chemicals that encompass the principal subclasses and a broad range of physical-chemical properties. In recent years, excellent reports and reviews regarding various aspects of these emerging contaminants have been published (Reemtsma and Jekel 2006; Ternes and Joss 2006; Petrovic and Barcelo 2007). This utility guide will provide key information from published and on-going studies regarding the analysis (chemical and bioassay), occurrence, treatment, health relevance, and communication of representative EDCs and PPCPs.

## EDCs

EDCs are chemical agents that interfere with the functioning of natural hormones in the body. They are also variously known as endocrine disruptors, hormonally-active agents, endocrine-active agents, endocrine modulating substances, and other similar variants of these terms. It is difficult to determine which chemicals should or should not be classified as EDCs. Hundreds of chemicals have been purported to be EDCs or potential EDCs based on a variety of different criteria, ranging from hormone receptor binding activity to cellular responses *in vitro* to stronger evidence of endocrine activity *in vivo*. For example, the United Kingdom's Institute for Environment and Health (IEH 2005) produced a consolidated listing of 966 chemicals or elements that have been suggested in the published literature to be potential endocrine disrupters. There currently is no consensus within the scientific community on a strategy to determine definitively whether a chemical is or is not an EDC, though several national and international agencies and organizations are making efforts toward this goal. Organizations, agencies, and expert groups use different terms and definitions to describe EDCs and employ various standards to identify specific chemicals as EDCs.

Participants in a U.S. EPA-sponsored workshop defined endocrine disrupting chemicals as "exogenous agents that interfere with the production, release, transport, metabolism, binding, action, or elimination of the natural hormones in the body responsible for the maintenance of homeostasis and the regulation of developmental processes" (Kavlock et al. 1996). The

International Programme on Chemical Safety (IPCS) (WHO 2002) adopted the following definition: "An endocrine disruptor is an exogenous substance or mixture that alters function(s) of the endocrine system and consequently causes adverse health effects in an intact organism, or its progeny, or (sub)populations. A potential endocrine disruptor is an exogenous substance or mixture that possesses properties that might be expected to lead to endocrine disruption in an intact organism, or its progeny, or (sub)populations." IPCS requires not only an effect, but an adverse effect, and the effect must occur in an intact organism rather than *in vitro*. Following much debate, the Endocrine Disruptor Screening and Testing Advisory Committee (EDSTAC) to the U.S. EPA Endocrine Disruptor Screening Program (EDSP) decided, "The EDSTAC describes an endocrine disruptor as an exogenous chemical substance or mixture that alters the structure or function(s) of the endocrine system and causes adverse effects at the level of the organism, its progeny, populations, or subpopulations of organisms, based on scientific principles, data, weight-of-evidence, and the precautionary principle" (EDSTAC 1998 Endocrine Disruptor).

For a chemical to be classified as an EDC, the U.S. EPA (Timm 2007) and the European Commission (2002) further stipulate that its critical effect must be one that results from a primary assault on the endocrine system rather than secondary to other more conventional toxicologic effects. That is, the endocrine-mediated effect should occur at a dose smaller than that required to cause general systemic toxicity. Likewise, according to the GWRC (2003), "An endocrine disrupter could be defined as 'a substance that causes perturbation of hormonal/endocrine status at dose levels which are not cytotoxic.' In other words, substances that cause changes in endocrine status only at doses that also cause tissue damage would not be considered."

Many chemicals might be expected to cause endocrine disruption based on results of screening-level tests such as quantitative structure-activity relationships (QSAR) or effects in some *in vitro* testing systems. Although this preliminary evidence might indicate endocrine activity, the more reliable and relevant *in vivo* tests do not always result in the conclusion that a chemical is indeed an EDC. In comparison with the number of chemicals that have been screened for endocrine activity, relatively few chemicals have been subjected to the more definitive *in vivo* testing. Also, while a positive result for endocrine activity in an *in vivo* test provides strong evidence that a chemical is an EDC, a negative result does not preclude the possibility that a chemical might have endocrine activity in a different test system.

Under the 1996 amendments to the Safe Drinking Water Act (P.L. 104–182), the U.S. EPA was tasked with developing methods to screen chemicals that have the propensity to contaminate water for endocrine disruptive effects. To meet the requirements of this legislation, the U.S. EPA convened the Endocrine Disruptor Screening and Testing Advisory Committee (EDSTAC) to provide recommendations on a conceptual framework, priority setting, screening and testing methodologies, and communication and outreach programs. The EDSTAC group consisted of various stakeholders and experts in reproductive toxicology. The committee began deliberations October of 1996 and issued a final report in July of 1998 recommending that human and wildlife impacts be considered and that estrogen, androgen, and thyroid (EAT) endpoints be examined (EDSTAC 1998). The conceptual framework devised by EDSTAC consists of an initial sorting, prioritization, Tier 1 & 2 testing, and a hazard assessment of an estimated 87,000 chemicals. In addition to discrete chemicals, EDSTAC recommended the evaluation of mixtures of chemicals in breast milk, baby formulas, hazardous waste sites, pesticides and fertilizers, drinking water DBPs, and gasoline. In response, the U.S. EPA established the Endocrine Disruptor Screening Program (EDSP) to develop screening methods and toxicity testing strategies that can be used to definitively determine whether a chemical is, or is not, an EDC. The U.S. EPA is cooperating with the European Organisation for Economic Co-operation and Development (OECD), which also currently is developing methods to identify EDCs. However, this is an arduous task and is not expected to be completed for several years. No scientific consensus has yet been reached within the EDSP or other expert groups regarding the specific suite of tests that should be used to determine whether a contaminant is or is not an EDC.

Despite the uncertainty surrounding the assignment of the term EDC to specific chemicals, some naturally-occurring and manmade chemicals are widely considered to be EDCs, including certain pharmaceuticals, pesticides, industrial chemicals, combustion byproducts, phytoestrogens (naturally occurring estrogens from plants), and hormones excreted by animals and humans. There are many other chemicals for which there is limited, incomplete evidence of potential endocrine activity or for which the evidence of endocrine activity is controversial. An even greater number of chemicals have not yet been tested for potential endocrine activity using any of the available methods.

### *Examples of Suspected EDCs*

***Steroid Hormones.*** With several exceptions, most steroid hormones are non-polar or weakly polar compounds derived from cholesterol, all bearing the same cyclopentanophenanthrene ring structure as cholesterol (Table 1.1). Androgen, estrogen, and progestagen are the three subcategories. Testosterone is the primary and also the most widely studied androgen. Estradiol, estriol, and estrone are three of the major endogenous estrogens. Progesterone is the only natural progestagen, exhibiting antiestrogenic and antigonadotropic properties. Hormones in water were first reported in the U.S. by Harvard University in 1965. The U.S. Department of the Interior published a study entitled "Steroids Hormones as Water Pollutants" in 1970 (Tabak and Bunch 1970). It is remarkable that the detection of trace steroids in water are considered "emerging contaminants" considering the existence of published studies (including from the U.S. Federal Government) more than three decades ago. Hormones are excreted from the body primarily as conjugated hormones; however, these are readily deconjugated during wastewater treatment to the original form (Johnson, Belfroid et al. 2000; Gentili, Perret et al. 2002; D'Ascenzo, Di Corcia et al. 2003; Isobe, Shiraishi et al. 2003). Natural and synthetic homones have been linked to estrogenic impacts of wastewater effluents (Routledge, Sheahan et al. 1998; Snyder, Villeneuve et al. 2001; Folmar, Hemmer et al. 2002; Kidd, Blanchfield et al. 2007).

***Alkylphenols.*** Alkylphenols have been used in various industrial and municipal applications for more than 40 years and have been reported to bioaccumulate and cause estrogenic effects to aquatic organisms (Giesy, Pierens et al. 2000; Snyder, Keith et al. 2001; Snyder, Keith et al. 2001; Tsuda, Takino et al. 2001). Some typical physical and chemical properties of compounds in this group are pH dependent due to the dissociation of proton from phenolic hydroxide group. Of particular interest in this group are octylphenols and nonylphenols, which occur primarily as degradation products from widely used detergents. Most alkylphenols exist as mixtures with different substitution locations and branched side chains. Bisphenol A is plasticizer that has been identified as an environmental contaminant (Welshons et al. 2006). As the name implies, bisphenol A contains two phenolic rings linked by an alkyl group (Table 1.1).

***Phthalates.*** Phthalates are dialkyl or alkyl aryl esters of 1,2-benzenedicarboxylic acid and are mainly used as plasticizers to increase the flexibility of hard plastics. They have been in use for about 50 years with global production estimated to be several million tons per year. Even though used in plastics, they are not covalently bound to the long polyvinyl molecules and can gradually migrate out of plastic products and into the environment. Several phthalates have been listed as suspected EDCs. The European Union has banned the use of three types of phthalates, including diethylhexyl phthalate (DEHP) in certain children products. DEHP is the mostly widely used plasticizer for polyvinyl products in the world due to its low cost. Structures of three common phthalates are provided in Table 1.1.

***Phytoestrogens.*** Phytoestrogens are naturally occurring compounds present in plants and fruits that act like estrogens in the body. Naturally occurring phytoestrogens have drawn attention several decades ago when a positive relation between cow mastitis and phytoestrogens was discovered (Brookbanks, Welch et al. 1969). The majority of phytoestrogens are from the flavonoid group, which is subdivided into flavone, flavonol, flavanone, flavanol, and isoflavonoid categories according to the presence and position of carbonyl and hydroxyl groups (Table 1.1). Genistein and daidzein are two of the most abundant and well-studied isoflavonoids that have shown some health benefits due to their anticarcinogenic, antiatherogenic, and antioxidative potentials. The other two main groups are lignans (found in flaxseed) and coumestans (derived from sprouting plants like alfalfa). Legumes (such as soybean) constitute a significant source for human phytoestrogen dietary intake of both genistein and daidzein. Phytoestrogens are not discussed in detail in this literature review; however, they are important to consider when evaluating holistic human exposure to estrogenic substances.

**Table 1.1**
**Structures of suspected EDCs**

| | | |
|---|---|---|
| **Steroid Hormones** | | |
| 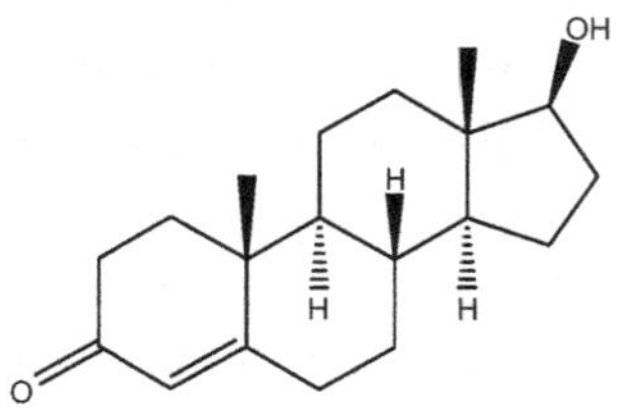 Testosterone | 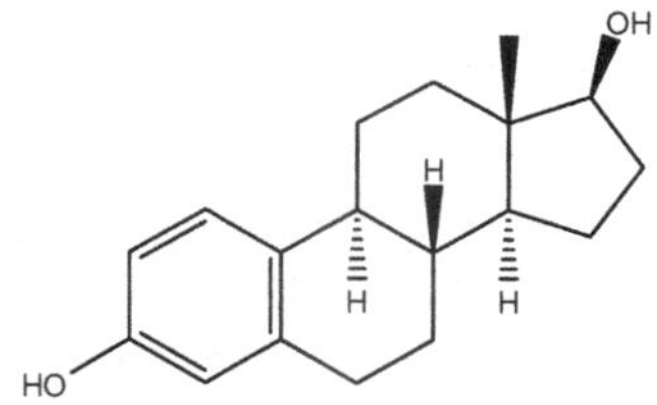 Estradiol | 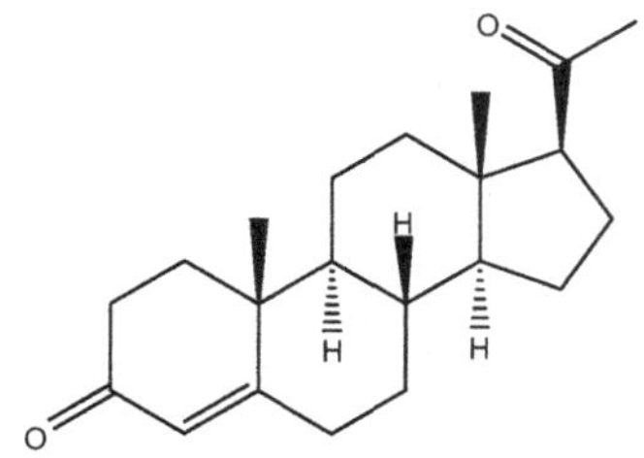 Progesterone |
| **Phenols** | | |
| 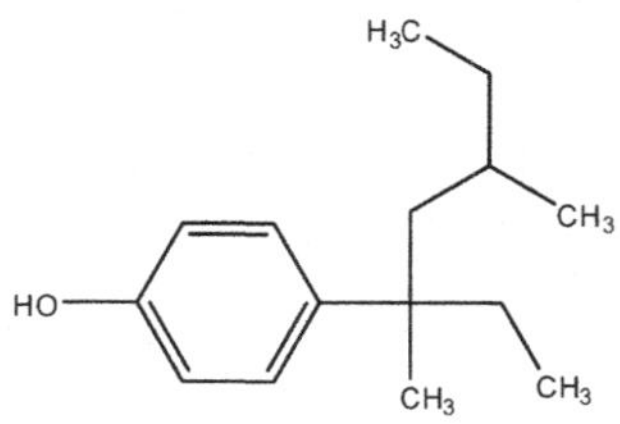 Nonylphenol | 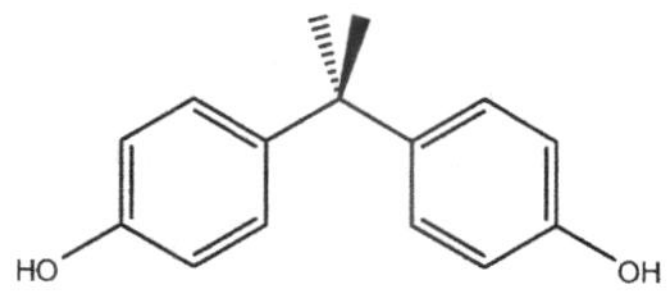 Bisphenol A | 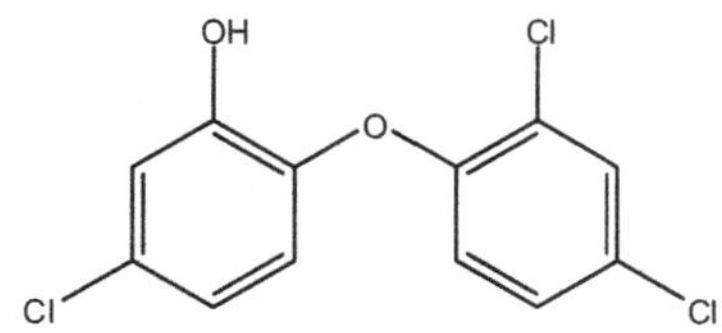 Triclosan |
| **Phthalates** | | |
| 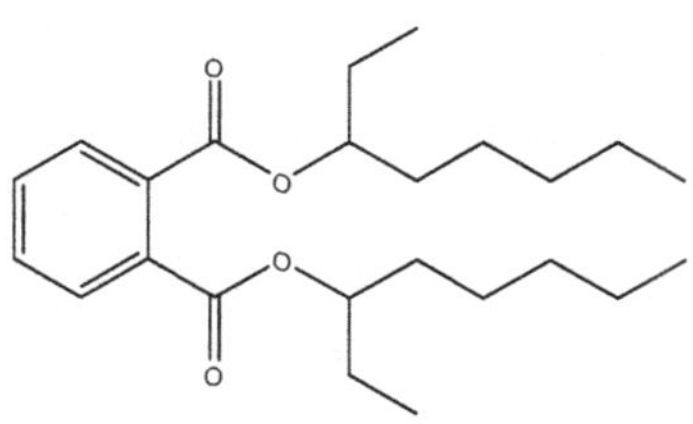 Diethylhexyl phthalate | 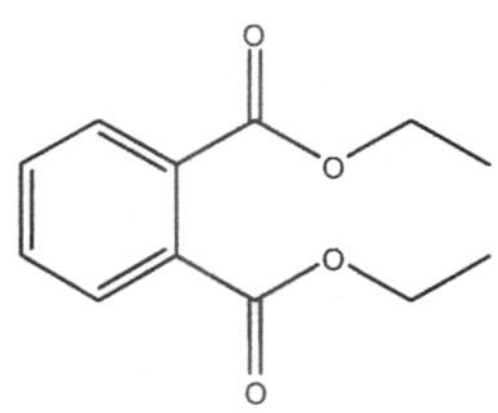 Diethyl phthalate | 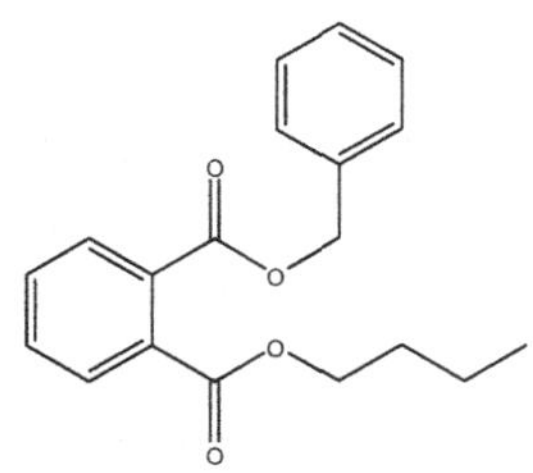 Benzylbutyl phthalate |
| **Phytoestrogens** | | |
| 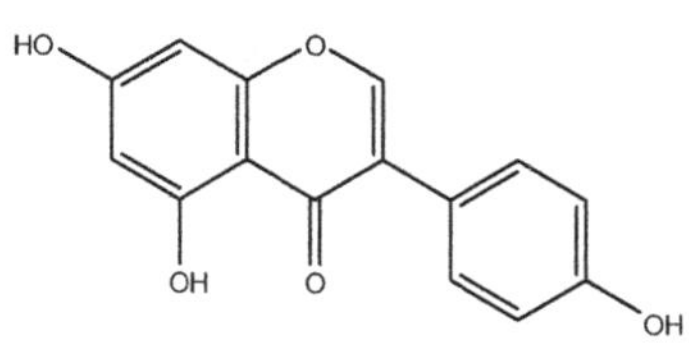 Genistein | 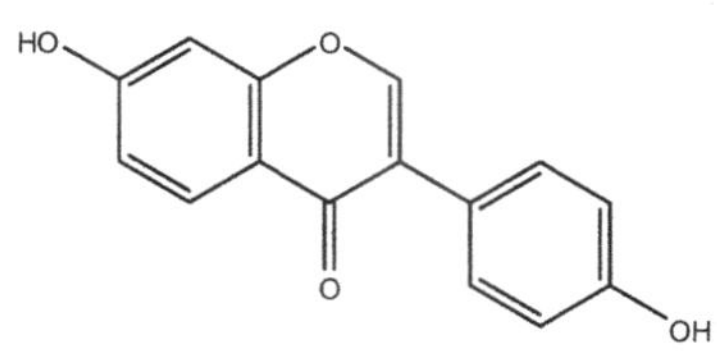 Daidzein | 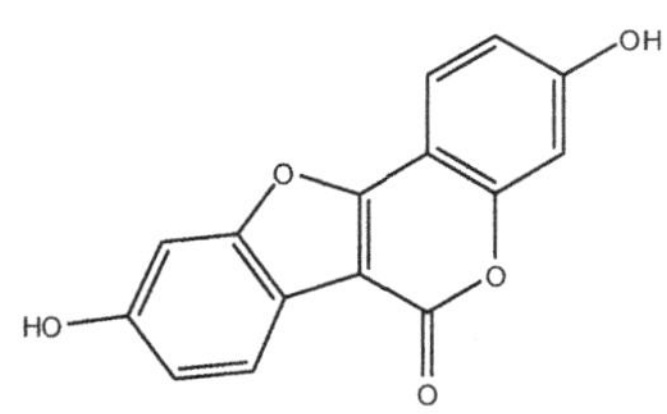 Coumestrol |

## Pharmaceuticals

Pharmaceuticals have been detected in WWTP effluents globally. The first published report demonstrating the existence of pharmaceuticals in the environment was published by the U.S. EPA in 1976 (Garrison, Pope et al. 1976). In 1977, a report was published documenting the detection of pharmaceuticals in wastewater outfalls in Kansas City, Missouri, USA (Hignite and Azarnoff 1977). Similarly to steroid hormones, it may be inappropriate to consider pharmaceuticals as "emerging contaminants" considering that the initial reports were published more than 30 years ago. Pharmaceuticals include prescription and non-prescription human drugs and veterinary medications and present a broad diversity of chemical structure and behavior (Table 1.2). Some pharmaceuticals are not readily removed by conventional water treatment processes and have been detected in finished drinking water (Snyder et al. 2007). Despite relatively small therapeutic doses of pharmaceuticals (generally mg or less), the cumulative use of certain pharmaceuticals can be quite large. For instance, the estimated usage of carbamazepine in France was 38 tons in 1998 alone (Bruchet, Hochereau et al. 2005). Pharmaceuticals generally have more comprehensive toxicological data due to rigorous testing required for registration. These data include clinical trials, thus human health data are available.

## Personal Care Products

Personal care products (PCPs) can include a wide range of consumer products (both active and inert ingredients) including: fragrances, sunscreens, insect repellants, over the counter medications, and detergents. PCPs could easily include thousands of chemicals in commerce. In some cases, products used as PCPs clearly result in orders of magnitude greater exposure from their direct use than from drinking water. For instance, triclosan is used as an active ingredient in some toothpaste, with certain products containing 0.3% triclosan by weight (0.003 g/g). It is obvious that human exposure to triclosan in toothpaste and hand sanitizer containing percentages of triclosan will be orders of magnitude greater than the nanograms of triclosan found in some wastewaters. Another example is synthetic musks used as fragrances. The human exposure to fragrances is obvious, as smelling a fragrance assures exposure. Regardless, it is clear that using modern analytical methods, nearly any chemical in commerce will be detectable in wastewater, and to a lesser extent in source waters.

**Table 1.2**
**Structures of some PPCPs**

| | | |
|---|---|---|
| **Analgesics** | | |
| 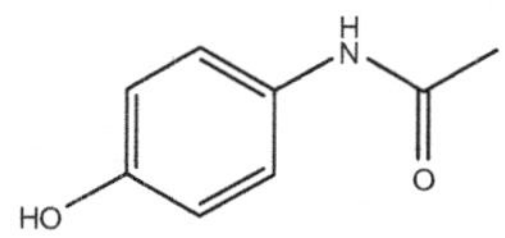 | 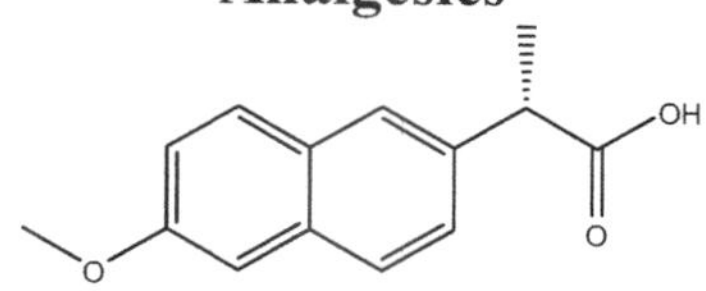 | 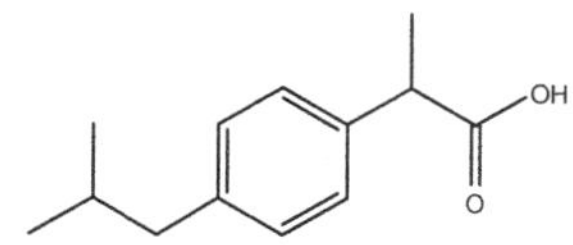 |
| Acetaminophen | Naproxen | Ibuprofen |
| **Antibiotics** | | |
| 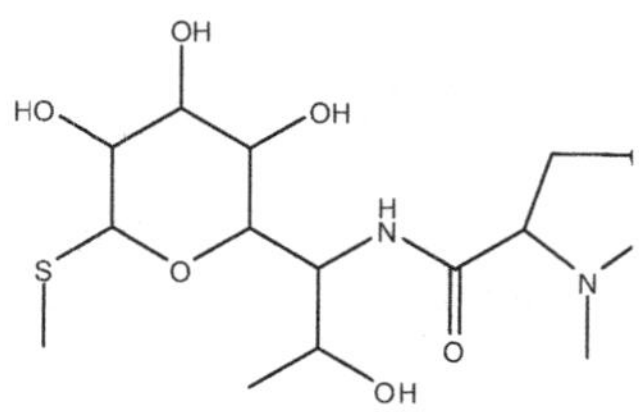 | 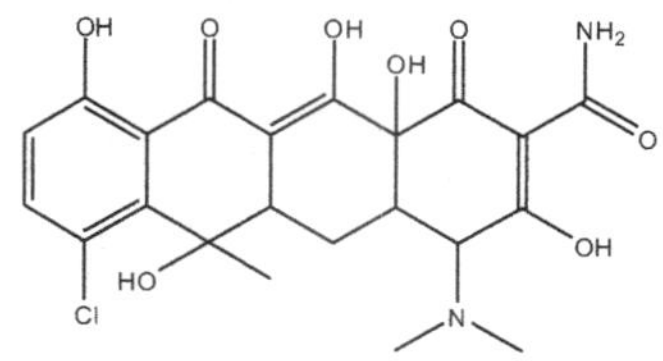 | 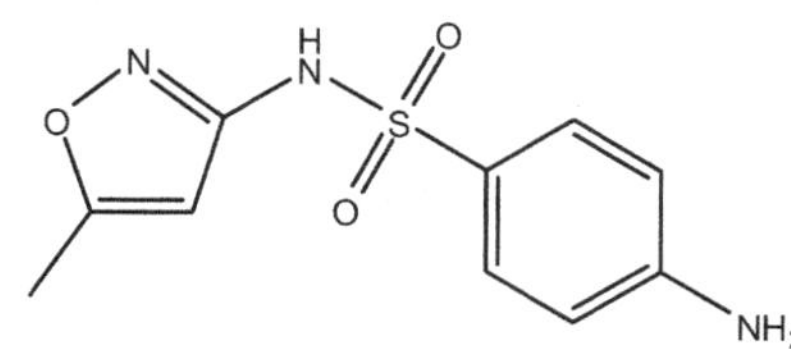 |
| Lincomycin | Chlortetracycline | Sulfamethoxazole |
| **Lipid Regulators** | | |
| 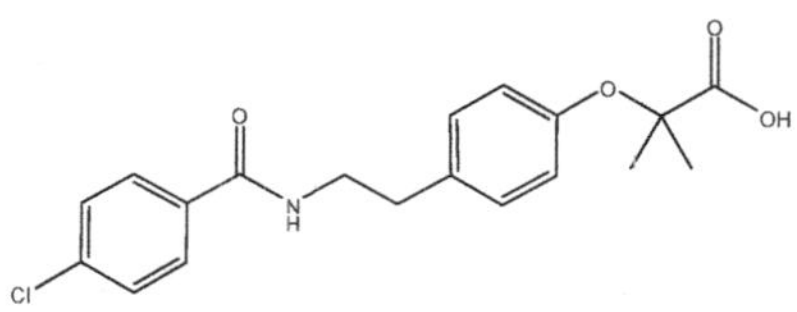 | 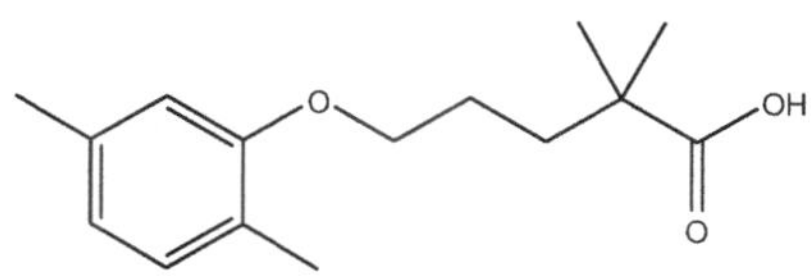 | 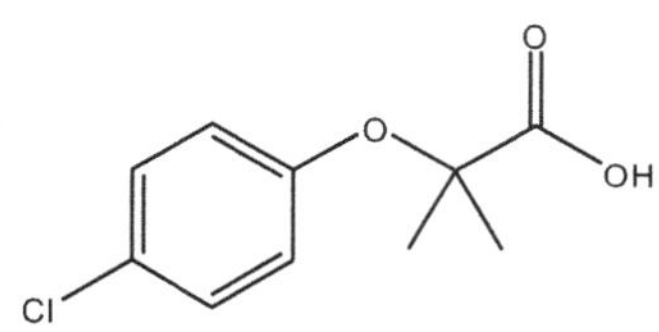 |
| Bezafibrate | Gemfibrozil | Clofibric acid |
| **β-Blocker** | **Anti-hypertensive** | **Insecticide** |
| 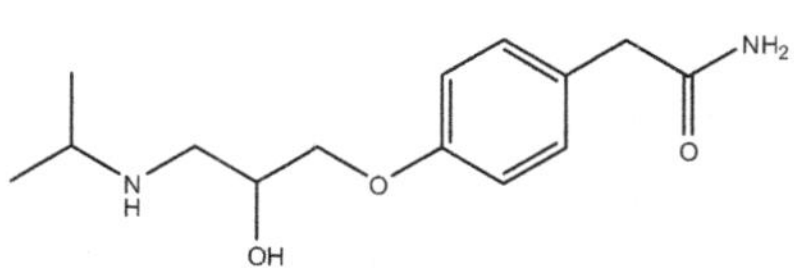 | 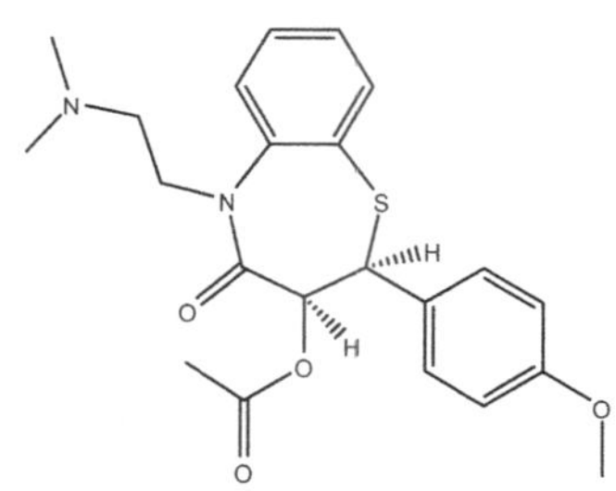 | 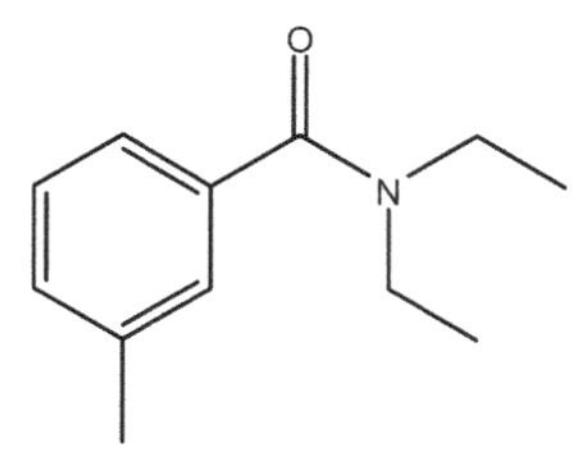 |
| Atenolol | Diltiazem | DEET |
| **X-ray contrast media** | **Anti-convulsant** | **Flame Retardant** |
| 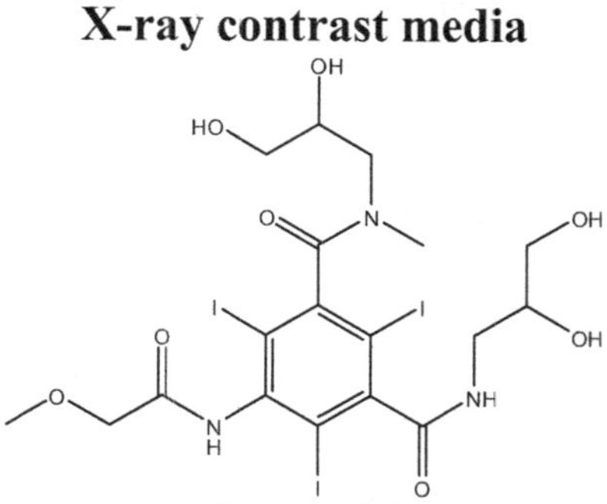 | 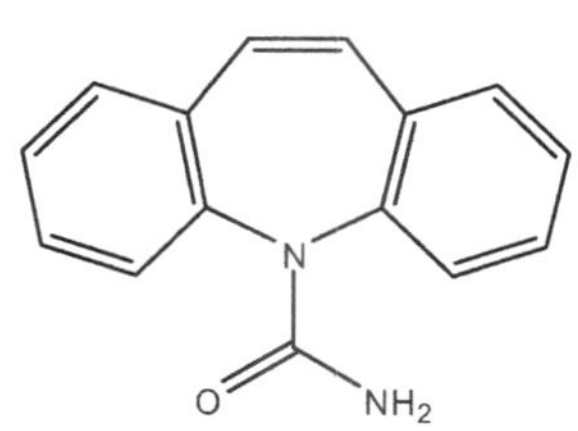 | 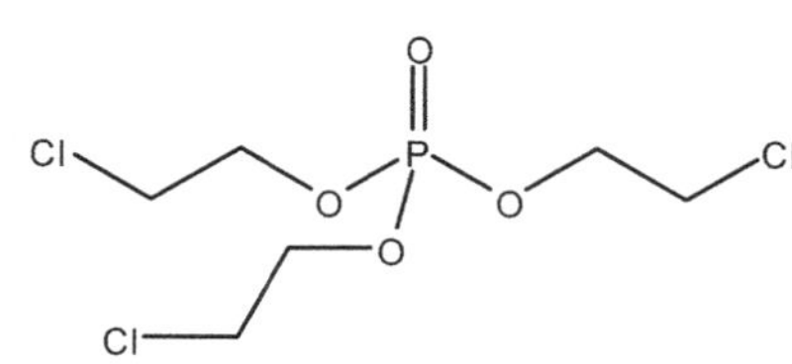 |
| Iopromide | Carbamazepine | TCEP |

## SELECTION OF INDICATOR EDCS AND PPCPS

There is a vast and diverse amount of information available regarding EDCs and PPCPs in potential drinking water sources. For the purpose of this utility guide, indicator chemicals were chosen as representatives of the nearly endless number of EDCs and PPCPs. These chemicals were chosen based upon structural diversity, relevancy (e.g., suspected EDC), ability to represent key subclasses, and availability of published information on occurrence in potential drinking water sources. The list of compounds is shown in Table 1.3 along with their physicochemical properties. This list of chemicals will be used through the report.

**Table 1.3**
**Physicochemical properties of selected EDCs and PPCPs**

| Compounds | CAS | MW | Solubility (mg/L) | $\log K_{ow}$ | pKa |
|---|---|---|---|---|---|
| **Pharmaceuticals** | | | | | |
| **Analgesics** | | | | | |
| Acetaminophen | 103-90-2 | 151.2 | $1.40 \times 10^4$ | 0.46 | 9.38 |
| Diclofenac | 15307-86-5 | 296.2 | 2.37 | 4.51 | 4.15 |
| Ibuprofen | 15687-27-1 | 206.3 | 21 | 3.97 | 4.91 |
| Ketoprofen | 22071-15-4 | 254.3 | 51 | 3.12 | 4.45 |
| Naproxen | 22204-53-1 | 230.3 | 15.9 | 3.18 | 4.15 |
| **Antibiotics** | | | | | |
| Chlortetracycline | 57-62-5 | 478.9 | 630 | -0.62 | 3.3 |
| Ciprofloxacin | 85721-33-1 | 331.3 | $3.00 \times 10^4$ | 0.28 | 6.09 |
| Clarithromycin | 81103-11-9 | 748.0 | 0.342 (est) | 3.16 | 8.99 |
| Erythromycin | 114-07-8 | 733.9 | 1.44 (est) | 3.06 | 8.88 |
| Lincomycin | 154-21-2 | 406.5 | 927 (est) | 0.56 | 7.8* |
| Roxithromycin | 80214-83-1 | 837.0 | na | 2.75 (est) | 9.2* |
| Sulfadiazine | 68-35-9 | 250.3 | 77 | -0.09 | 6.38 |
| Sulfadimethoxine | 122-11-2 | 310.3 | 343 | 1.63 | 6.91 (est) |
| Sulfamethazine | 57-68-1 | 278.3 | 1500 | 0.89 | 7.59 |
| Sulfamethoxazole | 723-46-6 | 253.3 | 610 | 0.89 | 5.5* |
| Trimethoprim | 738-70-5 | 290.3 | 400 | 0.91 | 7.12 |
| **β-Blockers** | | | | | |
| Atenolol | 29122-68-7 | 266.3 | $1.33 \times 10^4$ | 0.16 | 9.2‡‡ |
| Metoprolol | 37350-58-6 | 267.4 | $1.69 \times 10^4$ | 1.88 | 9.87 (est) |
| Propranolol | 525-66-6 | 259.3 | 61.7 | 3.48 | 9.42 |

(continued)

**Table 1.3 (Continued)**

| | | | | | |
|---|---|---|---|---|---|
| **Lipid regulators** | | | | | |
| Bezafibrate | 41859-67-0 | 361.8 | na | 4.25 (est) | 3.83 (est) |
| Clofibric acid | 882-09-7 | 214.6 | 583 (est) | 2.57 | 3.37 (est) |
| Gemfibrozil | 25812-30-0 | 250.3 | na | 4.77 (est) | 4.42 |
| **Anti-hypertensive** | | | | | |
| Diltiazem | 42399-41-7 | 414.5 | 465 | 2.79 (est) | 8.43 (est) |
| **Anti-convulsant** | | | | | |
| Carbamazepine | 298-46-4 | 236.3 | 18§§ | 2.45 | 13.9‡‡ |
| Primidone | 125-33-7 | 218.3 | 500 | 0.91 | 11.1 (est), 12.2 (est) |
| **X-ray contrast media** | | | | | |
| Diatrizoate | 117-96-4 | 613.9 | na | na | na |
| Iopromide | 73334-07-3 | 791.1 | $1.34 \times 10^5$*** | -2.05 | 10.25 (est) |
| **Muscle relaxant** | | | | | |
| Diazepam | 439-14-5 | 284.7 | 50 | 2.82 | 3.4 |
| **Steroids** | | | | | |
| **Estrogens** | | | | | |
| 17α-Estradiol | 57-91-0 | 272.4 | 3.9 | 3.94 (est) | 9.85 (est) |
| 17β-Estradiol | 50-28-2 | 272.4 | 3.6 | 4.01 | 10.4† |
| Estrone | 53-16-7 | 270.4 | 30 | 3.13 | 10.4† |
| 17α-Ethynylestradiol | 57-63-6 | 296.4 | 11.3 | 3.67 | 10.4‡ |
| **Androgens** | | | | | |
| Testosterone | 58-22-0 | 288.4 | 23.4 | 3.32 | na |
| **Personal Care Products** | | | | | |
| **Antimicrobials** | | | | | |
| Triclocarban | 101-20-2 | 315.6 | 2.0 – 4.6§ | 4.90 | 10.57 (est) |
| Triclosan | 3380-34-5 | 289.5 | 10 | 4.76 | 7.98 (est) |
| **Fragrances** | | | | | |
| Galaxolide (HHCB) | 1222-05-5 | 258.4 | 1.75†† | 5.9†† | na |
| Tonalide (AHTN) | 1506-02-1 | 258.4 | 1.2§§ | 4.6 – 6.4§§ | na |

(continued)

**Table 1.3 (Continued)**

| | | | | | |
|---|---|---|---|---|---|
| **Surfactant** | | | | | |
| Nonylphenol | 104-40-5 | 220.4 | 7 | 5.76 | 10.7 |
| **Antioxidants** | | | | | |
| Butylated hydroxyanisole (BHT) | 25013-16-5 | 181.3 | 213 (est) | 3.50 (est) | 9.94 (est) |
| **Insecticides** | | | | | |
| DEET | 134-62-3 | 191.3 | 912 (est) | 2.18 | 0.7 (est) |
| **Others** | | | | | |
| **Plasticizers** | | | | | |
| Bisphenol A | 80-05-7 | 228.3 | 120 | 3.32 | 9.6, 10.2** |
| Di-n-butyl phthalate | 84-74-2 | 278.3 | 11.2 | 4.5 | na |
| **Flame retardants** | | | | | |
| TECP | 115-96-8 | 285.5 | 7000 | 1.44 | na |
| TCPP | 13674-84-5 | 327.6 | 1200 | 2.59 | na |
| **Stimulants** | | | | | |
| Caffeine | 58-08-2 | 194.2 | 2.16 × 104 | -0.07 | 10.4 |

Note: Other than indicated explicitly, all are experimental values obtained from Environmental Science Database SRC PhysProp
*(Dodd, Buffle et al. 2006)
†(Deborde, Rabouan et al. 2005)
‡(Huber, Canonica et al. 2003)
§(Halden and Paull 2005)
**(Kosky, Silva et al. 1991)
††(Balk and Ford 1999)
‡‡(Jones, Voulvoulis et al. 2002)
§§(Carballa, Omil et al. 2005)
***(Liebig, Moltmann et al. 2006)

## SOURCES

EDCs and PPCPs can potentially originate from numerous sources and enter the environment by many routes. However, effluents from municipal WWTPs have been implicated as a major source to surface waters (Daughton 2001; Lee, Barber et al. 2004). WWTPs receive EDCs originating from many sources including plant material, personal care products, plastics, flame-retardant materials, cleaning products, pesticides, other household chemicals and consumer products, and hormones excreted by humans. PPCPs enter WWTPs when medicated people excrete pharmaceuticals or their metabolites, rinse them from their bodies during bathing, or flush unused medications down the sink or toilet. WWTPs also might receive hospital effluents containing pharmaceuticals, discarded pharmaceuticals or pharmaceutical-contaminated waste from human and veterinary medical facilities and pharmacies. WWTPs might also treat industrial effluents and stormwater runoff that contain EDCs and PPCPs from the same and additional sources. Although wastewater treatment processes remove some EDCs and PPCPs to varying degrees, chemicals that resist treatment may remain in effluents discharged to surface water. In surface water, the contaminants may be diluted, sequestered (e.g., in sediment), or degraded by physical or biological processes, but some persist in the environment or are detected due to relatively constant loading.

Although WWTP effluents constitute major sources and routes for EDCs and PPCPs to enter surface water and groundwater, there are many others (see Tables 1.4 and 1.5). For example, combined sewer overflows (CSO) as well as septic systems have been identified to release EDCs and PPCPs into the environment. In addition, untreated stormwater flows and urban runoff can contain a multitude of microcontaminants, including EDCs and PPCPs. It has been hypothesized that urban runoff and stormwater could contain pharmaceuticals excreted by medicated pets (Daughton 2001). EDCs and PPCPs excreted by pets also could potentially enter stormwater and urban runoff as well.

**Table 1.4**
**Sources of pharmaceuticals to surface water or groundwater**

| Source to groundwater or surface water | Route to surface water or groundwater | Source of Pharmaceuticals |
|---|---|---|
| Municipal WWTPs | ▪ Discharge of treated WWTP effluent to surface water<br>▪ Leakage from damaged sewage system infrastructure to groundwater<br>▪ Overflow of untreated sewage to surface water during storm events or system failures<br>▪ Use of recycled water for irrigation or for other domestic uses (source to groundwater or surface water)<br>▪ Use of recycled water for aquifer (groundwater) recharge<br>▪ Application of sewage biosolids to land | ▪ Pharmaceuticals, conjugates, and metabolites excreted from medicated people or washed from body surfaces<br>▪ Unused medications from households flushed down sink or toilet<br>▪ Unused medications and medical waste from pharmacies and human and veterinary medical facilities discarded to WWTPs<br>▪ Hospital effluents discharged to WWTPs |
| Septic systems | ▪ Leachate from septic systems to groundwater<br>▪ Leakage from damaged septic systems to groundwater<br>▪ Ponding and subsequent run-off from septic systems | ▪ Pharmaceuticals, conjugates, and metabolites excreted from medicated people<br>▪ Unused medications from households |
| Direct (untreated) releases to water | ▪ Releases from humans during recreation (e.g., swimming, bathing, boating)<br>▪ Untreated stormwater and urban runoff<br>▪ Untreated sewage discharges to surface water ("straight-piping") | ▪ Topically applied pharmaceuticals washed from bodies or ingested pharmaceuticals excreted into water during recreation<br>▪ Human waste illegally released from boats<br>▪ Wastes from medicated pets washed into water during storms and landscape irrigation<br>▪ Medications disposed/excreted to sink, toilet, bath, or shower |

(continued)

**Table 1.4 (Continued)**

| Source to groundwater or surface water | Route to surface water or groundwater | Source of Pharmaceuticals |
|---|---|---|
| Landfills | ▪ Leachate from landfills to groundwater | ▪ Disposal of unused pharmaceuticals and other medical wastes<br>▪ Disposal of medicated animal carcasses |
| Agriculture and aquaculture | ▪ Untreated runoff from CAFOs<br>▪ Soil amendment with animal manure<br>▪ Irrigation of crops with recycled water<br>▪ | ▪ Medications administered to livestock<br>▪ Medicated feed and excreta from farmed aquatic organisms<br>▪ Certain pharmaceuticals (e.g., antiobiotics) applied to crops or otherwise used as pest control agents |
| Pest control – non-agricultural | ▪ Contaminated carcasses washed into surface water or disposed in landfills<br>▪ Excess poisons flushed to municipal sewage | ▪ Certain pharmaceuticals used as pest control agents (e.g., anticoagulant warfarin used as a rat poison) |
| Industrial discharges | ▪ Discharge of regulated industrial manufacturing waste effluents | ▪ Pharmaceuticals, production intermediates, and metabolites/degradation products generated during pharmaceutical manufacture |

Source: Daughton 2001b
WWTP - wastewater treatment plant; CAFOs - confined animal feeding operations

**Table 1.5**
**Examples of sources and types of putative EDCs**

| Type | Sources and Routes to Water | Example Compounds |
|---|---|---|
| Hormones excreted by humans and animals | ▪ Municipal WWTP effluents, recycled/reuse water, septic systems<br>▪ Runoff from CAFOs<br>▪ Aquaculture<br>▪ Application of sewage sludge or manure to land | 17β-Estradiol<br>Estrone<br>Various conjugated estrogens |
| Phytoestrogens and other plant steroids, mycoestrogens (from fungi) | ▪ Phystoestrogens in foods (e.g., soy) and other plant material and in human waste discarded to sewage<br>▪ Plant steroids in pulp and paper mill effluents | Genistein<br>Daidzein<br>Coumestrol<br>β-Sitosterol |
| Ionic compounds | ▪ Runoff and leaching of manure and synthetic nitrate-based fertilizers<br>▪ Industrial (production of road flares, matches, automobile air bags) and military (munitions, aerospace) use and disposal of synthetic perchlorate<br>▪ Land application of fertilizers mined from Chilean caliche containing naturally-occurring perchlorate | Nitrate<br>Perchlorate<br>Thiocyanate |
| Metals | ▪ Mining<br>▪ Industrial applications<br>▪ Disposal of consumer products (e.g., batteries, cell phones) in landfills<br>▪ Incinerators<br>▪ Combustion of fossil fuels | Arsenic<br>Cadmium<br>Lead<br>Mercury |
| Human pharmaceuticals | ▪ Municipal WWTP effluents<br>▪ Recycled water | 17α-Ethynylestradiol<br>Conjugated estrogens |

(continued)

**Table 1.5 (Continued)**

| Type | Sources/Routes to Water | Example Compounds |
|---|---|---|
| Veterinary pharmaceuticals | ▪ Runoff from CAFOs | Trenbolone |
| Dioxins, Furans, PAHs | ▪ Forest fires (natural or human-caused)<br>▪ Volcanic activity<br>▪ Incinerators<br>▪ Industrial processes<br>▪ Combustion of fossil fuels | TCDD<br>Benzo[a]pyrene |
| Industrial chemicals and their degradation products | ▪ Industrial effluents and landfilled wastes<br>▪ Releases from consumer products | PCBs<br>Nonyphenol<br>Octylphenol<br>Bisphenol A |
| Biocides | ▪ Application of pesticides and herbicides used in agriculture, landscaping, homes, and businesses followed by runoff or leaching to groundwater or surface water<br>▪ Disposal of unused biocides to municipal WWTP or landfills | DDE<br>Lindane<br>Atrazine<br>Linuron<br>Vinclozolin<br>Tributyltin |

CAFOs - confined animal feeding operations; DDE - dichlorodiphenyl dichloroethene, degradation product of DDT (dichlorodiphenyltrichloroethane); PAHs – polycyclic aromatic hydrocarbons; PCBs – polychlorinated biphenyls; TCDD - tetrachlordibenzo-*p*-dioxin; WWTP - wastewater treatment plant

## OCCURRENCE AND EXPOSURE

The public and regulatory agencies are concerned about whether EDCs and PPCPs that might be ingested in drinking water pose a risk to consumers' health. To address these concerns, it is necessary to characterize potential exposure to and effects of these chemicals. Occurrence data for trace concentrations of EDCs and PPCPs in source water are sparse, and even fewer data are available for finished drinking water. A limited number of laboratories are capable of conducting these trace analyses, the costs can be substantial, and analytical methods have not been developed for all of the contaminants of interest that may occur in source water or drinking water.

Further development of analytical methods should be encouraged, but because the analytical costs and difficulty tend to increase substantially with decreasing detection limit requirements, the least expensive and simplest method that can quantitate concentrations below thresholds for health effects (if these are known) for humans and aquatic life should be used. Because data on occurrence, routes of exposure, and toxicity at environmentally relevant levels are lacking for many of the potential contaminants of concern, selection of contaminants for monitoring and treatment is a challenge. Efforts are underway to identify the EDCs and PPCPs in drinking water that are toxicologically most important or that might serve as surrogates or indicators for other contaminants in monitoring and treatment studies.

Humans are exposed to EDCs and PPCPs through routes other than drinking water. Exposure routes include diet, inhalation of airborne chemicals, medical treatments (PPCPs), and dermal absorption. Consequently, the contribution of drinking water to total exposure and its relative importance should be considered in risk assessments for these contaminants.

## HEALTH EFFECTS

Several studies have associated reproductive disorders in fish with EDCs in wastewater (Purdom, Hardiman et al. 1994; Bevans, Goodbred et al. 1996; Folmar, Denslow et al. 1996; Jobling, Noylan et al. 1998; Lazorchak, Hemming et al. 2001; Jobling and Tyler 2003; Snyder, Snyder et al. 2004). Although effects of EDCs in wastewater on fish have raised concerns for people who consume drinking water taken from wastewater-influenced source waters, there are substantial differences in exposure render fish more susceptible. Fish and other aquatic organisms are expected to receive much greater exposure than humans to EDCs and PPCPs in the aquatic environment. Fish are constantly immersed in water and can bioconcentrate waterborne contaminants by direct uptake from water across the gill. Fish also can be exposed to lipophilic contaminants through diet or exposure to contaminated sediment or suspended particulate material. Humans are likely to receive much less exposure to wastewater-associated contaminants by these routes. In oligotrophic to mesotrophic water bodies, fish often congregate near municipal WWTP discharges where increased nutrient input can result in greater availability of food and plant cover. This behavior also results in greater exposure to contaminants in wastewater. While people are exposed intermittently to waterborne contaminants, fish may be exposed continuously and possibly for an entire lifetime and are more likely to be exposed during critical periods of development when organisms are most sensitive to the effects of contaminants.

The adverse effects of the estrogenic pharmaceutical diethylstilbestrol (DES), a medication once prescribed to pregnant women, on children exposed to the drug during gestation have clearly shown that humans are susceptible to the effects of EDCs. Human and animal studies of the effects of long-term exposure to environmentally relevant doses are lacking for most putative EDCs. Results of available studies with laboratory animals indicate that certain EDCs can cause effects at low doses, and some studies on fish and wildlife provide strong evidence that they can be affected by EDCs in the environment. Industrial chemicals, organochlorine pesticides, and naturally occurring chemicals considered to be EDCs have been implicated as potential causes for a suite of reported effects in humans, including sperm count declines; increased incidences of cancers of the breast, ovary, and prostate; endometriosis; premature menopause; malformations of the male reproductive organs, and early puberty (WHO 2002). However, a specific chemical cause has not been established, and the existence of the effects has been disputed in some cases. To date, cause-effect relationships between low-level environmental exposures to EDCs and human health effects have not been established (WHO 2002). Toxicological risk assessments for EDCs are complicated by atypical dose-response relationships, multiple routes of exposure, and potential additive toxicity or interactions among chemicals within mixtures to which people are commonly exposed (including mixtures occurring in drinking water).

While a wealth of toxicological information may be available for pharmaceuticals, most testing for pharmaceuticals is conducted with theraepeutic doses, which are much greater than doses that would be expected from drinking water consumption. The effects of unintended chronic exposure to subtherapeutic doses that could occur via consumption of drinking water often are not known, particularly for sensitive subopulations such as pregnant women, infants, children, those who are elderly or who have compromised health. Based on information about the activity and potential toxicity of pharmaceuticals, possible health effects of long-term exposure to pharmaceuticals could include endocrine disruption, induction of antibiotic resistance in human pathogens, genotoxicity, carcinogenicity, allergic reactions, and reproductive or developmental effects. Risk assessments conducted to date do not indicate that the trace concentrations of pharmaceuticals detected in drinking water pose a health risk to consumers, but most of these assessments are based on comparisons of environmental concentrations or concentrations in drinking water with therapeutic doses, which are much greater than doses that could be attained through contaminated drinking water.

## CURRENT REGULATIONS

Regulatory agencies in the U.S. have not yet developed health risk-based standards for EDCs or PPCPs in drinking water. The Safe Drinking Water Act and amendments authorize the U.S. Environmental Protection Agency (U.S. EPA) to set drinking water standards for contaminants in drinking water in the U.S. Although the levels of certain contaminants that might be considered to be EDCs already are regulated in drinking water, these are not regulated on the basis of their potential effects on the endocrine system. One exception is perchlorate, an EDC that acts on the thyroid. The Massachusetts Department of Environmental Protection promulgated the first drinking water standard (2 ppb, or 2 μg/L) for perchlorate in the United States (Massachusetts DEP 2006) on the basis of effects mediated by the thyroid. There are no federal regulations for perchlorate in drinking water.

No federal regulatory limits have been set specifically for EDCs or PPCPs in wastewater in the U.S. However, the U.S. Food and Drug Administration (FDA) requires pharmaceutical manufacturers to present environmental risk assessments with New Drug Applications for pharmaceuticals that are expected to occur at concentrations greater than 1 ppb (1 μg/L) at the point of entry for effluents released to the aquatic environment (Velagaleti, Burns et al. 2002)(FDA 1998, FDA 2000).

## FUTURE ALTERNATIVES FOR CONTROL

### Water Treatment

Recent research indicates that many EDCs and PPCPs are removed to varying degrees through conventional and advanced wastewater and drinking water treatment processes. Because EDCs and PPCPs with widely varying properties might occur in the environment, a single treatment process is unlikely to be effective and feasible for all contaminants of potential concern present at a single utility. Each of the available treatment options has advantages and disadvantages in terms of effectiveness, feasibility, and cost. Little is known about the occurrence and potential toxicity of degradation products of EDCs and PPCPs that might result from treatment processes such as oxidation that alter chemical structures rather than removing chemicals from water. Conventional water treatment processes employing coagulation/flocculation and filtration are rather ineffective in removing EDCs and PPCPs from water. Various advanced treatment processes (e.g., ozone, advanced oxidation processes, granular activated carbon) are likely to achieve substantial removal of the majority of EDCs and PPCPs. Reverse osmosis is highly effective for removal of most of these contaminants, but it is not destructive and very costly. In many locations, the disposal of the brine is a significant problem. Riverbank filtration (RBF) has been identified as an efficient pre-treatment process to remove a wide range of EDCs and PPCPs from impaired surface water sources.

### Source Water Protection

Advanced drinking water treatment processes are likely to substantially reduce exposure to humans but will provide no protection for aquatic life. Source water protection (SWP) measures to reduce the release of EDCs and PPCPs to the environment should be considered, particularly when dealing with contaminants that are recalcitrant to environmental degradation and drinking water treatment processes. Detection and removal of contaminants may be less complicated at point sources, where they are likely to occur at elevated concentrations prior to dilution in source waters. Source water protection might potentially provide ancillary benefits to the environment and particularly to aquatic organisms, which are likely to receive much greater exposure than humans to EDCs and PPCPs in water and have demonstrated adverse effects. Although reports of endocrine effects in fish exposed to WWTP effluents have generated public concern about the potential for similar effects on humans exposed to drinking water obtained from sources influenced by these effluents, exposures to and effects of EDCs and PPCPs in fish cannot be readily equated to the same in humans, and efforts to prevent adverse effects in fish may or may not result in benefits for humans. Source water protection efforts should seek to identify sources of endocrine disruptors to the aquatic environment and to assess the relative importance of various sources to determine where expended effort and funds might produce the greatest benefits (e.g., improved wastewater treatment versus control of agricultural runoff).

## Regulations

Strong concerns voiced by members of the public and environmental groups have prompted proposals to set analytical detection limits as regulatory levels for the concentrations of pharmaceuticals and EDCs in wastewater, recycled/reuse water, and drinking water. While regulations might provide some level of comfort, this approach invites criticism for several reasons. First, analytical detection methods are improving at such a rapid rate that they are out-pacing improvements in treatment technologies. If analytical costs are not a consideration, it is practically impossible to remove all EDCs and PPCPs in water to levels below achievable detection limits. Second, analytical detection limits have no relationship to health-based standards. As analytical methods continue to improve, it is likely that detection limits for EDCs and PPCPs will more frequently fall below levels that produce any known biological effect. Consequently, striving to achieve "complete" removal will necessitate the use of increasingly expensive treatment technologies with no appreciable health benefit. Ideally, drinking water and wastewater treatment goals should be set for concentrations of contaminants that are safe and can be achieved at reasonable costs. However, estimating "safe" exposure levels is a daunting task. The potential risks associated with long-term, low level exposure to EDCs and PPCPs in drinking water generally are not well-studied, and limitations of toxicity testing make it likely that this will continue to be true for years to come.

Alternatively, to avoid the problem of decreasing detection limits, some have advanced the idea of setting analytical detection limits and levels of concern or regulations at arbitrary levels below which they believe there will be no (or negligible) effects or at levels below which analytical detection becomes difficult using laboratory equipment commonly available to public utilities ((UBA) 2003). This approach is more practical, but it still is based more on policy than on science and, if it is used at all, should be employed as an interim measure until more defensible criteria can be developed.

## Pharmaceutical Take-back Programs

Pharmaceutical take-back programs have been proposed as a measure to reduce the disposal of unused pharmaceuticals to municipal sewage systems and private septic systems. Proposed programs include voluntary or compulsory return of unused or expired pharmaceuticals to a collection point such as a pharmacy or municipal authority for disposal. In theory, preventing pharmaceuticals from entering water should be easier than removing trace concentrations of these chemicals from wastewater, recycled/reuse water, and drinking water to levels below constantly decreasing analytical detection limits.

However, there are some potential problems that must be assessed to determine whether these programs are feasible and cost-effective. First, prescription pharmaceuticals are regulated, so the receiver must have the legal authority to handle, transport, and dispose of these drugs. Security could become a problem at all stages from collection to transport and disposal for drugs that are desirable in illegal markets. Second, the method of disposal must be cost-effective. That is, the disposal method should result in less serious environmental consequences than the disposal of pharmaceuticals to sewage systems, and the cost per unit measure of benefit should be less expensive than using wastewater or drinking water treatment to remove pharmaceuticals from water. After pharmaceuticals are returned to a collection point, they could be returned to the manufacturer, incinerated, or deposited in a landfill. Manufacturers can be expected to increase drug prices to cover any costs related to disposal of returned pharmaceuticals.

Incineration has costs that must be considered, as well as environmental consequences such as air pollution. Landfill disposal also has costs, and future releases of pharmaceuticals back to groundwater through landfill leachate are a possibility. Because the risks associated with trace concentrations of most pharmaceuticals in the aquatic environment and in drinking water are largely undefined, it is difficult to characterize the potential benefits of reducing their releases to groundwater or surface water.

**Uncertainties, Costs, and Benefits**

Utilities and their customers will have to decide whether to wait for more information that can be used to guide action or to take proactive measures to remove EDCs and PPCPs from wastewater and drinking water despite limited information and uncertain benefits. Of particular concern are uncertainties related to selection of contaminants for monitoring and treatment and unknown benefits for humans and aquatic organisms. Consideration should be given to the willingness of the public to pay for potentially costly mitigation efforts, the importance of EDCs and PPCPs in surface waters and drinking water relative to other public health and environmental concerns, and potential losses or gains related to waiting for more information or taking action despite uncertainties. Effective communication among utilities, regulators, and the public is needed to address these issues.

## REFERENCES


(UBA), U. (2003). "Bewertung der Anwesenheit teil- oder nicht bewertbarer Stoffe im Trinkwasser aus gesundheitlicher Sicht." Bundesgesundheitsblatt-Gesundheitsforsch-Gesundheitsschutz **46**: 249-251.

Ahrer, W., E. Scherwenk, et al. (2001). "Determination of drug residues in water by the combination of liquid chromatography or capillary electrophoresis with electrospray mass spectrometry." Journal of Chromatography A **910**: 69-78.

Balk, F. and R. Ford (1999). "Environmental risk assessment for the polycyclic musks, AHTN and HHCB. II. Effect assessment and risk characterisation." Toxicology Letters **111**(1-2): 81-94.

Bevans, H. E., S. L. Goodbred, et al. (1996). "Synthetic organic compounds and carp endocrinology and histology in Las Vegas Wash and Las Vegas and Callville Bays of Lake Mead, Nevada, 1992 and 1995." Water-Resources Investigations Report 96-4266

Brookbanks, E. O., R. A. H. Welch, et al. (1969). "Estrogens in pasture and a possible relation with mastitis." New Zealand Veterinary Journal **17**: 159-160.

Bruchet, A., C. Hochereau, et al. (2005). "Analysis of drugs and personal care products in French source and drinking waters: the analytical challenge and examples of application." Water Science and Technology **52**: 53-61.

Bruchet, A., C. Prompsy, et al. (2002). "A broad spectrum analytical scheme for the screening of endocrine disruptors (EDs), pharmaceuticals and personal care products in wastewaters and natural waters." Water Science and Technology **46**(3): 97-104.

Carballa, M., F. Omil, et al. (2005). "Removal of cosmetic ingredients and pharmaceuticals in sewage primary treatment." Water Research **39**: 4790-4796.

D'Ascenzo, G., A. Di Corcia, et al. (2003). "Fate of natural estrogen conjugates in municipal sewage transport and treatment facilities." Science Of The Total Environment **302**(1-3): 199-209.

Daughton, C. G. (2001). Pharmaceuticals in the Environment: Overarching Issues and Overview. Pharmaceuticals and Personal Care Products in the Environment: Scientific and Regulatory Issues. C. G. Daughton and T. Jones-Lepp. Washington, D.C., American Chemical Society. **791:** 2-38.

Daughton, C. G. (2001, June 2001). "PPCPs as Environmental Pollutants." from http://www.epa.gov/esd/chemistry/pharma/faq.htm.

Deborde, M., S. Rabouan, et al. (2005). "Kinetics of Aqueous Ozone-Induced Oxidation of Some Endocrine Disruptors." Environmental Science and Technology **39**(16): 6086-6092.

Dodd, M., M. Buffle, et al. (2006). "Oxidation of antibacterial molecules by aqueous ozone: moiety-specific reaction kinetics and application to ozone-based wastewater treatment." Environmental Science & Technology **40**(6): 1969-77.

Drewes, J. E. and L. S. Shore (2001). "Concerns about pharmaceuticals in water reuse, groundwater recharge, and animal waste." ACS Symposium Series **791**: 206-228.

EDSTAC (Endocrine Disruptor Screening and Testing Advisory Committee). 1998. Endocrine Disruptor Screening and Testing Advisory Committee (EDSTAC) Final Report. U.S. Environmental Protection Agency. August. http://www.epa.gov/scipoly/oscpendo/edspoverview/finalrpt.htm

European Commission. 2002. Study on the Scientific Evaluation of 12 Substances in the Context of Endocrine Disrupter Priority List of Actions. Report No. UC 6052. November. Buckinghamshire, UK: European Commission. Available: http://ec.europa.eu/environment/endocrine/documents/wrc_report.pdf

FDA (United States Food and Drug Administration). 1998. United States Food and Drug Administration. Available: http://www.fda.gov/Cder/guidance/1730fnl.pdf.

FDA (United States Food and Drug Administration). 2000. Regulations in 21 CFR part 25 United States Food and Drug Administration. Available: http://www.fda.gov/cvm/Guidance/guide89.PDF

Folmar, L. C., N. D. Denslow, et al. (1996). "Vitellogenin induction and reduced serun testosterone concentrations in feral male carp (*Cyprinus carpio*) captured near a major metropolitan sewage treatment plant." Environmental Health Perspectives **104**(10): 1096-1101.

Folmar, L. C., M. J. Hemmer, et al. (2002). "A comparison of the estrogenic potencies of estradiol, ethynylestradiol, diethylstilbestrol, nonylphenol and methoxychlor *in vivo* and *in vitro*." Aquatic Toxicology **60**(1-2): 101-110.

Garrison, A. W., J. D. Pope, et al. (1975). GC/MS analysis of organic compounds in domestic wastewaters. Chem. Congr. North Am. Cont., Ann Arbor Science.

Gentili, A., D. Perret, et al. (2002). "Analysis of free estrogens and their conjugates in sewage and river waters by solid-phase extraction then liquid chromatography-electrospray-tandem mass spectrometry." Chromatographia **56**(1/2): 25-32.

Giesy, J. P., S. L. Pierens, et al. (2000). "Effects of 4-nonylphenol on fecundity and biomarkers of estrogenicity in fathead minnows (*Pimephales promelas*)." Environmental Toxicology and Chemistry **19**(5): 1368-1377.

GWRC (Global Water Research Coalition). 2003. Endocrine Disrupting Compounds: An Overview of Sources and Biological Methods for Measuring EDCs. September. Global Water Research Coalition (GWRC). 16 p.

Halden, R. U. and D. H. Paull (2005). "Co-Occurrence of Triclocarban and Triclosan in U.S. Water Resources." Environmental Science and Technology **39**(6): 1420-1426.

Huber, M. M., S. Canonica, et al. (2003). "Oxidation of pharmaceuticals during ozonation and advanced oxidation processes." Environmental Science & Technology **37**(5): 1016-1024.

Isobe, T., H. Shiraishi, et al. (2003). "Determination of estrogens and their conjugates in water using solid-phase extraction followed by liquid chromatography-tandem mass spectrometry." Journal of Chromatography A **984**(2): 195-202.

Jobling, S., M. Noylan, et al. (1998). "Widespread sexual disruption in wild fish." Environmental Science & Technology **32**: 2498-2506.

Jobling, S. and C. R. Tyler (2003). "Endocrine disruption in wild freshwater fish." Pure and Applied Chemistry **75**(11-12): 2219-2234.

Johnson, A. C., A. C. Belfroid, et al. (2000). "Estimating steroid oestrogen inputs into activated sludge treatment works and observations on their removal from the effluent." The Science of the Total Environment **256**(2-3): 163-173.

Jones, O. A. H., N. Voulvoulis, et al. (2002). "Aquatic environmental assessment of the top 25 English prescription pharmaceuticals." Water Research **36**: 5013-5022.

Kavlock, R.J., G.P. Daston, C. Derosa, P. Fenner-Crisp, L.E. Gray, S. Kaattari, G. Lucier, M. Luster, M.J. Mac, C. Maczka, R. Miller, J. Moore, R. Rolland, G. Scott, D.M. Sheehan, T. Sinks, and H.A. Tilson. 1996. Research needs for the risk assessment of health and environmental effects of endocrine disruptors: A report of the U.S. EPA-sponsored workshop. *Environmental Health Perspectives*, 104(Supplement 4):715-740.

Kidd, K., P. J. Blanchfield, et al. (2007). Collapse of a fish population after exposure to a synthetic estrogen. National Academy of Sciences of the United States of America, National Academy of Sciences of the United States of America.

Kim, S.-C. and K. Carlson (2006). "Occurrence of ionophore antibiotics in water and sediments of a mixed-landscape watershed." Water Research **40**: 2549-2560.

Kosky, P. G., J. M. Silva, et al. (1991). "The aqueous phase in the interfacial synthesis of polycarbonates. Part 1. Ionic equilibria and experimental solubilities in the BPA-sodium hydroxide-water system." Industrial & Engineering Chemistry Research **30**(3): 462-7.

Laganá, A., A. Bacaloni, et al. (2000). "Trace analysis of estrogenic chemicals in sewage effluent using liquid chromatography combined with tandem mass spectrometry." Rapid Commun. Mass Spectrom. **14**: 401-407.

Lazorchak, J. M., J. M. Hemming, et al. (2001). Evaluation of a wastewater discharge using vitellogenin gene expression and plasma protein levels in fathead minnows. SETAC, Baltimore, MD.

Lee, K. E., L. B. Barber, et al. (2004). Presence and Distribution of Organic Wastewater Compounds in Wastewater, Surface, Ground, and Drinking Waters, Minnesota, 2000-02. Reston, Virginia, United States Geological Survey: 47.

Liebig, M., J. F. Moltmann, et al. (2006). "Evaluation of measured and predicted environmental concentrations of selected human pharmaceuticals and personal care products." Environmental Science and Pollution Research **13**: 110-119.

Massachusetts DEP. 2006. Perchlorate Fact Sheet. July. Available: http://mass.gov/dep/toxics/pchlorqa.pdf.

Petrovic, M. and D. Barcelo (2007). Analysis, fate and removal of pharmaceuticals in the water cycle. Amsterdam, The Netherlands, Elsevier Science.

Purdom, C. E., P. A. Hardiman, et al. (1994). "Estrogenic effects of effluents from sewage treament works." Chemistry and Ecology **8**: 275-285.

Reemtsma, T. and M. Jekel (2006). Organic Pollutants in the Water Cycle: Properties, Occurrence, Analysis and Environmental Relevance of Polar Compounds. Weinheim, Wiley-VCH.

Routledge, E. J., D. Sheahan, et al. (1998). "Identification of estrogenic chemicals in STW effluent. 2. *In vivo* responses in trout and roach." Environmental Science & Technology **32**(11): 1559-1565.

Snyder, E. M., S. A. Snyder, et al. (2004). "Reproductive responses of common carp (*Cyprinus carpio*) exposed in cages to influent of the Las Vegas Wash in Lake Mead, Nevada from late winter to early spring." Environmental Science & Technology **38**(23): 6385-6395.

Snyder, S. A., T. L. Keith, et al. (2001). "Identification and quantification method for nonylphenol and lower oligomer nonylphenol ethoxylates in fish tissues." Environmental Toxicology and Chemistry **20**(9): 1870-1873.

Snyder, S. A., T. L. Keith, et al. (2001). "Bioconcentration of nonylphenol in fathead minnows (*Pimephalas promelas*)." Chemosphere **44**(8): 1697-1702.

Snyder, S. A., T. L. Keith, et al. (1999). "Analytical methods for detection of selected estrogenic compounds in aqueous mixtures." Environmental Science & Technology **33**(16): 2814-2820.

Snyder, S. A., B. J. Vanderford, et al. (2003). "Analytical methods used to measure endocrine disrupting compounds in water." Practice Periodical of Hazardous,Toxic, and Radioactive Waste Management **7**(4): 224-234.

Snyder, S. A., D. L. Villeneuve, et al. (2001). "Identification and quantification of estrogen receptor agonists in wastewater effluents." Environmental Science & Technology **35**(18): 3620-3625.

Snyder, S. A., E. C. Wert, H. Lei, P. Westerhoff, and Y. Yoon. 2007. Removal of EDCs and Pharmaceuticals in Drinking Water and Reuse Treatment Processes, American Water Works Association: Denver, Colorado.

Stumm-Zollinger, E. and G. M. Fair (1965). "Biodegradation of steroid hormones." Journal of the Water Pollution Control Federation **37**: 1506-1510.

Ternes, T., M. Bonerz, et al. (2001). "Determination of neutral pharmaceuticals in wastewater and rivers by liquid chromatography-electrospray tandem mass spectrometry." Journal of Chromatography a **938**(1-2): 175-185.

Ternes, T. and A. Joss (2006). Human pharmaceuticals, hormones and fragrances - the challenge of micropollutants in urban water management. London, IWA Publishing.

Ternes, T. A. (2001). "Analytical methods for the determination of pharmaceuticals in aqueous environmental samples." Trends in Analytical Chemistry **20**(8): 419-434.

Ternes, T. A., R. Hirsch, et al. (1998). "Methods for the determination of neutral drugs as well as betablockers and $\beta_2$-sympathomimetics in aqueous matrices using GC/MS and LC/MS/MS." Fresenius' Journal of Analytical Chemistry **362**: 329-340.

Tsuda, T., A. Takino, et al. (2001). "Evaluation of 4-nonylphenols and 4-tert-octylphenol contamination of fish in rivers by laboratory accumulation and excretion experiments." Water Research **35**(7): 1786-1792.

Velagaleti, R., P. K. Burns, et al. (2002). "Impact of current good manufacturing practices and emission regulations and guidances on the discharge of pharmaceutical chemicals into the environment from manufacturing, use, and disposal." Environmental Health Perspectives **110**(3): 213-220.

Welshons, W.V., S.C. Nagel, and F.S. Vom Saal. 2006. Large effects from small exposures. III. Endocrine mechanisms mediating effects of bisphenol A at levels of human exposure. *Endocrinology*, 147(6):S56-S69.

WHO. 2002. *Global Assessment of the State-of-the-Science of Endocrine Disruptors*. WHO/PCS/EDC/02.2. Edited by T Damstra, S Barlow, A Bergman, R Kavlock, and G Van Der Kraak: World Health Organization (WHO). http://www.who.int/ipcs/publications/new_issues/endocrine_disruptors/en/print.html

Zwiener, C. and F. H. Frimmel (2004). "LC-MS analysis in the aquatic environment and in water treatment - a critical review. Part I: Instrumentation and general aspects of analysis and detection." Analytical and Bioanalytical Chemistry **378**(4): 851-861.

Zwiener, C. and F. H. Frimmel (2004). "LC-MS analysis in the aquatic environment and in water treatment technology - a critical review. Part II: Applications for emerging contaminants and related pollutants, microorganisms and humic acids." Analytical and Bioanalytical Chemistry **378**(4): 862-874.

# CHAPTER 2
# ANALYTICAL METHODS

## INTRODUCTION

Although the term "emerging contaminants" has often been applied to chemicals that have been recently detected in the environment, the analysis of endocrine disrupting compounds (EDCs) and pharmaceuticals and personal care products (PPCPs) has been ongoing for decades. In the 1960s, Stumm-Zollinger and Fair (1965) used UV absorbance to study the biodegradation of steroid hormones in wastewater. In the 1970s, Tabak and Bunch (1970) studied steroid hormones as water pollutants and Hignite and Azarnoff (1977) studied the presence of chlorophenoxyisobutyrate, a biologically active metabolite of clofibrate, and salicylic acid, commonly known as aspirin, in wastewater using gas chromatography-mass spectrometry (GC-MS). Despite these reports of EDCs and pharmaceuticals in the environment, they received little attention until researchers in the United Kingdom and United States linked the occurrence of trace steroids to biological activity in fish and cellular bioassays (Desbrow, Routledge et al. 1998; Routledge, Sheahan et al. 1998; Snyder, Keith et al. 1999; Snyder, Villeneuve et al. 2001). These developments, along with advances in analytical instrumentation, led to a rapid increase in the number of analytical techniques used to study steroid hormones and other exogenous agents such as PPCPs in water. These techniques have increased the sensitivity and accuracy of EDC and PPCP analysis, allowing ultra-trace levels of a wide variety of contaminants to be identified and quantified.

Because EDCs and PPCPs represent an extremely broad spectrum of compounds, developing a single all-encompassing technique for their analysis can be quite challenging. EDCs and PPCPs vary widely in their physico-chemical properties (e.g., polarity, molecular weight, $pK_a$, water solubility, etc.) making analysis by traditional analytical techniques difficult at times. Additionally, the concentration of EDCs and PPCPs in the environment can be quite low, typically sub-μg/L, which further increases the complexity of analysis by necessitating extraction and concentration steps. In general, however, a method for the analysis of target EDCs and PPCPs uses the following sequence: extraction, to separate them from the matrix and remove interferences; concentration, to increase the concentration of the target analyte to a sufficient level to allow for quantification; chromatography, to separate target analytes from background interferences and from each other; and quantification, to determine the concentration of the target analyte in solution.

The purpose of this chapter is to familiarize the reader with analytical techniques commonly used for the detection and quantification of EDCs and PPCPs in water. It should be noted, however, that an exhaustive review of every analytical technique available for the analysis of EDCs and PPCPs is beyond the scope of this chapter. For that purpose, many excellent reviews have already been written (Lopez de Alda and Barcelo 2001; Ternes 2001; Richardson 2002; Richardson 2004; Koester and Moulik 2005; Petrovic, Hernando et al. 2005; Richardson and Ternes 2005; Richardson 2006). Rather, this chapter focuses on the fundamentals of techniques that are frequently encountered in this field. In addition, this chapter provides an update on methods recently developed for the compound list introduced in Chapter 1. This update can be found in Table A.1. Table A.1 should be viewed as a guide to aid in the development of a strategy to analyze target EDCs and/or PPCPs and used as a starting point for a more in-depth literature review.

## SAMPLE COLLECTION/PRESERVATION

### Sample Collection

Due to the common use of pharmaceuticals, the ubiquitous nature of personal care products and the extremely low concentrations expected in the environment, great care must be taken to avoid contamination of samples by samplers, sampling equipment and laboratory personnel. Communication between the laboratory and those collecting samples regarding the list of target compounds is important to help prevent contamination by identifying and eliminating possible undesired sources of target analytes. In general, nitrile gloves should be worn at all times during the collection and handling of samples to prevent contamination with personal care products applied directly to the skin, such as triclosan, DEET, and various sunscreen agents. Similarly, smoking and handling or ingesting pharmaceuticals or caffeinated beverages should be avoided shortly before and during sampling. To monitor background levels of the EDCs and PPCPs, travel blanks should be used.

Samples should be collected in amber, glass bottles to prevent analyte loss due to photodegradation, contamination with various plasticizers, and adsorption to the walls of plastic sampling bottles. It is also recommended that bottles be silanized (Ahrer, Scherwenk et al. 2001) to prevent the possibility of analyte interaction with silanol groups on the glass surface. Furthermore, sample bottles should be cleaned thoroughly with applicable solvents (e.g. water, methanol, acetone, dichloromethane, hexane) to ensure the cleanliness of the bottles prior to sampling. In addition, sampling equipment should be composed of materials such as stainless steel or Teflon that will not leach target EDCs/PPCPs and should be cleaned with solvent between sample locations to prevent cross contamination.

### Preservation

At the time of collection, samples are generally preserved to reduce microbial degradation, hydrolysis, and adsorption of the target analytes. This is typically accomplished through lowered temperature and/or chemical preservatives. However, it must be noted that preservative selection depends greatly on the EDCs/PPCPs selected for analysis. For example, some EDCs/PPCPs may have an adverse reaction with a chemical preservative; therefore, it is advisable to test the target analytes with the selected preservative in a controlled experiment before using it in the field.

After samples are collected, they should be cooled to prevent analyte degradation. This usually involves placing the sample in a cooler with ice while other samples are taken and when they are transported back to the laboratory. If samples are to be transported over long distances, it is recommended that blue ice be used to maintain sample temperature during shipment. Once samples have been received by the laboratory, they may then be stored at 4°C or less until analysis.

Chemical preservatives are often used to prevent analyte degradation. Several have been commonly used including reducing the sample pH to 2 or below using either sulfuric (Vanderford, Pearson et al. 2003) or hydrochloric acid (Hernando, Heath et al. 2006), adding formaldehyde to a final concentration between 1 – 4 percent (Baronti, Curini et al. 2000; Ferguson, Iden et al. 2001), or adding sodium azide to a final concentration of 1 g/L (Vanderford and Snyder 2006). As stated above, care must be taken to ensure the preservative of choice does not interfere with the target analytes. For example, formaldehyde has been extensively used to preserve samples for steroid analysis; however, Vanderford et al. (2003) reported that using formaldehyde for the preservation of pharmaceuticals resulted in significant changes in their concentrations over time.

Although sample preservatives can reduce the amount of degradation that occurs before the samples are extracted and/or analyzed, it is recommended that samples be extracted as soon as possible after they are received. Long wait times can result in sample adsorption to both the bottle and suspended/dissolved organic matter in the sample. Typical holding times range from 24 to 7 days (Miao and Metcalfe 2003; Hernando, Heath et al. 2006; Moldovan 2006; Vieno, Tuhkanen et al. 2006).

### Residual Oxidant Quenching

When collecting and analyzing samples from drinking water-treatment or potable reuse facilities, it is important to know whether residual oxidants, such as free chlorine, may be present. If the residual oxidants are not quenched, target analytes will be exposed to chlorine until the samples are extracted. Analytes that are susceptible to oxidation will be further degraded due to this increase in contact time, leading to misinterpretation of analyte concentrations present at the time of sampling. Therefore, residual oxidants must be quenched using suitable chemical agents.

Commonly used quenching agents include sodium thiosulfate (Acero, Rodriguez et al. 2005), sodium sulfite (Ho, Onstad et al. 2006), ammonium chloride (Pepich, Domino et al. 2004), and ascorbic acid (Ye, Weinberg et al. 2007). However, researchers have found that some quenching agents react adversely with various target analytes (Trenholm, Vanderford et al. 2006; Ye, Weinberg et al. 2007). Therefore, it is essential that, like the preservation agents, tests are performed to ensure the selected quenching agent does not interfere with the target analytes. Furthermore, adverse reactions between preservatives and quenching agents should be explored, especially with regard to safety.

## EXTRACTION/CONCENTRATION

Detection of emerging contaminants in water is often desired at trace levels (sub-μg/L) since some compounds have been found to have aquatic impacts at these concentrations (Routledge, Sheahan et al. 1998; van Aerle, Pounds et al. 2002; Segner, Navas et al. 2003). However, most analytical instruments are not able to directly detect compounds at these levels. Therefore, an extraction step is used to concentrate the target compounds to a detectable level. Extraction can vary in degree of selectivity, speed, and convenience and depends not only on the approach and conditions used but on the geometric configurations of the extraction phase. Conventional extraction techniques such as liquid-liquid (Yook, Hong et al. 1994; Holm, Rügge et al. 1995; Romero, Ventura et al. 2002), Soxhlet (Bennie, Sullivan et al. 1997; Pryor, Hay et al. 2002; Fatoki and Awofolu 2003), and steam distillation (Kubeck and Naylor 1990; Fowler, Haviland et al. 1998; Snyder, Keith et al. 2001) have been used to extract organic compounds from water in the past; however, more recent techniques have been applied to address the need for reduction of solvent use, automation, and miniaturization and further improve the selectivity necessary to separate trace quantities of EDCs and PPCPs from challenging matrices.

Solid phase microextraction (SPME), first developed by Pawliszyn and co-workers (Arthur and Pawliszyn 1990; Arthur, Killam et al. 1992), uses a glass fiber coated with a desired stationary phase. Target analytes adsorb to the fiber after it has been placed directly into the sample or in the headspace of a heated sample vial. The fiber can then be directly inserted into a GC for desorption and analysis. The technique has been applied to analyze various EDCs/PPCPs (Cai, Jiang et al. 2004; Lamas, Salgado-Petinal et al. 2004; Rodríguez, Carpinteiro et al. 2004). Because of its reliance on equilibria, this technique can suffer from reproducibility problems and it can be difficult to use in the presence of high amounts of organic matter.

Semipermeable membrane devices (SPMD) have also been used recently for EDC/PPCP analysis (Koester and Moulik 2005). SPMDs are passive sampling devices that are typically composed of low density polyethylene tubing containing a thin film of triolein. The device is deployed into a body of water and nonpolar organic compounds adsorb to the film, thereby concentrating them. The film can then be extracted with organic solvent and the extract analyzed for trace contaminants. Similarly, a passive sampling device for more polar analytes called the polar organic chemical integrative sampler (POCIS) has been developed (Alvarez, Petty et al. 2004) and applied to the extraction of various polar EDCs/PPCPs (Alvarez, Stackelberg et al. 2005; Vermeirssen, Koerner et al. 2005). The POCIS consists of a sorbent contained between two microporous polyethersulfone membranes. The membranes allow water and dissolved chemicals to pass through to the sorbent where the chemicals are trapped. The chemicals can then be extracted and analyzed.

Solid-phase extraction (SPE), however, is by far the most common technique employed for sample enrichment. An SPE cartridge or disk is packed with a sorbent that has an affinity for the compounds of interest. The sorbent binds the target compounds as an aqueous sample is passed through the cartridge or disk. The sorbent can then be washed with an aqueous solution and/or organic solvents to remove unwanted interferences that have also been retained by the sorbent. Subsequently, the sorbent is dried with nitrogen or air. Target compounds are then eluted from the sorbent using organic solvents. The solvents are chosen so that the target analytes have a higher affinity for the solvent than for the SPE sorbent. The resulting extract is generally concentrated by evaporation to a volume of one milliliter or less. Depending on the quality of the extract, interferences may be removed using various cleanup procedures. Common cleanup techniques use gel permeation/size exclusion chromatography to remove interferences by size and/or silica gel/alumina columns that separate interferences based on polarity (Khim, Kannan et al. 1999; Khim, Villeneuve et al. 1999; Khim, Villeneuve et al. 1999; Snyder, Keith et al. 1999; Snyder, Kelly et al. 2001; Snyder, Villeneuve et al. 2001).

Many different SPE sorbents have been manufactured to extract organic pollutants with a variety of physico-chemical properties. The most commonly used sorbents for environmental analysis include standard C18, Oasis HLB, Isolut ENV+, Lichrolut EN, and Strata-X. Due to the combination of lipophilic divinylbenzene and hydrophilic N-vinylpyrrolidone polymers, the Oasis HLB sorbent is preferred because it is able to extract acidic, neutral, and basic compounds at a wide range of pHs, including neutral pH (Gros, Petrovic et al. 2006). It has been used to extract various multiple classes of EDCs and pharmaceuticals from an array of different matrices (Vanderford, Pearson et al. 2003; Diaz-Cruz and Barcelo 2005; Gros, Petrovic et al. 2006; Nikolai, McClure et al. 2006; Vanderford and Snyder 2006).

## QUANTIFICATION

### Chromatography

Since the use of the term "chromatography" was published over one hundred years ago by Mikhail Tswett (1906), the field of chromatography has grown tremendously. The separation of various chemicals to aid in their analysis has been crucial to the identification, detection and quantification of environmental pollutants. Although there are many variations on the theme of chromatography, such as paper chromatography, thin layer chromatography, ion chromatography, size-exclusion chromatography, immunoaffinity chromatography, gel permeation chromatography, and capillary electrophoresis, by far the most common separation techniques in use today for the analysis of environmental contaminants are gas and liquid chromatography.

#### *Gas Chromatography (GC)*

GC has been used for many years for chemical analysis. Traditionally, GC works by injecting a liquid sample into a heated injection chamber, in which all components of the sample are quickly vaporized and swept by a carrier gas, usually helium, into a long, thin silica chromatography column (i.e., 30 m long by 0.25 μm diameter) internally coated with a stationary phase. The chromatography column is positioned in an electronically controlled oven, which can be programmed to either maintain a constant temperature (isocratic) or gradually increase the temperature over time (gradient). The individual compounds will separate as they flow through the column based on their boiling point, affinity for the stationary phase, and temperature profile of the GC oven. A detector then monitors the compounds as they leave the column. There are many types of detectors including flame-ionization, electron capture, nitrogen-phosphorus, flame-photometric, and mass spectrometric (MS). Each detector has its own advantages and disadvantages with regard to sensitivity, selectivity and cost.

In general, GC is amenable to non-polar, non-thermally labile, volatile compounds due to the temperatures involved and the need for the compounds to be in the gaseous state. This has made GC the standard instrument for measuring non-polar compounds such as PCBs, DDT, lindane and PAHs and volatile compounds such as benzene, toluene, acetone, and trihalomethanes. To analyze more polar compounds, such as many EDCs and PPCPs, derivatization of the compounds is often necessary. This renders the compounds sufficiently volatile so that they can be eluted from the column at reasonable temperatures without thermal degradation or molecular rearrangement. Derivatization can also increase thermal stability and decrease polarity. Often this involves silylation, alkylation or acylation of the compound.

#### *Liquid Chromatography*

Typical LC systems consist of a pump, sample injector, analytical column, and a detector. In general, analytes in the liquid phase are injected into an aqueous and/or organic solvent-containing mobile phase that is being pumped through an analytical column. The stainless steel column has been packed with a solid stationary phase that will separate compounds based on their physico-chemical properties and differences in partitioning behavior between the liquid mobile phase and the solid stationary phase. Although there are several types of liquid chromatography, reverse-phase liquid chromatography (RPLC) is by far the most widespread. RPLC involves the use of a non-polar stationary phase and a polar liquid phase. The LC mobile phase composition can either be held at a constant ratio of aqueous/organic content (isocratic elution) or the mobile phase can be gradually ramped from high aqueous/low organic content to low aqueous/high organic content to elute the analytes from the column in a more efficient manner (gradient elution). Common organic solvents used are methanol and acetonitrile and aqueous mobile phases generally include a buffer, salt and/or ion-pairing reagent to assist in the separation of the target analytes. After the analytes elute from the column, they are detected by various methods including refractive index, UV absorbance, fluorescence, and mass spectrometry.

LC is a suitable separation technique for many types of compounds including those with a wide range of polarities, molecular weights, functional groups, and acidities. For this reason, LC has rapidly become the separation technique of choice for analyzing EDCs and PPCPs in aqueous samples.

## Detection Techniques

### *Non-Selective Detectors*

Non-selective detection has been used for the analysis of pharmaceuticals and EDCs. Non-selective detectors rely on the detection of physical or chemical properties that are not unique to a particular compound, but may be representative of a large group or class of compounds (e.g. UV absorbance at 254 nm, electrical conductivity, fluorescence, flame ionization, etc.). They are generally used to simply detect a target analyte rather than to discriminate between analytes. To differentiate between analytes, it is imperative to couple non-selective detectors with some form of chromatographic separation. EDCs such as human steroid hormones have been analyzed in the past using non-selective detection techniques such as fluorescence (Ying, Kookana et al. 2002), UV and electrochemical detection (Penalver, Pocurull et al. 2002), and diode array detection (de Alda and Barcelo 2001). Likewise, pharmaceuticals have also been analyzed by UV (Vervoort, Ruyter et al. 2001) and diode array/fluorescence detectors (Santos, Aparicio et al. 2005).

However, due to the low concentrations of EDCs and pharmaceuticals in the environment, it is often difficult for non-selective detectors to achieve environmentally relevant detection limits. In addition, the complexity of many environmental samples leads to high background levels and higher signal to noise ratios because of the inability of non-selective detectors to differentiate between target analytes and compounds in the sample with similar physico-chemical properties. As a result, mass spectrometry has become the standard for the analysis of EDCs and pharmaceuticals in the environment.

### *Mass Spectrometry*

Mass spectrometry has been used to identify unknown compounds, quantify known compounds, and elucidate the structures and chemical properties of molecules since its invention in the early part of the 20$^{th}$ century. Because of its selectivity and sensitivity, it has also been the preferred method of analysis for environmental pollutants since the 1970s. More recently, mass spectrometry has been used to study unregulated emerging organic contaminants like EDCs and PPCPs in the environment at trace levels (Richardson 2000; Richardson 2002; Richardson 2004;

Richardson 2006). The most common mass spectrometers used for environmental applications are discussed below.

***Quadrupole-Based.*** Quadrupole mass spectrometers increased in popularity with the advent of relatively inexpensive gas chromatography-mass spectrometry (GC-MS) systems. A quadrupole mass spectrometer consists of four parallel rods equally spaced around a central axis. These mass spectrometers separate ions based on an oscillating electric field created by radio frequencies and direct current voltages. At a particular radio frequency and voltage, only ions of a single mass to charge ratio (m/z) can pass through the quadrupoles to the detector. Quadrupole based instruments are considered rugged and have relatively large linear calibration ranges; however, they are generally low-resolution instruments. Single quadrupole instruments have excellent sensitivity but limited selectivity since co-eluting compounds are difficult to differentiate. Often single quadrupole systems utilize in-source collision-induced dissociation (CID) to fragment the analyte resulting in increased selectivity.

Triple-quadrupole (MS/MS) mass spectrometers are more costly than single quadrupole systems, but offer many advantages for environmental analysis. The triple-quadrupole system consists of three main quadrupoles: an initial scanning quadrupole (Q1), followed by an RF-only collision chamber (Q2), and a final scanning quadrupole (Q3). This allows the mass spectrometer to be used for a variety of different scan modes, including precursor, product and neutral loss scans. In terms of environmental analysis of target analytes, the most often used technique is selected reaction monitoring (SRM). This mode allows a precursor ion to be selected in Q1, fragmented in Q2, and a single product ion (charged precursor ion fragment) to be selected in Q3. The triple quadrupole mass spectrometer exhibits excellent sensitivity and selectivity, but is not of great use for the identification of unknowns. Due to their sensitivity and selectivity, triple quadrupole mass spectrometers are rapidly becoming the technique of choice for the targeted analysis of environmental pollutants.

A quadrupole ion trap (QIT) mass spectrometer is a variation of the quadrupole mass spectrometer in which ions are "trapped" in a 3-dimensional quadrupole field. Ions with a range of *m/z* ratios are stored in a field that is created with a fixed-frequency RF applied to a cylinder ring electrode. End cap electrodes are positioned on either side of this ring electrode. As the amplitude of this fixed-frequency RF is increased, ions are "ejected" from the trap through the end caps and are recorded at the detector. Ion trap mass spectrometers can also be used for $MS^n$ measurements by constructing waveforms to isolate an ion, induce its fragmentation, isolate one of the products, induce its fragmentation, etc. This function can be used to generate precursor/product ion scans similar to a triple quadrupole mass spectrometer. Ion trap mass spectrometers have good sensitivity, especially related to full scans of the mass spectrometer, but are low-resolution instruments. They are generally less expensive than time of flight (TOF) and triple-quadrupole instruments, and are similar in price to single quadrupole instruments. Ion traps traditionally suffer from matrix issues, as the ion trap can only hold a finite amount of ions. In very difficult matrices, sensitivity can rapidly decrease. The key advantages are low cost and $MS^n$ spectrometric measurements, which can be used for structural elucidation and SRM experiments.

Two-dimensional linear ion traps have also been developed. In these ion traps, ions can be stored along the entire length of the quadrupole, resulting in greater storage capacity. In addition, hybrid triple quadrupole instruments that combine standard quadrupoles with linear ion traps have been used more recently for environmental analysis (Nikolai, McClure et al. 2006; Zhao, Boyd et al. 2006).

***Time of Flight (TOF) and Quadrupole-TOF (QTOF).*** Time of flight (TOF) mass spectrometers were first developed in the 1950s and were almost entirely replaced by double-focusing mass spectrometers by the 1980s. During the 1990s, rapid data acquisition techniques and improved reflective ion mirrors (reflectrons) resulted in a resurgence of TOF instruments. A TOF instrument uses a flight tube in which ions are accelerated using an electric field. This acceleration results in the separation of the ions based on mass because the $m/z$ ratio is proportional to the travel time in the flight tube. A longer flight tube results in greater ion separation and, therefore, greater mass resolution. Since very long flight tubes are not practical, ion mirrors are used to increase flight times and result in higher mass resolution measurements without significantly increasing the size of the instrument. Because of their high resolution and mass accuracy capabilities, TOF instruments can be used for identification of unknown compounds by providing a molecular formula and confirming or denying a suggested structure (Reemtsma 2003). In addition, the ability of the TOF to provide full scan spectra with relatively high sensitivity makes this instrument an interesting choice for qualitative analysis. Furthermore, TOF instruments are well suited to screening purposes. Hybrid quadrupole-TOF (QTOF) systems also show great promise, as high-resolution MS/MS measurements are possible. These measurements provide more structural information and add selectivity by allowing the selection of a parent ion (Reemtsma 2003). Moreover, QTOF systems are robust for environmental applications as they can handle difficult matrices while maintaining high resolution for confirmation and unknown identification.

### *Hyphenated Techniques*

The combination of techniques of chromatographic separation and detection, often called hyphenated techniques, is the standard for detection of environmental contaminants. The two most powerful and most widely used hyphenated techniques are gas chromatography-mass spectrometry (GC-MS) and liquid chromatography-mass spectrometry (LC-MS) due to their robustness, sensitivity and selectivity. Both of these techniques are discussed below.

***GC-MS.*** Because MS separates and detects an analyte based on mass to charge ratio ($m/z$), the compound must be charged (ionized) before it enters the MS. In GC-MS, there are two common ionization techniques: electron ionization (EI) and chemical ionization (CI). In EI, the GC column eluent is directed through a beam of electrons created by a filament that produces electrons having an energy of 70 eV. The electron interacts with analyte molecules in the gas phase, resulting in the loss or gain of an electron by the analyte creating a positively or negatively charged molecule, respectively. This type of ionization results in molecular fragmentation, which is related to the structural properties of the compound. Thus, each compound has a unique MS "fingerprint" that allows for the identification of the compound based on its fragmentation pattern. However, because the analyte is fragmented before it reaches the detector, a loss in sensitivity results. In CI, a gas (typically methane or ammonia) is first ionized and then interacts with the analyte, resulting in the gain (positive ionization) or loss (negative ionization) of a proton. CI is considered a "soft" ionization process because it generally results in less fragmentation than EI. Therefore, even though CI can increase analyte sensitivity, it often provides less structural information. Once the analyte is ionized, it enters the MS and is selected and detected, as discussed above.

***LC-MS.*** Unlike GC-MS, the separation of analytes in LC-MS occurs in the liquid phase. Thus, analytes reach the MS as dissolved solutes in the liquid phase rather than in the gaseous phase. Because ionization in the GC-MS occurs via the use of a filament at well below atmospheric pressure, a similar type of ionization for LC-MS is not applicable. Therefore, ionization in LC-MS occurs in an entirely different manner. There are three common ionization techniques: electrospray ionization (ESI), atmospheric chemical ionization (APCI) and atmospheric pressure photoionization (APPI).

In ESI, the eluent from the LC column containing the target analytes is directed into a capillary tube, onto which a large voltage is applied. This induces a charge in the liquid and, as it exits the capillary tube, a fine mist of charged droplets forms. Typically, with the aid of a stream of nitrogen gas and heaters in the source, the droplets reduce in size. When the charge repulsion on the surface of the droplet exceeds the surface tension, the droplet disintegrates into smaller droplets. The process is repeated until the charged analyte escapes the droplet (ion desorption) or the solvent has evaporated to leave the charged analyte in the gas phase. This process can form positive and negative ions when positive and negative voltages, respectively, are applied to the capillary. ESI is by far the most utilized form of LC-MS ionization.

In APCI, the eluent from the LC column is nebulized and directed into a heated chamber, in which the solvent and analytes are almost completely vaporized. As the eluent exits the chamber, the solvent becomes a reagent gas when it is charged by a corona discharge needle. The reagent gas then transfers the charge to the target analyte. APPI is very similar to APCI, except a photon-emitting krypton lamp is used to directly ionize the target analyte. These techniques are generally only used when compounds are not found to be amenable to ionization by ESI because they tend to be less sensitive.

### *GC-MS versus LC-MS*

In general, GC-MS is more amenable to volatile, thermally stable, less polar compounds. Therefore, it has been the method of choice in the past for legacy pollutants such as PCBs, PAHs, DDT, etc. However, newly discovered contaminants, such as pharmaceuticals and personal care products, are often polar and non-volatile. This has, in part, led to the surge of popularity for the use of LC-MS to monitor emerging contaminants. For many of these compounds to be monitored by GC-MS, they need to be derivatized prior to analysis. This process can be painstaking, labor-intensive and ineffective. On the other hand, LC-MS has the ability to analyze a wide variety of compounds without the need for derivatization.

However, the ESI process that is most frequently used during LC-MS analysis can be susceptible to matrix effects. During the ionization process, non-target analytes at a greater concentration and/or that have a higher affinity for becoming charged will exhaust the available charge and leave target analytes uncharged. If uncorrected, matrix effects may result in improper data interpretation because the effects can vary substantially between matrices and lead to the reporting of artificially lessened concentrations. Researchers have tried to minimize matrix effects using various extraction, cleanup, and elution techniques (Reemtsma 2003; Quintana, Rodil et al. 2004; Kloepfer, Quintana et al. 2005) or compensate for them using different calibration techniques (Ferguson, Iden et al. 2000; Lindsey, Meyer et al. 2001). However, most become problematic when applied to the simultaneous analysis of a broad range of compounds that encompass many different classes and structures in matrices having varying degrees of suppression and enhancement (Vanderford and Snyder 2006). Perhaps the most promising method to date is the use of isotope dilution to correct for matrix effects. In this method, isotopically labeled versions of each analyte are added to all samples prior to SPE. Results obtained for unlabeled target analytes are corrected for matrix effects based on the recovery of the labeled version. This method has shown promise when applied to the analysis of a varied group of pharmaceuticals, personal care products, pesticides and EDCs (Vanderford and Snyder 2006).

### Quality Assurance/Quality Control (QA/QC)

For any environmental analysis method, an extensive plan for QA/QC is advisable. In the case of emerging contaminants such as EDCs and pharmaceuticals, it is even more critical. However, a detailed plan for QA/QC is beyond the scope of this document. Rather, listed below are a series of suggestions that specifically relate to the analysis of EDCs and pharmaceuticals in water.

Because the concentrations of these compounds are often less than 100 ng/L, extensive care must be taken to prevent accidental contamination by sampling and laboratory personnel. As discussed above, the common use of many target pharmaceuticals and the ubiquitous nature of personal care products make contamination of samples and equipment a common problem. In a report following their national reconnaissance of U.S. streams, the USGS reported a significant number of episodes of blank contamination (Barnes, Kolpin et al. 2002), suggesting that even the most experienced laboratories encounter blank-related issues. For these reasons, blanks should be an integral portion of every analysis to ensure that reported concentrations are present in the environment and, if contamination is suspected, to help determine a source of contamination. Therefore, frequent travel blanks, extraction blanks, and instrument blanks are recommended.

Due to the common occurrence of detrimental effects on the concentration of target analytes (e.g. matrix effects, sample preparation losses, improper internal standard selection, etc.), matrix spikes are suggested to monitor the accuracy of the method. In this manner, deficiencies in the method can be revealed and corrected by techniques such as isotope dilution (Vanderford and Snyder 2006).

## CONCLUSIONS

In conclusion, a variety of methods exist for the detection and quantification of EDCs and PPCPs. Due to their physico-chemical diversity, EDCs and PPCPs often require multiple methods and techniques for their analysis. EDCs and PPCPs are typically extracted using SPE and analyzed by GC-MS or LC-MS, depending on the characteristics of the target compounds. EDCs and PPCPs tend to be polar and non-volatile and, therefore, are amenable to analysis by GC-MS with derivatization or LC-MS. Quality assurance/control and sampling handling procedures, including preservation, hold times, and quenching, are very important to maintain the accuracy and precision of analytical results for these trace contaminants.

## REFERENCES


Abuin, S., R. Codony, R. Compano, M. Granados, and M.D. Prat. 2006. Analysis of macrolide antibiotics in river water by solid-phase extraction and liquid chromatography-mass spectrometry. *Journal of Chromatography A*, 1114:73-81.

Acero, J.L., E. Rodriguez, and J. Meriluoto. 2005. Kinetics of reactions between chlorine and the cyanobacterial toxins microcystins. *Water Research*, 39:1628-1638.

Ahrer, W., E. Scherwenk, and W. Buchberger. 2001. Determination of drug residues in water by the combination of liquid chromatography or capillary electrophoresis with electrospray mass spectrometry. *Journal of Chromatography A*, 910:69-78.

Alvarez, D.A., J.D. Petty, J.N. Huckins, T.L. Jones-Lepp, D.T. Getting, J.P. Goddard, and S.E. Manahan. 2004. Development of a passive, in situ, integrative sampler for hydrophilic organic contaminants in aquatic environments. *Environmental Toxicology and Chemistry*, 23(7):1640-1648.

Alvarez, D.A., P.E. Stackelberg, J.D. Petty, J.N. Huckins, E.T. Furlong, S.D. Zaugg, and M.T. Meyer. 2005. Comparison of a novel passive sampler to standard water-column sampling for organic contaminants associated with wastewater effluents entering a New Jersey stream. *Chemosphere*, 61:610-622.

Andresen, J.A., A. Grundmann, and K. Bester. 2004. Organophosphorus flame retardants and plasticizers in surface waters. *Science Of The Total Environment*, 332:155-166.

Arthur, C.L., and J. Pawliszyn. 1990. Solid phase microextraction with thermal desorption using fused silica optical fibers. *Analytical Chemistry*, 62:2145-2148.

Arthur, C.L., L.M. Killam, M. Lim, D.W. Potter, and J. Pawliszyn. 1992. Analysis of Substituted Benzene Compounds in Groundwater Using Solid-Phase Microextraction. *Environmental Science & Technology*, 26(5).

Balakrishnan, V.K., K.A. Terry, and J. Toito. 2006. Determination of sulfonamide antibiotics in wastewater: A comparison of solid phase microextraction and solid phase extraction methods. *Journal of Chromatography A*, 1131(1-2):1-10.

Barnes, K.L., D.W. Kolpin, M.T. Meyer, E.M. Thurman, E.T. Furlong, S.D. Zaugg, and L.B. Barber. 2002. Water-Quality Data for Pharmaceuticals, Hormones, and Other Organic Wastewater Contaminants in U.S. Streams, 1999-2000 Edited by Iowa City, Iowa: USGS.

Baronti, C., R. Curini, G. D'ascenzo, A. Di Corcia, A. Gentili, and R. Samperi. 2000. Monitoring natural and synthetic estrogens at activated sludge sewage treatment plants and in a receiving river water. *Environmental Science & Technology*, 34(24):5059-5066.

Batt, A.L., and D.S. Aga. 2005. Simultaneous analysis of multiple classes of antibiotics by ion trap LC/MS/MS for assessing surface water and groundwater contamination. *Analytical Chemistry*, 77(9):2940-2947.

Bennie, D.T., C.A. Sullivan, H.B. Lee, T.E. Peart, and R.J. Maguire. 1997. Occurence of alkylphenols and alkylphenol mono- and diethoxylates in natural waters of the Laurentian Great Lakes basin and the upper St. Lawrence River. *The Science of the Total Environment*, 193:263-275.

Benotti, M., and B.J. Brownawell. 2007. Distributions of Pharmaceuticals in an Urban Estuary during both Dry- and Wet-Weather Conditions. *Environmental Science & Technology*, 41:5795-5802.

Buerge, I.J., H.-R. Buser, M.D. Mueller, and T. Poiger. 2003. Behavior of the polycyclic musks HHCB and AHTN in lakes, two potential anthropogenic markers for domestic wastewater in surface waters. *Environmental Science & Technology*, 37:5636-5644.

Cahill, J.D., E.T. Furlong, M.R. Burkhardt, D. Kolpin, and L.G. Anderson. 2004. Determination of pharmaceutical compounds in surface- and ground-water samples by solid-phase extraction and high-performance liquid chromatography-electrospray ionization mass spectrometry. *Journal of Chromatography A*, 1041(1-2):171-180.

Cai, Y.Q., G.B. Jiang, J.F. Liu, X. Liang, Z.W. Yao, J.M. Liu, J.Y. Liu, and Q.X. Zhou. 2004. Solid-phase microextraction coupled with high performance liquid chromatography-fluorimetric detection for the determination of bisphenol A, 4-n-nonylphenol, and 4-tert-octylphenol in environmental water samples. *Analytical Letters*, 37(4):739-753.

Castiglioni, S., R. Bagnati, D. Calamari, R. Fanelli, and E. Zuccato. 2005. A multiresidue analytical method using solid-phase extraction and high-pressure liquid chromatography tandem mass spectrometry to measure pharmaceuticals of different therapeutic classes in urban wastewaters. *Journal of Chromatography A*, 1092:206-215.

De Alda, M.J.L., and D. Barcelo. 2001. Determination of steroid sex hormones and related synthetic compounds considered as endocrine disrupters in water by fully automated on-line solid-phase extraction-liquid chromatography-diode array detection. *Journal of Chromatography A*, 911:203-210.

Desbrow, C., E.J. Routledge, G.C. Brighty, J.P. Sumpter, and M. Waldock. 1998. Identification of Estrogenic Chemicals in STW Eflleunt. 1. Chemical Fractionation and *in vitro* Biological Screening. *Environmental Science & Technology*, 32(11):1549-1558.

Diaz-Cruz, M.S., and D. Barcelo. 2005. LC-MS2 trace analysis of antimicrobials in water, sediment and soil. *Trac-Trends in Analytical Chemistry*, 24(7):645-657.

Einsle, T., H. Paschke, K. Bruns, S. Schrader, P. Popp, and M. Moeder. 2006. Membrane-assisted liquid-liquid extraction coupled with gas chromatography-mass spectrometry for determination of selected polycyclic musk compounds and drugs in water samples. *Journal of Chromatography A*, 1124:196-204.

Fatoki, O.S., and R.O. Awofolu. 2003. Methods for selective determination of persistent organochlorine pesticide residues in water and sediments by capillary gas chromatography and electron-capture detection. *Journal of Chromatography A*, 983(1-2):225-236.

Ferguson, P.L., C.R. Iden, and B.J. Brownawell. 2000. Analysis of alkylphenol ethoxylate metabolites in the aquatic environment using liquid chromatography-electrospray mass spectrometry. *Analytical Chemistry*, 72(18):4322-4330.

Ferguson, P.L., C.R. Iden, A.E. Mcelroy, and B.J. Brownawell. 2001. Determineation of steroid estrogens in wastewater by immunoaffinity extraction coupled with HPLC-electrospray-MS. *Analytical Chemistry*, 73:3890-3895.

Fono, L.J., and D.L. Sedlak. 2005. Use of the Chiral Pharmaceutical Propranolol to Identify Sewage Discharges into Surface Waters. *Environmental Science & Technology*, 39:9244-9252.

Fono, L.J., E.P. Kolodziej, and D.L. Sedlak. 2006. Attenuation of wastewater-derived contaminants in an effluent-dominated river. *Environmental Science & Technology*, 40:7257-7262.

Fowler, B.R., L. Haviland, and S. Kennedy. 1998. Analysis of alkylphenols in Fraser river water and suspended sediments. *poster*.

Gatidou, G., N.S. Thomaidis, A.S. Stasinakis, and T.D. Lekkas. 2007. Simultaneous determination of the endocrine disrupting compounds nonylphenol, nonylphenol ethoxylates, triclosan and bisphenol A in wastewater and sewage sludge by gas chromatography-mass spectrometry. *Journal of Chromatography A*, 1138(1-2):32-41.

Gobel, A., C.S. Mcardell, M.J.F. Suter, and W. Giger. 2004. Trace determination of macrolide and sulfonamide antimicrobials, a human sulfonamide metabolite, and trimethoprim in wastewater using liquid chromatography coupled to electrospray tandem mass spectrometry. *Analytical Chemistry*, 76:4756-4764.

Gomez, M.J., M. Petrovic, A.R. Fernandez-Alba, and D. Barcelo. 2006. Determination of pharmaceuticals of various therapeutic classes by solid-phase extraction and liquid chromatography-tandem mass spectrometry analysis in hospital effluent wastewaters. *Journal of Chromatography A*, 1114:224-233.

Gros, M., M. Petrovic, and D. Barcelo. 2006a. Multi-residue analytical methods using LC-tandem MS for the determination of pharmaceuticals in environmental and wastewater samples: a review. *Analytical and Bioanalytical Chemistry*, 386(4):941-952.

Gros, M., M. Petrovic, and D. Barcelo. 2006b. Development of a multi-residue analytical methodology based on liquid chromatography-tandem mass spectrometry (LC-MS/MS) for screening and trace level determination of pharmaceuticals in surface and wastewaters. *Talanta*, 70(4):678-690.

Halden, R., and D. Paull. 2004. Analysis of triclocarban in aquatic samples by liquid chromatography electrospray ionization mass spectrometry. *Environmental Science & Technology*, 38(18):4849-4855.

Hao, C., L. Lissemore, B. Nguyen, S. Kleywegt, P. Yang, and K.R. Solomon. 2006. Determination of pharmaceuticals in environmental waters by liquid chromatography/electrospray ionization/tandem mass spectrometry. *Analytical and Bioanalytical Chemistry*, 384:505-513.

Hernando, M.D., E. Heath, M. Petrovic, and D. Barcelo. 2006. Trace-level determination of pharmaceutical residues by LC-MS/MS in natural and treated waters. A pilot-survey study. *Analytical and Bioanalytical Chemistry*, 385:985-991.

Hignite, C., and D.L. Azarnoff. 1977. Drugs and drug metabolites as environmental contaminants: chlorophenoxyisobutyrate and salicylic acid in sewage water effluent. *Life Sciences*, 20(2):337-341.

Ho, L., G. Onstad, U. Von Gunten, S. Rinck-Pfeiffer, K. Craig, and G. Newcombe. 2006. Differences in the chlorine reactivity of four microcystin analogues. *Water Research*, 40:1200-1209.

Holm, J.V., K. Rügge, P.J. Bjerg, and T.H. Christensen. 1995. Occurence and distribution of pharmaceutical organic compounds in the groundwater downgradient of a landfill (Grindsted, Denmark). *Environmental Science & Technology*, 29(5):1415-1420.

Huerta-Fontela, M., M.T. Galceran, and F. Ventura. 2007. Ultraperformance liquid chromatography-tandem mass spectrometry analysis of stimulatory drugs of abuse in wastewater and surface waters. *Analytical Chemistry*, 79:3821-3829.

Hummel, D., D. Loeffler, G. Fink, and T.A. Ternes. 2006. Simultaneous Determination of Psychoactive Drugs and Their Metabolites in Aqueous Matrices by Liquid Chromatography Mass Spectrometry. *Environmental Science & Technology*, 40:7321-7328.

Isobe, T., H. Shiraishi, M. Yasuda, A. Shinoda, H. Suzuki, and M. Morita. 2003. Determination of estrogens and their conjugates in water using solid-phase extraction followed by liquid chromatography-tandem mass spectrometry. *Journal of Chromatography A*, 984(2):195-202.

Khim, J.S., K. Kannan, D.L. Villeneuve, C.H. Koh, and J.P. Giesy. 1999a. Characterization and distribution of trace organic contaminants in sediment from Masan Bay, Korea. 1. Instrumental analyses. *Environmental Science & Technology*, 33(23):4199-4205.

Khim, J.S., D.L. Villeneuve, K. Kannan, and J.P. Giesy. 1999b. Characterization and distribution of trace organic contaminants in sediment from Masan Bay, Korea. 2. *In vitro* gene expression assays. *Environmental Science & Technology*, 33(23):4206-4211.

Khim, J.S., D.L. Villeneuve, K. Kannan, K.T. Lee, S.A. Snyder, C.H. Koh, and J.P. Giesy. 1999c. Alkylphenols, polycyclic aromatic hydrocarbons, and organochlorines in sediment from Lake Shihwa, Korea: Instrumental and bioanalytical characterization. *Environmental Toxicology and Chemistry*, 18(11):2424-2432.

Kloepfer, A., J. Quintana, and T. Reemtsma. 2005. Operational options to reduce matrix effects in liquid chromatography-electrospray ionization-mass spectrometry analysis of aqueous environmental samples. *Journal of Chromatography A*, 1067(1-2):153-160.

Koester, C.J., and A. Moulik. 2005. Trends in environmental analysis. *Analytical Chemistry*, 77:3737-3754.

Kolodziej, E.P., and D.L. Sedlak. 2007. Rangeland grazing as a source of steroid hormones to surface waters. *Environmental Science & Technology*, 41:3514-3520.

Kreuzinger, N., M. Clara, B. Strenn, and B. Vogel. 2004. Investigation on the behaviour of selected pharmaceuticals in the groundwater after infiltration of treated wastewater. *Water Science and Technology*, 50(2):221-228.

Kubeck, E., and C.G. Naylor. 1990. Trace analysis of alkylphenol ethoxylates. *JAOCS*, 67(6):400-405.

Lamas, J.P., C. Salgado-Petinal, C. García-Jares, M. Llompart, R. Cela, and M. Gómez. 2004. Solid-phase microexraction-gas chromatography-mass spectrometry for the analysis of selective serotonin reuptake inhibitors in environmental water. *Journal of Chromatography A*, 1046:241-247.

Lindsey, M.E., M. Meyer, and E.M. Thurman. 2001. Analysis of trace levels of sulfonamide and tetracycline antimicrobials in groundwater and surface water using solid-phase extraction and liquid chromatography/mass spectrometry. *Analytical Chemistry*, 73(19):4640-4646.

Loos, R., J. Wollgast, T. Huber, and G. Hanke. 2007. Polar herbicides, pharmaceutical products, perfluorooctanesulfonate (PFOS), perfluorooctanoate (PFOA), and nonylphenol and its carboxylates and ethoxylates in surface and tap waters around Lake Maggiore in Northern Italy. *Analytical and Bioanalytical Chemistry*, 387(4):1469-1478.

Lopez De Alda, M.J., and D. Barcelo. 2001. Review of analytical methods for the determination of estrogens and progestogens in waste waters. *Fresenius Journal of Analytical Chemistry*, 371:437-447.

Loraine, G.A., and M.E. Pettigrove. 2006. Seasonal variations in concentrations of pharmaceuticals and personal care products in drinking water and reclaimed wastewater in southern California. *Environmental Science & Technology*, 40:687-695.

Marklund, A., B. Andersson, and P. Haglund. 2005. Organophosphorus flame retardants and plasticizers in Swedish sewage treatment plants. *Environmental Science & Technology*, 39:7423-7429.

Meyer, J., and K. Bester. 2004. Organophosphate flame retardants and plasticizers in wastewater treatment plants. *Journal of Environmental Monitoring*, 6:599-605.

Miao, X.-S., and C.D. Metcalfe. 2003. Determination of carbamazepine and its metabolites in aqueous samples using liquid chromatography-electrospray tandem mass spectrometry. *Analytical Chemistry*, 75:3731-3738.

Miao, X.-S., F. Bishay, M. Chen, and C.D. Metcalfe. 2004. Occurrence of antimicrobials in the final effluents of wastewater treatment plants in Canada. *Environmental Science & Technology*, 38(13):3533-3541.

Moldovan, Z. 2006. Occurrences of pharmaceutical and personal care products as micropollutants in rivers from Romania. *Chemosphere*, 2006:1808-1817.

Nikolai, L.N., E.L. Mcclure, S.L. Macleod, and C.S. Wong. 2006. Stereoisomer quantification of the .beta.-blocker drugs atenolol, metoprolol, and propranolol in wastewaters by chiral high-performance liquid chromatography-tandem mass spectrometry. *Journal of Chromatography A*, 1131(1-2):103-109.

Noppe, H., K. De Wasch, S. Poelmans, N. Van Hoof, T. Verslycke, C.R. Janssen, and H.F. De Brabander. 2005. Development and validation of an analytical method for detection of estrogens in water. *Analytical and Bioanalytical Chemistry*, 382(1):91-98.

Pedersen, J.A., M. Soliman, and I.H. Suffet. 2005. Human Pharmaceuticals, Hormones, and Personal Care Product Ingredients in Runoff from Agricultural Fields Irrigated with Treated Wastewater. *Journal of Agricultural and Food Chemistry*, 53:1625-1632.

Penalver, A., E. Pocurull, F. Borrull, and R.M. Marce. 2002. Method based on solid-phase microextraction-high-performance liquid chromatography with UV and electrochemical detection to determine estrogenic compounds in water samples. *Journal of Chromatography A*, 964(1-2):153-160.

Peng, X., Z. Wang, W. Kuang, J. Tan, and K. Li. 2006. A preliminary study on the occurrence and behavior of sulfonamides, ofloxacin and chloramphenicol antimicrobials in wastewaters of two sewage treatment plants in Guangzhou, China. *Science Of The Total Environment*, 371:314-322.

Pepich, B.V., M.M. Domino, T.A. Dattilio, P.S. Fair, and D.J. Munch. 2004. Validating sample preservation techniques and holding times for the approved compliance monitoring methods for haloacetic acids under the US EPA's stage 1 D/DBP rule. *Water Research*, 38(4):895-902.

Petrovic, M., M.D. Hernando, M.S. Diaz-Cruz, D. Barcelo, W.M.A. Niessen, and U.A.T. Brinkman. 2005. Liquid chromatography-tandem mass spectrometry for the analysis of pharmaceutical residues in environmental samples: a review. *Journal of Chromatography A*, 1067(1-2):1-14.

Polo, M., M. Llompart, C. Garcia-Jares, and R. Cela. 2005. Multivariate optimization of a solid-phase microextraction method for the analysis of phthalate esters in environmental waters. *Journal of Chromatography A*, 1072:63-72.

Pryor, S.W., A.G. Hay, and L.P. Walker. 2002. Nonylphenol in anaerobically digested sewage sludge from New York state. *Environmental Science & Technology*, 36(17):3678-3682.

Quintana, J.B., R. Rodil, and T. Reemtsma. 2004. Suitability of hollow fibre liquid-phase microextraction for the determination of acidic pharmaceuticals in wastewater by liquid chromatography-electrospray tandem mass spectrometry without matrix effects. *Journal of Chromatography A*, 1061:19-26.

Rabiet, M., A. Togola, F. Brissaud, J.L. Seidel, H. Budzinski, and F. Elbaz-Poulichet. 2006. Consequences of Treated Water Recycling as Regards Pharmaceuticals and Drugs in Surface and Ground Waters of a Medium-sized Mediterranean Catchment. *Environmental Science & Technology*, 40:5282-5288.

Reemtsma, T. 2003. Liquid-chromatography-mass spectrometry and strategies for trace-level analysis of polar organic pollutants. *Journal of Chromatography A*, 1000:477-501.

Richardson, S.D. 2000. Environmental Mass Spectrometry. *Analytical Chemistry*, 72:4477-4496.

Richardson, S.D. 2002. Environmental Mass Spectrometry: Emerging Contaminants and Current Issues. *Analytical Chemistry*, 74(12):2719-2742.

Richardson, S.D. 2004. Environmental mass spectrometry: Emerging contaminants and current issues. *Analytical Chemistry*, 76(12):3337-3364.

Richardson, S.D., and T.A. Ternes. 2005. Water analysis: emerging contaminants and current issues. *Analytical Chemistry*, 77:3807-3838.

Richardson, S.D. 2006. Environmental mass spectrometry: Emerging contaminants and current issues. *Analytical Chemistry*, 78:4021-4045.

Rodríguez, I., J. Carpinteiro, J.B. Quintana, A.M. Carro, R.A. Lorenzo, and R. Cela. 2004. Solid-phase microextraction with on-fiber derivatization for the analysis of anti-inflammatory drugs in water samples. *Journal of Chromatography A*, 1024(1-2):1-8.

Romero, J., F. Ventura, and M. Gomez. 2002. Characterization of paint samples used in drinking water reservoirs: Identification of endocrine disruptor compounds. *Journal of Chromatographic Science*, 40(4):191-197.

Routledge, E.J., D. Sheahan, C. Desbrow, G.C. Brighty, M. Waldock, and J.P. Sumpter. 1998. Identification of estrogenic chemicals in STW effluent. 2. *In vivo* responses in trout and roach. *Environmental Science & Technology*, 32(11):1559-1565.

Sandstrom, M.W., D.W. Kolpin, E.M. Thurman, and S.D. Zaugg. 2005. Widespread detection of N,N-diethyl-m-toluamide in U.S. Streams: Comparison with concentrations of pesticides, personal care products, and other organic wastewater compounds. *Environmental Toxicology and Chemistry*, 24(5):1029-1034.

Santos, J.L., I. Aparicio, E. Alonso, and M. Callejon. 2005. Simultaneous determination of pharmaceutically active compounds in wastewater samples by solid phase extraction and high-performance liquid chromatography with diode array and fluorescence detectors. *Analytica Chimica Acta*, 550:116-122.

Sapkota, A., J. Heidler, and R.U. Halden. 2007. Detection of triclocarban and two co-contaminating chlorocarbanilides in US aquatic environments using isotope dilution liquid chromatography tandem mass spectrometry. *Environmental Research*, 102:21-29.

Segner, H., J.M. Navas, C. Schäfers, and A. Wenzel. 2003. Potencies of estrogenic compounds in *in vitro* screening assays and in life cycle tests with zebrafish *in vivo*. *Ecotoxicology and Environmental Safety*, 54(3):315-322.

Seitz, W., W.H. Weber, J.-Q. Jiang, B.J. Lloyd, M. Maier, D. Maier, and W. Schulz. 2006. Monitoring of iodinated X-ray contrast media in surface water. *Chemosphere*, 64(8):1318-1324.

Seitz, W., W. Schulz, and W.H. Weber. 2006. Novel applications of highly sensitive liquid chromatography/mass spectrometry/mass spectrometry for the direct detection of ultra-trace levels of contaminants in water. *Rapid Communications In Mass Spectrometry*, 20:2281-2285.

Serodio, P., and J.M.F. Nogueira. 2006. Considerations on ultra-trace analysis of phthalates in drinking water. *Water Research*, 40:2572-2582.

Snyder, S.A., T.L. Keith, D.A. Verbrugge, E.M. Snyder, T.S. Gross, K. Kannan, and J.P. Giesy. 1999. Analytical methods for detection of selected estrogenic compounds in aqueous mixtures. *Environmental Science & Technology*, 33(16):2814-2820.

Snyder, S.A., T.L. Keith, C.G. Naylor, C.A. Staples, and J.P. Giesy. 2001a. Identification and quantification method for nonylphenol and lower oligomer nonylphenol ethoxylates in fish tissues. *Environmental Toxicology and Chemistry*, 20(9):1870-1873.

Snyder, S.A., D.L. Villeneuve, E.M. Snyder, and J.P. Giesy. 2001b. Identification and quantification of estrogen receptor agonists in wastewater effluents. *Environmental Science & Technology*, 35(18):3620-3625.

Snyder, S.A., K.L. Kelly, A.H. Grange, G.W. Sovocool, E.M. Snyder, and J.P. Giesy. 2001c. Pharmaceuticals and personal care products in the waters of Lake Mead, Nevada. In *Pharmaceuticals and Personal Care Products in the Environment: Scientific and Regulatory Issues, Vol. Symposium Series 791*. Edited by C. G. Daughton and T. L. Jones-Lepp. Washington,D.C.: American Chemical Society.

Stumm-Zollinger, E., and G.M. Fair. 1965. Biodegradation of steroid hormones. *Journal of the Water Pollution Control Federation*, 37:1506-1510.

Tabak, H.H., and R.L. Bunch. 1970. Steroid hormones as water pollutants. I. Metabolism of natural and synthetic ovulation-inhibiting hormones by microorganisms of activated sludge and primary settled sewage. *Dev. Ind. Microbiol.*, 11:367-376.

Ternes, T.A. 2001. Analytical methods for the determination of pharmaceuticals in aqueous environmental samples. *Trends in Analytical Chemistry*, 20(8):419-434.

Trenholm, R.A., B.J. Vanderford, J.C. Holady, D.J. Rexing, and S.A. Snyder. 2006. Broad range analysis of endocrine disruptors and pharmaceuticals using gas chromatography and liquid chromatography tandem mass spectroscopy. *Chemosphere*, 65:1990-1998.

Tswett, M. 1906. Physical chemical studies on chlorophyll adsoprtions. *Berichte der Deutschen Botanischen Gesellschaft*, 24:316-323.

Van Aerle, R., N. Pounds, T.H. Hutchinson, S. Maddix, and C.R. Tyler. 2002. Window of sensitivity for the estrogenic effects of ethinylestradiol in early life-stages of fathead minnow, *Pimephales promelas*. *Ecotoxicology*, 11(6):423-434.

Vanderford, B.J., R.A. Pearson, D.J. Rexing, and S.A. Snyder. 2003. Analysis of endocrine disruptors, pharmaceuticals, and personal care products in water using liquid chromatography/tandem mass spectrometry. *Analytical Chemistry*, 75(22):6265-6274.

Vanderford, B.J., and S.A. Snyder. 2006. Analysis of pharmaceuticals in water by isotope dilution liquid chromatography/tandem mass spectrometry. *Environmental Science & Technology*, 40:7312-7320.

Vermeirssen, E.L.M., O. Koerner, R. Schoenenberger, M.J.-F. Suter, and P. Burkhardt-Holm. 2005. Characterization of environmental estrogens in river water using a three pronged approach: active and passive water sampling and the analysis of accumulated estrogens in the bile of caged fish. *Environmental Science and Technology*, 39:8191-8198.

Vervoort, R.J.M., E. Ruyter, A.J.J. Debets, H.A. Claessens, C.A. Cramers, and G.J. De Jong. 2001. Characterisation of reversed-phase liquid chromatography stationary phases for the analysis of basic pharmaceuticals: eluent properties and comparison of empirical test methods. *Journal of Chromatography A*, 931:67-79.

Vieno, N., T. Tuhkanen, and L. Kronberg. 2006a. Analysis of neutral and basic pharmaceuticals in sewage treatment plants and in recipient waters using solid phase extraction and liquid chromatography-tandem mass spectrometry detection. *Journal of Chromatography A*, 1134:101-111.

Vieno, N.M., T. Tuhkanen, and L. Kronberg. 2006b. Analysis of neutral and basic pharmaceuticals in sewage treatment plants and in recipient rivers using solid phase extraction and liquid chromatography-tandem mass spectrometry detection. *Journal of Chromatography A*, 1134(1-2):101-111.

Xu, W.-H., G. Zhang, S.-C. Zou, X.-D. Li, and Y.-C. Liu. 2007. Determination of selected antibiotics in the Victoria Harbour and the Pearl River, South China using high-performance liquid chromatoraphy-electrospray ionization tandem mass spectrometry. *Environmental Pollution*, 145:672-679.

Yamamoto, A., N. Kakutani, K. Yamamoto, T. Kamiura, and H. Miyakoda. 2006. Steroid hormone profiles of urban and tidal rivers using LC/MS/MS equipped with electrospray ionization and atmospheric pressure photoionization sources. *Environmental Science & Technology*, 40:4132-4137.

Yang, S., J. Cha, and K. Carlson. 2004. Quantitative determination of trace concentrations of tetracycline and sulfonamide antibiotics in surface water using solid-phase extraction and liquid chromatography/ion trap tandem mass spectrometry. *Rapid Communications In Mass Spectrometry*, 18(18):2131-2145.

Ye, Z., H.S. Weinberg, and M.T. Meyer. 2007. Trace analysis of trimethoprim and sulfonamide, macrolide, quinolone, and tetracycline antibiotics in chlorinated drinking water using liquid chromatography electrospray tandem mass spectrometry. *Analytical Chemistry*, 79(3):1135-1144.

Ying, G.-G., R.S. Kookana, and Z. Chen. 2002. On-line solid-phase extraction and fluorescence detection of selected endocrine disrupting chemicals in water by high-performance liquid chromatography. *J. Environ. Sci. Heal. B.* , 37(3):225-234.

Yook, K.-S., S.-M. Hong, and J.-H. Kim. 1994. Comparison of liquid-liquid extraction and solid-phase extraction coupled with GC/MS for determination of priority pollutants in water. *Analytical Science Technology*, 7(4):441-453.

Zhang, Z.L., A. Hibberd, and J.L. Zhou. 2006. Optimisation of derivatisation for the analysis of estrogenic compounds in water by solid-phase extraction gas chromatography-mass spectrometry. *Analytica Chimica Acta*, 577:52-61.

Zhao, Y.-Y., J. Boyd, S.E. Hrudey, and X.-F. Li. 2006. Characterization of new nitrosamines in drinking water using liquid chromatography tandem mass spectrometry. *Environmental Science & Technology*, 40:7636-7641.

# CHAPTER 3
# BIOLOGICALLY-BASED METHODS FOR DETECTION AND ANALYSIS

## INTRODUCTION

This chapter focuses on the use of biologically-based methods for detection and analysis of EDCs and pharmaceuticals in drinking water and characterization of the potential of these drinking water contaminants to impact human health. Other reviews also have addressed the use of biologically-based methods to detect EDCs and/or PPCPs in water. For example, the Global Water Research Coalition (GWRC 2003 Endocrine) published a brief overview of sources, occurrence, and biological methods for measuring EDCs in the aquatic environment. GWRC also published the results of a workshop on methods for chemical and biological analysis of EDCs in water systems (GWRC 2003 Methodologies). Snyder et al. (2000, 2003) reviewed analytical methods (including *in vitro* and *in vivo* bioassays) for measuring endocrine disrupting compounds in water. A review of pharmaceuticals in the environment published by the United Kingdom Environment Agency (Ayscough et al. 2000) focuses on human rather than veterinary drugs and does not assess potential impacts on human health through drinking water.

## BIOLOGICALLY-BASED METHODS FOR DETECTION AND ANALYSIS

Biologically-based methods for detection and analysis of chemical substances include immunochemical methods and biossays. Immunochemical methods include immunoextraction techniques used to extract and concentrate chemical substances from water, immunochromatography, and immunoassays used to detect and quantify waterborne chemicals. Immunochemical techniques are not intended to measure an effect on a biological system. A bioassay is a procedure that uses the response of an organism or its components as an analytical tool to detect the presence of a chemical or class of chemicals, to test for effects of chemicals, and/or to evaluate the potency of chemicals in comparison to a reference material. Bioassays can be used to detect and sometimes quantify contaminants in water on the basis of the effect of those contaminants in a biological system. Bioassays also can be used to test for endocrine activity or pharmaceutical bioactivity associated with individual contaminants that occur, or that might be expected to occur, in drinking water. Drinking water is a matrix that contains a mixture of regulated and unregulated contaminants, many of which have not been identified or characterized. Bioassays can be useful for assessing the integrated effects of contaminants that occur in a mixture of known or unknown chemical composition.

### Immunochemical Methods

#### *Immunoassays*

Immunoassays are "rapid" tests that can be used on-site or in the laboratory as a tool to detect specific molecules. Immunoassays rely on the ability of an antibody to bind with high specificity to the 3-dimensional structure of a molecule (antigen or analyte). Monoclonal and/or polyclonal antibodies are used in a variety of formats. Direct immunoassays detect the analyte itself, while indirect immunoassays detect the antibodies and indirectly reveal the presence and concentration of the analyte. Descriptions of four commonly used formats (monlyclonal-polyclonical sandwich assays, competitive inhibition assays, antigen-down immunoassays, and rapid assays) are presented elsewhere (Immunochemistry Technologies 2005). Immunoassays are sensitive because they rely on radioisotopes, fluorescent compounds, or the products of enzymatic reactions that can be detected in minute quantities. In an enzyme-linked immunosorbent assay (ELISA), antibodies or antigen are linked to an enzyme conjugate that catalyzes a reaction that can be monitored (e.g., through a color change). Immunofluorescence assays use a fluorescent compound as the conjugate, while radioimmunoassays (RIAs) use radiolabeled antibody or antigen (radiolabeled version of the analyte). Generally, RIA is a more sensitive technique than ELISA, but RIA is more costly and requires handling and disposal of radioactive materials (Snyder et al. 2003).

Immunoassay is an alternative or complementary method that is relatively more rapid and simple than analytical chemistry methods. Immunoassays are accurate and sensitive, cost-effective, and require only small sample volumes and thus are useful for screening large numbers of samples with high throughput (Shim et al. 2006). However, because antibodies bind to chemicals based on their structure, immunoassays can be subject to interference by chemicals in the sample that have a 3-dimensional structure similar to the analyte, and cross-reactivity to other chemicals similar to the analyte should be reported along with immunoassay results. Consequently, the results of immunoassays are considered to be less reliable than results of analytical chemistry techniques that allow for confirmation of the presence of the analyte.

### *Immunochromatography and Immunoextraction*

Despite the advantages of immunoassays, they often require long reaction times, involve multiple steps, and require equipment and conditions that might not be easily adapted to field work. Thus, immunoassays commonly are confined to laboratories equipped for this type of analysis and might still require skilled analysts. Immunochromatography techniques have been developed to maximize the convenience and speed of immunoassays. They involve immobilization of an analyte binding site on a membrane.

Immunoextraction is a method in which antibodies are covalently immobilized or adsorbed onto a chromatographic support for use in isolating a specific analyte or group of related chemicals from a complex matrix. There is growing interest in the use of on-line immunoextraction as a tool for removing and concentrating chemicals from many types of environmental samples including water, as well as serum and urine and extracts or portions of food, sediment, and sludge. When immunoextraction is combined with liquid chromatography, it is possible to separate and analyze structurally similar chemicals that cross-react with the same antibody. On-line immunoextraction has been coupled with reversed-phase liquid chromatography for analysis of atrazine in water samples. (Nelson et al. 2007)

Application of analytical chemistry methods and immunochemical methods for identification and quantification requires advance knowledge of the specific contaminants that might be present in a sample in order to select extraction and detection methods directed toward them. Analytical standards are needed to develop, validate, and conduct analytical chemistry and immunochemical methods, but analytical standards are not available for some environmental contaminants. For example, Bolz, Koerner and Hagenmaier (2000) reported that identification of the individual p-isomers of nonylphenol was not possible at that time because pure standards of the individual compounds were not available.

## Bioassays

A bioassay (or biological assay) is a procedure that uses the response of a living organism, a biological system, or its components (e.g., animals, cells or tissues in culture, hormone receptors) as an analytical tool to detect the presence of a chemical or class of chemicals, to test for the effects of chemicals, and/or to evaluate the potency of chemicals in comparison to a reference material. Biological systems and components of those systems at all levels of biological organization (molecule, cell, tissue, organ, organ system, organism, population, and community) can be employed in bioassays. Bioassays applied for the protection of human health to drinking water or to drinking water contaminants are restricted to biological systems in the molecule to organism levels, though they may be used as a part of clinical or epidemiology studies to investigate potential effects on a human population. Population- and community-level bioassays using fish and wildlife are applied to effluent or environmental water samples, for example, to investigate the potential impact of waterborne contaminants on populations of aquatic organisms or on their predator-prey interactions.

*In vivo* bioassays use living organisms (animals in this case) for biologically based testing. *In vitro*[1] (literally "in glass") bioassays exploit the responses of components of organisms (e.g., tissues, cells in culture, receptors) maintained artificially in a test tube, culture dish, or elsewhere in a laboratory. In general, *in vitro* assays are more rapid, simpler to perform, less expensive, and more reproducible than *in vivo* tests, and they also commonly require smaller amounts of test material (Murk et al. 2002, Snyder et al. 2003). *In vitro* tests are particularly useful for screening-level testing and for elucidating the MOA by which a chemical causes an effect *in vivo* (WHO 2002). Also, *in vitro* bioassays typically cannot fully account for all of the pharmacokinetic (uptake, distribution, excretion), metabolic, and homeostatic processes that influence exposure and toxicity *in vivo*. Consequently, the results of *in vitro* bioassays are not always predictive of effects *in vivo*.

Screening for contaminants based on biological activity is useful for evaluating the presence or absence of a particular class of compounds (e.g., estrogenic or androgenic), while the majority of instrumental techniques are used to identify and quantify specific pre-selected target chemicals (Snyder et al. (2003). In contrast to analytical chemistry and immunochemical techniques, bioassays for effects of chemicals in water can be conducted without prior knowledge of the specific chemicals in water samples. However, awareness of the specific contaminants that might be present in the sample can help the analyst to select bioassays with appropriate detection limits and to account for substances in the sample that might interfere with the chosen bioassays. Bioassays also do not require pure standards to detect and quantify the effects of contaminants in water, with the exception of one or more standard reference substances used to test the sensitivity and responsiveness of the assay and to quantify and characterize the response. Because bioassays can respond to multiple contaminants in a sample at the same time, they are not as specific as analytical chemistry techniques or immunoassays.

## BIOLOGICALLY-BASED METHODS FOR EDCS

The majority of biologically-based methods for EDCs, and particularly immunoassays and *in vitro* bioassays that can be applied to drinking water, are intended to detect estrogenic or anti-estrogenic chemicals or their effects. Commercial bioassays are available for estrogenic and

[1] A third type of bioassay is the ex vivo experiment, in which part of the experiment is carried out on a whole living organism, then part of the organism is removed for examination *in vitro*, or in which part of a subject is removed, manipulated externally, then returned to the organism.

androgenic EDCs, but to our knowledge, commercial bioassays for EDCs that act on the thyroid have not yet become available (Meulenberg and Marchesini 2006). Although a wide variety of bioassays directed at EDCs have been developed, standardized methods are lacking. Various laboratories have employed different test methods, resulting in disagreements over proper test species, conditions, endpoints, protocols, and interpretation of results. In the absence of standardized and widely accepted test methods and criteria, professional judgement is needed to determine whether there is sufficient information to conclude that an effect resulted from endocrine disruption.

Because EDCs can exert their effects through many different mechanisms, no single bioassay exists that can determine definitively whether a chemical is or is not an EDC. A weight-of-evidence approach that considers all available toxicological information (e.g., data arising from multiple toxicity tests, screening assays, models) is recommended for evaluating whether a chemical is an EDC (WHO 2002). Furthermore, reproductive endocrine systems differ between male and female animals such that sex-specific bioassays are needed to test for potential effects of EDCs. To date, there is no consensus within the scientific community on the battery of tests that should be used to identify EDCs.

The U.S. EPA's Endocrine Disrupter Screening Program (EDSP) (see below) is one of a number of initiatives in progress in different parts of the world aimed at developing standardized screening and testing methods (including bioassays) for EDCs. An overview and history of the Endocrine Disruptor Screening Program (EDSP) is available in the EDSP Endocrine Primer (EDSP 2007 Endocrine Primer). The U.S. EPA developed the Endocrine Disruptor Screening Program (EDSP) in response to a Congressional mandate in the Federal Food, Drug, and Cosmetic Act (FFDCA).

The EDSP is charged with identifying or developing screening methods and toxicity testing strategies for EDCs. The EDSP focuses exclusively on the estrogenic, androgenic, and thyroidal MOAs and has adopted a tiered testing strategy. Bioassays under consideraton for use in the EDSP are presented in Table 3.1

**Immunochemistry Methods**

Immunoassay methods (described above) for synthetic EDCs generally are developed and applied in the same manner as for other chemicals. Immunoassay methods (particularly RIAs) for hormones have been developed for clinical and endocrinology research applications and used in a similar fashion for environmental samples. However, lower detection limits might be required to detect hormones in environmental water samples or drinking water, and potentially interfering substances might differ between clinical and environmental samples.

ELISA has become an accepted method for analysis of steroid hormones and pesticide residues in environmental samples including water (Holland 2003). Although most laboratories are not capable of developing their own antibodies and immunoassays, commercial immunoassay kits are available for many EDCs (Holland 2003). As discussed above, an immunochromatography method was developed for detection of atrazine in water samples (Shim et al. 2006). Immunoaffinity columns can be used for efficient extraction and clean-up of environmental samples preceding instrumental analyses for EDCs (Holland 2003). This technology has been applied to water samples to extract EDCs including atrazine and its metabolites, other triazine herbicides, and PAHs. Immunoaffinity columns also have been used to extract 17ß- and 17α,19-nortestosterone from urine (Nelson 2007), demonstrating their usefulness for extraction of steroids from aqueous samples. Additional examples of studies that applied immunochemical methods for detection and analysis of EDCs to water samples are presented in Appendix D.

In clinical chemistry for the diagnosis of thyroid-related diseases and disorders, measurements of thyroid hormones and their binding proteins and receptors are commonly based on antibodies or radiolabeled tracers. Analagous radioligand binding assays have been developed and applied for assessment environmental contaminants for potential thyroidal activity (Meulenberg 2006, Meulenberg and Marchesini 2006). Meulenberg and Marchesini (2006) noted that there is a need for non-radioactive bioassays for the assessment of thyroidal activity of contaminants in aqueous samples, including drinking water. They developed a biosensor assay using TBG or TTR as binding molecules and T4 as a ligand. T4 was covalently linked to a sensor chip surface without changing its affinity for TBG or recombinant transthyretin (rTTR). Competition between analytes in the sample and immobilized T4 for binding sites on TBG or rTTR is assessed by measuring the change in mass on the surface of the sensor chip. Greater concentrations of analytes capable of binding to TBG or rTTR result in smaller amounts of those binding proteins bound to the T4 immobilized on the biosensor surface. Results are expressed as T4 equivalents. The biosensor assay cannot identify the specific contaminants responsible for the assay response and must be coupled with an analytical chemistry method for that purpose. This biosensor assay for thyroidal activity was applied to surface water and to water samples collected from intake sites for drinking water treatment plants.

**Bioassays**

*In vitro* and *in vivo* bioassays are intended to evaluate the potential for chemicals to cause a biologic response. Compared with *in vivo* bioassays, *in vitro* bioassays for EDCs generally are quicker, simpler to conduct, and less expensive, making them more attractive for screening purposes. Because *in vitro* bioassays for EDCs often are based on relatively well-characterized MOAs and use endpoints that are more easily interpreted and explained (Zacharewski 1997), they also are particularly useful for investigating the MOA of substances purported to be EDCs. However, *in vitro* data alone are insufficient for evaluating the potential human health effects of EDCs. At this time, only *in vivo* bioassays can account for the various pharmacokinetic, metabolic, and pharmacodynamic interactions that can occur in an animal (Zacharewski 1997). For example, *in vitro* bioassays have limited metabolic capabilities compared with *in vivo* bioassays. Of particular concern is the potential for false negative results in the *in vitro* bioassays for chemicals that require metabolic activation to cause an effect. *In vitro* bioassays also do not account for bioaccumulation or interactions with binding globulins that can affect steroid uptake and metabolism (Zacharewski 1997). Consequently, the results of *in vitro* bioassays typically are less biologically relevant than results derived from *in vivo* bioassays and have limited utility for assessing risk to human health (Zacharewski 1997), so *in vivo* data are required to serve as a point of departure (POD) for development of acceptable daily intakes or reference doses that are commonly used to set guidelines for drinking water contaminants.

Despite their necessity, *in vivo* bioassays for EDCs suffer from the same disadvantages as *in vivo* bioassays for other types of chemicals They are relatively time-consuming, laborious, complicated, and costly, making them less suitable than the *in vitro* assays for high throughput screening. *In vivo* bioassays for estrogenic EDCs also tend to exhibit rather poor sensitivity and only modest responsiveness (Zacharewski 1997). Complex responses such as uterine weight gain (considered to be the hallmark estrogen response) can be affected by mechanisms other than direct receptor binding and might not be selective for chemicals that interact with the ER

(Zacharewski 1997). For example, carbon tetrachloride has been purported to potentiate the action of E2 on the uterus by inhibiting metabolism of the endogenous hormone and thus can be considered to be an EDC, but carbon tetrachloride itself does not bind to the ER (Zacharewksi 1997). Thus, *in vivo* bioassays generally are more useful for assessing the potential for a substance to act through any of a number of possible MOAs to cause a particular effect, but may be less useful than *in vitro* bioassays for determining the exact mechanism by which the effect occurs. For this reason, data from *in vitro* bioassays are commonly used as a complement to *in vivo* bioassays to aid in investigating the mechanism by which an effect occurs *in vivo*.

### *In vitro Bioassays*

***In vitro Receptor Binding Assays.*** Some EDCs can bind to hormone receptors and either mimic the effect of a natural endogenous hormone or interfere with the binding of natural hormones to the receptors and thus block their activity (EDSP 2006 Assays). The androgen receptor (AR) is involved in the development of male sexual characteristics, and the ER is involved in female sexual maturation and reproductive function (EDSP 2006 Assays). Thyroid hormone receptors play a role in cellular metabolism and development as well as a number of other bodily processes (WHO 2002). *In vitro* receptor competitive binding assays are cell-free tests in which a test chemical competes for binding at receptors with an endogenous hormone (natural ligand) or other strongly binding reference substance. Results for ER- and AR-binding assays usually are expressed in terms of estradiol equivalents (EEQ) or androgen equivalents (Aeq), respectively. Receptor binding assays for EDCs are useful for evaluating the ability of a chemical to bind to a hormone receptor and provide an indication the potential for a chemical to disrupt hormone receptor-mediated function *in vivo* (U.S. EPA 1998). Receptor binding assays also are useful for evaluating the relative strength of EDCs with respect to their ability to bind to hormone receptors. Receptor binding assays are amenable to high-throughput screening, and commercial assays are available (Zacharewski et al. 2002). The EDSP will use *in vitro* receptor binding assays to evaluate binding to nuclear ER and AR (U.S. EPA 1998) (Table 3.1).

As explained above, it is important to note that receptor binding is only an initial step in receptor-mediated MOAs. *In vitro* receptor binding assays provide no information about the ability of a putative EDC to propagate events that occur after receptor binding. Binding of a substance to the ER alone is not a sufficient predictor of estrogenicity. A chemical must remain bound to the receptor and be capable of triggering a cascade of events that lead to an ultimate adverse effect (Zacharewki 1997). Many estrogenic EDCs do not induce ER-mediated gene expression in reporter gene assays to the extent that would be expected based on their affinity for the ER (Zacharewski et al. 2002). Changes in endocrine function or reproductive fitness might occur only with continued accumulation of occupied nuclear receptors (Zacharewski 1997). Receptor binding assays cannot distinguish between receptor agonists (EDCs that mimic the effects of hormones) and antagonists (EDCs that block the effects of hormones) and can be subject to non-specific binding (Zacharewski 1997, Zacharewski et al. 2002). These cell-free assays also do not account for the ability of cell membranes to exclude chemicals. Thus, receptor binding assays applied to water samples might tend to overestimate the potential for *in vivo* activity.

Receptor binding assays have been applied to water samples and extracts or concentrates of water samples including raw and treated wastewater (sheep estrogen receptor binding assay (Leusch et al. 2005), Murk et al. 2002), river water (Murk et al. 2002, Oh et al. 2006, Quanrud et al. 2004), groundwater (Quanrud et al. 2004) and simulated drinking water (Hu et al. 2002, rat estrogen receptor competitive binding assay (Huber et al. 2004)). Hormone receptors from fish have been used in ER and AR binding assays (Leusch et al. 2005, 2006a, 2006b), primarily to predict effects of wastewater or surface water contaminants on fish.

Cell-based bioassays for EDCs evaluate not only receptor binding but also the ability of an EDC to produce a subsequent biologic effect. Cells in culture provide for a more realistic expsosure scenario than cell-free assays because they can represent the effects of cell membranes on uptake of chemicals and sometimes can metabolize chemicals to a limited extent. In some test systems, a metabolic activation system consisting of enzymes that metabolize chemicals is used to better represent *in vivo* metabolic capabilities. Some cellular bioassays can distinguish between hormone agonists and antagonists. Cell-based bioassays for EDCs commonly use cultured cells stably transfected with transcriptional activation reporter systems. In these assays, a reporter gene is inserted into a region of the cellular DNA that undergoes transcription in response to binding by a hormone-receptor complex.

***Vertebrate Cell Proliferation Assays.*** Cell proliferation assays for estrogenic EDCs are based on the responses of ER-containing, estrogen-responsive cells such as the MCF-7 or T47-D human breast cancer cell lines (Zacharewski 1997). The E-screen, a type of cell proliferation assay, uses cells cultured in medium supplemented with serum stripped of naturally-occurring steroids. The cells proliferate in response to exposure to estrogenic substances. After a 6-day incubation period, the cells are counted. The test is based on the assumption that there are unidentified factors in human serum that inhibit the proliferation of the cells and that estrogenic substances overcome these inhibitory factors to induce cell proliferation (Zacharewski 1997). Cell proliferation assays are relatively simple to perform. The MCF-7 E-screen produces few false positives and, with a reported detection limit of 10 pg E2/mL (30 pM E2), is one of the most sensitive *in vitro* bioassays for xenoestrogens (Zacharewski 1997). Because the E-screen uses only endogenous cellular components, responses in the E-screen are probably more biologically relevant than responses obtained with bioassays based on transformed cell lines, which contain non-native components (Zacharewski et al. 2002).

However, cell proliferation assays have several drawbacks. They are subject to interference by mitogens that can induce the cells to grow in the absence of estrogenicity. When levels of estrogen are very low in the media, MCF-7 cells can enhance their sensitivity to E2. Differences in cell lines, culture conditions, and serum lots can have a significant effect on the assessment of relative potency for the same substance evaluated in separate assays. Cells are enumerated using inconsistent methods ranging from counting cells or cell nuclei by Coulter Counter or hemacytometer, to measuring incorporation of tritiated thymidine, to using metabolizable dyes in spectrophotometric detection methods. In some cases, a chemical can be negative in a test with one cell line and positive with another. Cell proliferation assays also are only modestly responsive, limiting the evaluation of both estrogenic and antiestrogenic substances, and require relatively long incubation times. There is a lack of studies demonstrating that E-screen results are predictive of *in vivo* responses. Finally, MCF-7 cells express receptors for other types of hormones that can suppress E2-induced cell proliferation. (Zacharewski 1997)

The E-screen has been applied to treated landfill leachate (Behnisch et al. 2001), WWTP influent and effluent (Körner et al. 1999), surface water (Matsuoka et al. 2005, Oh et al. 2006, Soto et al. 2004), and simulated drinking water (Alum et al. 2004).

Soto et al (2004) reported on the use of the A-screen on water samples. This assay uses

MCF7-AR1 cells, which are stable transfectants of MCF7 cells expressing the wild-type human androgen receptor. The cells proliferate in response to estrogens and respond to androgens by decreasing their proliferation rate. The synthetic, nonmetabolizable androgen R1881 was used as a positive control. The A-screen was applied to water samples from feedlot retention ponds and surface water as well as tap water.

***Post-confluent Cell Accumulation and Foci Formation.*** After MCF-7 cells become confluent (i.e., spread out in an even monolayer across the culture plate), they form muticellular clusters of overlapping cells (foci) on a monolayer background. This post-confluent cell accumulation and foci formation is thought to be an ER-mediated response that can be exploited in a bioassay to evaluate the estrogenicity of test substances. Assays based on post-confluent cell accumulation and foci formation appear to be selective for estrogens versus other types of hormones (androgens, progestagens), demonstrate equivalent or better sensitivity than some other types of *in vitro* bioassays, and exhibit ~20-fold maximum induction (suitable for assessment of antiestrogenic effects). However, it has not been well-characterized with regard to its potential to be complementary or redundant with the cell proliferation assays or its ability to predict *in vivo* effects, and it requires relatively longer incubation times. (Zacharewski 1997)

***Gene or Protein Expression and Enzyme Activity Assays.*** Increased expression of proteins (e.g., hormone receptors, sex hormone binding globulin) and induction of enzyme activities can be used to evaluate the estrogenic potency of test substances. In some cases, expression of these proteins and enzyme activities is specific to certain cell lines or tissues and might not be relevant to other tissues or species (Zacharewski 1997). Measurement of these endpoints commonly involves labor-intensive methods such as northern blotting, western blotting, and ELISA (Zacharewski 1997) that are not amenable to high-throughput screening and more recently reverse-transcription polymerase chain reaction (RT-PCR) (Zacharewski et al. 2002) to detect native products. However, there is growing interest in the use of microarray technology to simplify the measurement of endpoints in protein expression and enzyme activity assays (see below) (Zacharewski et al. 2002). Expression of some proteins and enzyme activities may be modulated by non-receptor-mediated interactions that increase the likelihood of false positives. Many protein expression assays and enzyme activity assays are quite sensitive ($EC_{50}$[2] values in the picomolar range) and are strongly inducible.

Microarray technology is gaining attention for its potential to facilitate rapid, sensitive, inexpensive screening of a large number of molecular biomarkers of exposure to EDCs, pharmaceuticals, and other xenobiotics. According to Zacharewski et al. (2002), "Biomarkers are mechanistically-based experimental endpoints that are designed to efficiently detect and characterize chemical exposures. Biomarkers have been frequently used to detect and characterize xenobiotics for hormone-like activity." Marker genes are useful as biomarkers because of their sensitivity, strong dose-dependence, and responsiveness. Microarrays can be used to assess the level of mRNA expression for hundreds to tens of thousands of genes simultaneously in a single experiment, while the more commonly used northern blotting and RT-PCR methods can quantify expression of only a few genes per experiment. Microarrays have been investigated for use in monitoring endogenous gene expression in MCF-7 cells used in the E-screen, as well as in other cell lines. Commonly used bioassays for EDCs are directed at characterizing substances that act through a certain limited MOA. Microarrays have the potential to enable the large-scale investigation of gene expression profiles of known or suspected EDCs without prior knowledge of their targets for toxicity or MOA. This feature is particularly useful for EDCs, can act through a wide range of MOAs. Gene expression profiles also can provide

[2] EC50- Effective concentration 50%; Concentration that produces 50% of the maximum response

insight into potential endocrine MOAs for substances that are not well-characterized. (Zacharewski et al. 2002)

Microarrays also have limitations. Human cells vary in their gene expression responses due to differences among cell types and genotype. Gene-based biomarkers lack specificity because genes often are regulated by more than one signalling pathway. Adverse effects are not an unavoidable consequence of gene expression, nor does an absence of a response in gene expression preclude an adverse effect. Changes in multiple biomarkers that are correlated with an adverse effect *in vivo* and support of microarray results with other experimental evidence *in vitro* and *in vivo* will allow greater confidence in the results of microarray analyses. (Zacharewski et al. 2002)

***Recombinant Receptor/Reporter Gene Assays.*** These assays can be broadly assigned to three categories: (1) endogenous promoter-regulated reporter genes, (2) response element-regulated reporter genes, and (3) chimeric receptor/response element-regulated reporter genes (Zacharewski 1997). Though estrogenic substances are the primary focus here, these assays can be applied to any response mediated through a nuclear hormone receptor (Zacharewski 1997). Some additional bioassays are described by Zacharewski et al. (2002).

Endogenous promoter-regulated reporter gene assays make use of recombinant reporter gene constructs consisting of endogenous promoters from hormone-responsive genes linked to reporter genes (Zacharewski 1997). Reporter genes encode for enzymes such as firefly luciferase, ß-galactosidase (LacZ), chloramphenicol acetyltransferase (CAT), alkaline phosphatase (Zacharewski 1997) and more recently jellyfish green fluorescent protein (GFP) (Zacharewski et al. 2002). Reporter genes are used rather than assessing endogenous hormone-inducible proteins because reporter genes can be measured more easily and quickly and results are quantitative (Zacharewski et al. 2002). A test substance must be able to bind to the hormone receptor *and* induce gene expression to cause a response. These assays have excellent responsiveness and sensitivity and can be used to assess the relative potency of receptor-mediated agonists and antagonists (Zacharewski 1997). Estrogen-inducible human and rat progesterone receptor promoter-regulated reporter genes are available for insertion into cultured cells, allowing for evaluation of species-specific responses using relevant cell lines derived from different tissues (e.g., breast and uterine cell lines) (Zacharewski 1997). However, commonly-used promoters are still susceptible to induction by mechanisms that do not involve ER (Zacharewski 1997). To investigate androgenic and antiandrogenic effects, human AR and androgen-regulated reporter genes have been transfected into various cell lines, including monkey kidney (CV1) cells, human prostate cancer (LNCaP) cells, mouse vas deferens epithelial cells, and Chinese hamster ovary (CHO) cells (Zacharewski et al. 2002).

Response element-regulated reporter gene assays have been developed in an attempt to circumvent problems with the endogenous promoter-regulated reporter genes. For example, some researchers have developed reporter genes regulated by EREs to ensure that induction of a response occurs only through the ERE. Unfortunately, these constructs are exquisitely sensitive to estrogens in serum used in culture medium, resulting in high background and reduced responsiveness in assays. Using serum stripped of steroids does not sufficiently alleviate the problem, as studies indicate that the remaining estrogen is still enough to produce a response in the *in vitro* bioassays for estrogenicity. Also, other types of receptors can interact with EREs and interfere with expression of the reporter gene. (Zacharewski 1997)

Chimeric receptor/reporter gene assays are more complicated and labor-intensive than protein expression or cell proliferation assays, but they demonstrate greater selectivity, responsiveness, and sensitivity, and they are less susceptible to interference by estrogens in serum. Because they contain exogenous, non-native proteins, their responses probably are less biologically relevant than those obtained with cell proliferation assays and endogenous gene expression/protein/enzyme activity assays (Zacharewski 1997).

The E2 Bioassay uses a chimeric receptor/reporter gene system for investigation of estrogenic substances. The system is comprised of two central components. The first component is a chimeric receptor comprised of the ligand binding domain of the ER linked to the DNA binding domain of the yeast transcription factor Gal4. Complementary DNA (cDNA) encoding for the chimeric receptor has been inserted into a mammalian expression vector which can be transfected into recipient cells such as MCF-7 cells, which then express the chimeric receptor. The second component is a Gal4-regulated luciferase reporter gene, which consists of firefly luciferase cDNA regulated by a rabbit protein promoter and multiple Gal4 response elements. The reporter gene is expressed when an estrogenic substance binds *and* activates the chimeric receptor. Both the chimeric receptor and the reporter gene constructs have been transfected into MCF-7 cells (transiently[3]) and HeLa cells [a cervical cancer cell line] (stably). An estrogenic substance administered to the cells binds to the ER ligand binding domain of the chimeric receptor and activates it. The activated chimeric receptor-ligand complex binds to the Gal4 response elements on the reporter gene and initiates expression of the firefly luciferase cDNA and induction of firefly luciferase activity. Luciferase activity in the E2 Bioassay is a measure of the estrogenic activity of the test substance.

The ER ligand binding domain in the E2 Bioassay chimeric receptor behaves like native ER. No mammalian proteins are known to bind and initiate the gene expression throught the response element used in this bioassay, so induction of rhe reporter gene occurs only through the chimeric receptor (i.e., there is no known interference by other mammalian proteins). However, chimeric receptor/reporter gene assays have disadvantages in addition to those mentioned above. They require specialized equipment and skilled analysts. There is some variability in responsiveness, possibly due to differences in the quality of DNA used to insert the constructs into cells, changes in culture conditions, and/or the number of cell passages (i.e., the number of times cells are grown and then transferred to a new culture dish). Responsiveness of the constructs also can vary with different cell lines. (Zacharewski 1997)

The estrogen receptor-mediated chemically activated luciferase gene expression (ER-CALUX) assay is an example of an example of a cell-based reporter gene assay that has been applied to water samples (Murk et al. 2002), including drinking water (Ghijsen and Hoogenboezem 2000). The bioassay uses a recombinant human T47-D breast adenocarcinoma cell line which contains endogenous ER linked to a stably transfected ER-mediated firefly luciferase gene. Because metabolism of a chemical *in vivo* can result in greater or lesser estrogenic activity, a metabolic transformation step can be added to the ER-CALUX assay by incubating test chemicals with liver microsomes obtained from rats (Legler et al. 2002). Ghijsen and Hoogenboezem (2000) used the ER-CALUX assay to assess the estrogenic potential of extracts of surface water samples collected from the Rhine and Meuse Rivers and extracts of finished drinking water samples taken from drinking water treatment plants in the Rhine and Meuse River basins. The authors reported an unusually strong bioassay response (0.201 ng/L EEQ) in single finished drinking water sample and no detectable response in any of other

[3] In bioassays using transient transfections, the construct must be reintroduced into the cell line each time the cells are grown for use because they cannot replicate the introduced DNA. Stably transfected cell lines replicate the introduced DNA internally so that reintroduction of the construct is not necessary. (Zacharewski et al. 2002)

samples collected in three rounds of sampling. They measured concentrations of known EDCs in the extracts using analytical chemistry methods and predicted responses that they could be expected to produce in the bioassay based on bioassay results for the individual chemicals. Then they compared the predicted bioassay response with the observed response of the bioassay to the extracts to evaluate whether the measured EDCs could account for all of the bioactivity observed in the ER-CALUX assay. In the single drinking water sample that induced a detectable response in the bioassay, the concentrations of selected EDCs were insufficient to account for the response.

The MVLN bioasay for estrogenicity uses human breast cancer cells stably transfected with an estrogen-responsive reporter gene encoding for firefly luciferase. The MVLN bioassay has been used successfully to study estrogenicity of complex matrices such as municipal wastewater effluents (Coors et al. 2003, Snyder et al. 2001), municipal landfill leachate, sludge, and sediments (Coors et al. 2003) and should be applicable to drinking water samples as well. In their report on the use of the MVLN bioassay to characterize municipal landfill leachate, Coors et al. (2003) observed, "Because of the detection limit of about 10 pM E2 in the bioassay (corresponding to about 0.3 ng E2 in 1 L of effluent sample), very low estrogenic activities may not be found with this method."

The Ishikawa cell-alkaline phosphatase (Ishikawa cell-ALP) bioassay uses human endometrial cancer cells which possess estrogen receptors. Alkaline phosphatase activity increases in response to estrogen exposure. Matsuoka et al. (2005) used the Ishikawa cell-ALP bioassay to detect and quantify estrogenicity of river water samples and compared the responses in that bioassay with responses obtained with E-screen and YES. Although there were some differences between responses of the E-screen and the Ishikawa cell-ALP bioassay, results generally were similar. In comparison, YES responses were low or below detection limits for the bioassay. The authors concluded that YES is better suited to analysis of samples containing relatively high levels of estrogenic substances.

***Yeast-based Assays.*** The yeast strain *Saccharomyces cerevisiae* has been used extensively in place of vertebrate cells in bioassays for EDCs (Zacharewski 1997). These assays use transformed yeast cells containing a hormone receptor of interest and a hormone response element linked to an easily measured reporter gene (Zacharewski et al. 2002). Yeast has several advantages over other types of cell lines, including lack of known endogenous receptors that can confound interpretation of the results, media that does not contain steroids, and genetics that are amenable to insertion of mammalian proteins and reporter genes (Zacharewski et al. 2002, Zacharewski 1997). Yeast has limited metabolic capability, so if metabolites are available, they can be studied in a yeast-based assay to assess whether the parent chemical and/or its metabolites are the toxic form (Zacharewski et al. 2002). Yeast-based assays also are attractive for high-throughput screening.

There currently are two variations of yeast-based assays used to investigate estrogenic EDCs. The more common version uses yeast strains transformed with cDNA encoding for human ER and and ERE-regulated LacZ reporter gene that encodes for the enzyme ß-galactosidase. Although the $EC_50$ for these bioassays is higher than those for bioassays using mammalian cells, the yeast-based bioassay is more responsive, allowing for a lower detection limit (0.07 pM). The yeast cells also have been bioengineered to secrete the ß-galactosidase enzyme into the culture medium, improving ease of qualitative and quantitative assessment. The yeast estrogen screen (YES) is based on yeast cells that express human estrogen receptor (hER) and two estrogen response elements (ERE) linked to the lacZ reporter gene. The product ß-galactosidase indicates an estrogenic response. The yeast two-hybrid system assay is based on the ligand-dependent interaction of two proteins - human estrogen receptor (ERα) and the coactivator TIF2 - to induce transcriptional activation of a reporter gene (LacZ) that ultimately induces ß-galactosidase activity indicative of an estrogenic response (Hu et al. 2002).

The other variety of yeast-based bioassay for estrogenic EDCs is an ER-mediated phenotypic transactivation assay that uses the URA3 selectable marker as a reporter gene. The URA3 gene encodes for an enzyme involved in uracil synthesis (OMPdecase). Yeast cells that do not have this enzyme activity cannot grow on minimal media unless the media is supplemented with uracil. The recombinant yeast strain PL3 expresses human ER and contains the URA3 reporter gene under regulatory control of three EREs. The yeast can grow on selective media deficient in uracil only when a ligand binds to the ER and induces expression of the URA3 gene, resulting in translation of the enzyme required for uracil synthesis. In this assay, the $EC_{50}$ for E2 is 3 nM or greater. The response is highly inducible (2,500-fold) in the range of 10-1,000 nM E2.

Yeast-based bioassays have some signficant disadvantages. Ligand potencies assessed in receptor-mediated yeast-based assays do not always agree with those assessed in mammalian cells. Certain pharmaceutical and experimental agents that are strongly antiestrogenic *in vivo* in mammals and *in vitro* in mammalian cell-based assays do not block the ER-mediated response and even act as partial agonists in the yeast-based assays. Differences in permeability between yeast cell walls and mammalian cell membranes, differences in metabolic capabilities between yeast and animal cells, variations among yeast strains, differences in receptor levels and protein proteolysis mechanisms, non-receptor cell-specific factors, multi-drug resistance efflux pumps, and endogenous yeast binding proteins have been postulated as factors that might be responsible for discrepancies between results from yeast-based and mammalian cell-based bioassays. Consequently, the results of yeast-based assessments should be confirmed using other *in vitro* and *in vivo* bioassays. (Zacharewski et al. 2002, Zacharewski 1997)

The YES assay been applied to water samples including river water (Matusuoka et al. 2005, Murk et al. 2002), WWTP influent and effluent (Murk et al. 2002, Routledge 2003, Svenson et al. 2003), and drinking water or simulated drinking water (Fawell et al. 2000, Hu et al. 2003, Huber et al. 2004, Lenz et al. 2004). A yeast two-hybrid system has been applied to investigate estrogenic activity of bisphenol A and its chlorination byproducts (Hu et al. 2002) and estrogenic/antiestrogenic activity of 4-nonylphenol and its chlorination byproducts (Hu and Xie 2002) during simulated drinking water treatment. Antiestrogenic activity can be observed in the yeast two-hybrid system as inhibition of the bioassay response to E2 (Hu and Xie 2002).

***Steroidogenesis Assays.*** All of the *in vitro* bioassays discussed in detail so far have been designed for investigation of EDCs that act through receptor-mediated mechanisms. However, not all EDCs act through binding to a receptor. Certain EDCs exert their effects by interfering with enzymes involved in the synthesis of steroid hormones. *In vitro* steroidogenesis assays are designed to detect chemicals that inhibit production of male and female sex steroid hormones. U.S. EPA optimized a version of a steroidogenesis assay that uses testis slices as a source of steroidogenic enzymes, but due to concerns about its inability to distinguish between chemicals that inhibit steroidogenesis and chemicals that kill cells responsible for testosterone synthesis, further work on the validation of this assay was abandoned. The discontinued assay was replaced by a cell-based assay that employs the H295R human adrenocortical carcinoma cell line. This assay shows promise for ability to detect chemicals that induce enzymes responsible for steroid synthesis in addition to chemicals that inhibit it. (EDSP 2006 Assays)

Aromatase assays are a subcategory of steroidogenesis assays. Aromatase, an enzyme complex involved in estrogen biosynthesis, converts androgens to estrogens (E2 and estrone) (EDSP 2006, WHO 2002). A yeast-based screening assay has been developed for concurrent identification of AR ligands and aromatase inhibitors (Zacharewski et al. 2002). Oh et al. (2006) evaluated the effect of contaminants in river water extracts on aromatase activity in JEG-3 cells, a human choriocarcinoma-derived cell line. This same bioassay could be applied to drinking water extracts.

***In vivo Bioassays***

***Fish Bioassays.*** *In vivo* fish bioassays using vitellogenin induction as a response also have been used to characterize estrogenicity of drinking water (Fawell et al. 2001, Harries et al. 1996). Fish bioassays can provide an assessment of exposure integrated over time (Fawell et al. 2001), but the relevance of the vitellogenin induction endpoint and fish model to human health is not entirely clear at this time. Although there is considerable homology in the endocrinology of vertebrates, there are significant differences in endocrine function among some species. This warrants careful consideration during interspecies extrapolations (WHO 2002). There also are substantial differences between fish and humans with regard to exposure to and uptake of waterborne contaminants and metabolic and detoxification systems.

***Rodent Bioassays.*** Rodents are common test species in bioassays used to predict the potential for toxic effects of chemicals in humans. *In vivo* rodent bioassays are widely used to assess the estrogenic effects of substances using a variety of endpoints including effects on organ weights (e.g., uterus), cell differentiation, protein expression, and enzyme activities (Zacharewski 1997). Microarrays have been used to study hormonal regulation of genes *in vivo* and have been proposed as a tool to investigate changes in gene expression during *in vivo* bioassays for EDCs (Zacharewski et al. 2002). Several assays under consideration for use in the EDSP are rodent bioassays, including the Hershberger, pubertal male, pubertal female, uterotrophic, 15-day intact adult male, mammalian 2-generation, and in utero through lactation assays (EDSP 2006). For more details, see Table 3.1.

For various reasons, the appropriateness of rodents as models to assess the risk that EDCs pose to humans has been questioned. For example, unlike humans, rodents to not express sex SHBG after birth (Zacharewski 1998). Waring and Harris (2005) reported that humans have relatively active metabolic detoxification enzymes compared with many other species. They also stated that rodents are not particularly good animal models for reproductive toxicity in humans and suggested that rabbits or pigs would be better models. Primates also are considered to be suitable substitutes for humans in toxicity testing, but ethical considerations and costs for primate testing often are prohibitive. Despite their drawbacks, *in vivo* rodent bioassays remain an important component of assessments conducted on the potential effects of EDCs (Zacharewski 1997).

***Human Bioassays and Epidemiologic or Clinical Studies of Humans.*** In some cases, humans can be assessed for evidence of exposure to EDCs and/or endocrine disruption as a result of such exposure. People who received occupational exposure to industrial chemicals or pesticides have been evaluated for exposure to and effects of EDCs. Epidemiologic studies have been conducted on humans exposed unintentionally to contaminants in their environment, diet, and drinking water supplies. Clinical studies have been conducted on people exposed to endocrine-disrupting pharmaceutical agents. While these studies are useful, they are plagued by a host of problems. Most importantly, any ethical concerns associated with the use of human subjects must be addressed. Studies must control for effects of behavior and personal choices like smoking, drug and alcohol use, diet, and exercise. Long delays between exposure to EDCs and potential effects can confound attempts to identify an association or assign causality. Studies of endocrine disruption in humans and the limitations of human studies to assess the potential for chemicals to cause endocrine disruption are reviewed in more detail elsewhere (WHO 2002).

Much of the work that has been done to date on potential human health effects of EDCs in drinking water has suffered from poor characterization of exposure. It is particularly difficult to assess exposure in people because researchers can exert only limited control over human subjects and generally are restricted to the use of non-invasive or minimally invasive assessments of exposure and effects. Furthermore, wide-scale sampling and analyses for EDCs in drinking water can be laborious and expensive, particularly given that people might obtain their drinking water from multiple sources. Exposure to the chemical or chemicals of interest through routes other than drinking water also must be considered, as well as co-exposure to chemicals (e.g., other EDCs) that can affect the endpoints used for assessment of exposure and effects.

Particularly noteworthy for this review is a clinical study of the effects of the EDC perchlorate intentionally administered via drinking water to human volunteers (Greer et al. 2002). Groups of healthy men and women were given perchlorate in doses ranging from 0.007 to 0.5 mg/kg of body weight per day for 14 days. The assessment endpoint for the study was inhibition of iodide uptake into the thyroid, which is not considered to be an adverse effect but rather an early step that must occur in a series of events leading to subsequent adverse effects. The study identified a NOEL for inhibition of iodide uptake by the thyroid at 0.007 mg/kg per day. If inhibition of iodide uptake into the thyroid gland is sufficiently great in terms of magnitude and duration, thyroid hormone synthesis declines. When stored thyroid hormone is depleted, blood levels of thyroid hormones will decline below the normal physiologic range. No significant changes in blood thyroid hormone concentrations were observed at any dose in this study. The results agreed with those from other similar studies, one of which was 6 months in duration. It should be noted that perchlorate represents a somewhat unique case among EDCs in that its MOA is relatively well-studied and simple in comparison with many other known or potential EDCs. Perchlorate has only one known MOA - competitive inhibition of iodide uptake into the thyroid gland – while multiple MOAs are possible for many other EDCs. The U.S. EPA selected the NOEL taken from the study by Greer et al. (2002) as the POD for development of a reference dose for perchlorate. U.S. EPA's use of the inhibition of iodide uptake into the thyroid as the POD for development of a reference dose is unusual in that this endpoint does not represent an adverse effect. Selection of this endpoint is considered to be protective and conservative because inhibition of iodide uptake must precede any adverse effects, and prevention of this early biochemical change should protect against any subsequent adverse effects.

## BIOLOGICALLY-BASED METHODS FOR PHARMACEUTICALS

Traditional toxicity endpoints (such as reproductive and developmental toxicity) that can be used to assess the toxicity of pharmaceuticals *in vivo* are described in Chapter 6. Pharmaceuticals are useful for therapeutic purposes because they are biologically active and potent, but they also can be toxic and exert negative side-effects. Therapeutic doses are intended to maximize therapeutic benefits while minimizing the potential negative side-effects. People use pharmaceuticals only when the benefits outweigh the risks. The concern for adverse effects is magnified for unintended exposures such as those that occur through drinking water. This is particularly true for non-target groups for whom the pharmaceuticals were never intended. For example, certain drugs are not prescribed to pregnant women because of the risk of harm to the fetus.

Because drugs are intended to exert potent effects through a particular MOA, toxicity tests directed at identifying effects through the pharmaceutical MOA in non-target populations or through MOA that are responsible for major pharmaceutical side-effects might be more sensitive, resulting in more conservative estimates of health-protective levels in drinking water.

## DRINKING WATER AS A BIOANALYTICAL MATRIX

An important consideration when using biologically-based methods to analyze drinking water is the occurrence of residual disinfectants and disinfection byproducts. This is particularly important for *in vitro* cellular bioassays, in which residual chlorine in drinking water can be toxic to cells in culture. To prevent toxicity caused by residual chlorine, water samples can be aerated prior to subjecting them to a bioassay, provided that the sample holding time is not exceeded for the target analytes. However, in cases where the bioassay is intended to characterize the effect of a mixture of unknown chemicals that might be present in a drinking water sample, aeration might result in the loss of volatile chemicals other than chlorine, resulting in a reduced response in the bioassay and underestimation of the potential for toxicity. Also, some drinking water contaminants (e.g., E2) can be degraded by residual disinfectants during sample transport and storage such that oxidant quenching is required to maintain the integrity of the target analytes and their associated bioactivity. Depending on sample holding conditions, long holding times also might result in degradation of the target analytes by other processes, e.g., by photodegradation or biodegradation. For these reasons, chemical quenching of an oxidant in the water sample is an attractive option, but quenching agents must be carefully selected to avoid additional toxic effects related to sample pre-treatment. Bioassay controls should be run with and without quenching agent to ensure that the added chemical does not affect the results of the bioassay. Alum et al. (2004) added sodium thiosulfate to quench residual chlorine in drinking water samples prior to subjecting them to the E-screen, but the quenching agent produced detrimental effects in the controls. The investigators found that ammonium chloride adequately quenched the residual chlorine without interfering with the bioassay.

Addition of chemicals to water samples for sample preservation also must be carefully considered if the samples are intended for use in a bioassay. For example, if water samples are collected for bioassay-directed fractionation and analysis, it might be necessary to apply measures to maintain the bioactivity of the sample. However, any chemical preservatives that are added to samples must be non-toxic in the bioassay.

**Table 3.1**
**Bioassays current under consideration for use in the Endocrine Disruptor Screening Program (EDSP)***

| Bioassay | Tier* | Description |
|---|---|---|
| Amphibian (Frog) Metamorphosis | 1 | *In vivo* bioassay that uses tadpoles to determine whether chemicals affect the thyroid during metamorphosis and consequently result in developmental effects. |
| Androgen Receptor (AR) Binding | 1 | Two *in vitro* assays under consideration to examine the ability of a test chemical to bind with androgen receptors, one using rat prostate cytosol and one using recombinant rat androgen receptor. |
| Aromatase *In vitro* | 1 | *In vitro* assay that detects substances that inhibit the activity of aromatase, an enzyme complex responsible for estrogen biosynthesis that converts androgens into estrogens (estradiol and estrone). |
| Estrogen Receptor (ER) Binding | 1 | Two *in vitro* assays under consideration to examine the ability of a test chemical to bind with estrogen receptors, one using rat uterine cytosol and one using the alpha isoform (ER-α) of the human recombinant estrogen receptor. |
| Fish Screen | 1 | *In vivo* bioassay used to screen for estrogenic and androgenic effects in fish exposed to test chemicals. Examines abnormalities associated with survival, reproductive behavior, secondary sex characteristics, histopathology, and fecundity (i.e., number of spawning events, number of eggs per spawning event, fertility, and development of offspring). |
| Hershberger | 1 | *In vivo* rat bioassay that detects androgenic and anti-androgenic effects. Accessory sex gland weights, including several androgen-dependent tissues, are measured in castrated or immature male rats exposed to test chemicals. |
| Pubertal Female | 1 | *In vivo* rat bioassay used to screen for estrogenic and thyroid activity in females during sexual maturation. Examines abnormalities associated with sex organs, puberty markers, and thyroid tissue. |
| Pubertal Male | 1 | *In vivo* rat bioassay used to screen for androgenic, anti-androgenic, and thyroid activitiy in males during sexual maturation. Examines abnormalities associated with sex organs, puberty markers, and thyroid tissue. |

(continued)

**Table 3.1 (Continued)**

| | | |
|---|---|---|
| Steroidogenesis – Cell-based H295R | 1 | *In vitro* cell-based assay that uses the H295R human adrenocortical carcinoma cell line to detect chemicals that inhibit steroidogenesis. Shows promise for being able to to detect chemicals that induce steroidogenic enzymes in addition to those that inhibit them. |
| Uterotrophic | 1 | *In vivo* female rat bioassay used to screen for estrogenic effects. Uterine weight changes are measured in ovariectomized or immature female rats. |
| 15-Day Intact Adult Male Rat | 1 | *In vivo* male rate bioassay used to screen primarily for anti-androgenic and thyroid activity. Screens for abnormalities associated with primary and secondary sex organs, systemic hormone concentrations, and thyroid. |
| Amphibian 2-Generation (Development, Reproduction | 2 | *In vivo* frog bioassay used to characterize dos-response characteristics and adverse reproductive and developmental effects. |
| Avian 2-Generation | 2 | *In vivo* bioassay uses Japanese quail to characterize dose-response characteristics and adverse reproductive and developmental effects. |
| Fish Lifecycle | 2 | *In vivo* bioassay uses fish to characterize dose-response characteristics and adverse reproductive and developmental effects. |
| Invertebrate (Mysid) Lifecycle | 2 | *In vivo* bioassay uses mysid shrimp to characterize dose-response characteristics and adverse reproductive and developmental effects. |
| Mammalian 2-Generation | 2 | *In vivo* bioassay uses rats to characterize dose-response characteristics and adverse reproductive and developmental effects. |
| In Utero through Lactation | TBD | Pregnant rats are used assess post-natal development of the neonate after in utero and lactational exposure. |

*Source:* EDSP 2006
TBD – to be determined
*Tier 1 assays are screening-level bioassays designd to identify the potential for chemicals to interact with the estrogen, androgen, and thyroid systems; they are not designed to be used for prediction of *in vivo* toxicity. Tier 2 assays are intended to identify adverse effects resulting from interference with the endocrine system and to provide a quantitative estimate of the dose necessary to cause an adverse response; they will be accepted as definitive for risk assessments.

## REFERENCES


Aherne, G.W., J. English, and V. Marks. 1985. The role of immunoassay in the analysis of microcontaminants in water samples. *Ecotoxicology and Environmental Safety*, 9:79-83.

Alum, A., Y. Yoon, P. Westerhoff, and M. Abbaszadegan. 2004. Oxidation of bisphenol A, 17ß-estradiol, and 17α-ethynyl estradiol and byproduct estrogenicity. *Environmental Toxicology*, 19(3):257-264.

Ayscough, N.J., J.Fawell, G.Franklin, and W.Young. 2000. *Review of Human Pharmaceuticals in the Environment.* R&D Tech. Report P390. WRc Report No. EA4761. WRc-NSF Ltd, Buckinghamshire. Bristol: Environment Agency.

Barton, H.A. 2003. Endocrine active substances and dose response for individuals and populations. *Pure & Applied Chemistry*, 75(11):2159-2166.

Behnisch, P.A., K. Fujii, K. Shiozaki, I. Kawakami, and S. Sakai. Estrogenic and dioxin-like potency in each step of a controlled landfill leachate treatment in Japan. *Chemosphere*, 43(4-7):977-984.

Bolz, U., W. Koerner, and H. Hagenmaier. 2000. Development and validation of a GC/MS method for determination of phenolic xenoestrogens in aquatic samples. *Chemosphere*, 40(9-11):929-935.

Coors, A., P.D. Jones, J.P. Giesy, and H. T. Batte. 2003. Removal of Estrogenic Activity from Municipal Waste Landfill Leachate Assessed with a Bioassay Based on Reporter Gene Expression. *Environmental Science and Technology*, 37(15):3430-3434.

Daughton, C.G., and T.A. Ternes. 1999. Pharmaceuticals and personal care products in the environment: agents of subtle change? *Environmental Health Perspectives*, 107(Supplement 6):907-938.

Degitz, S.J., G.W. Holcombe, K.M. Flynn, P.A. Kosian, J.J. Korte, and J.E. Tietge. 2005. Progress towards development of an amphibian-based thyroid screening assay asing *Xenopus laevis*. Organismal and thyroidal responses to the model compounds 6-propylthiouracil, methimazole, and thyroxine. *Toxicological Sciences*, 87(2):353-364.

EDSP (Endocrine Disruptor Screening Program). 2007. Endocrine Primer [website]. Endocrine Disruptor Screening Program (EDSP), U.S. Environmental Protection Agency. Updated June 18, 2007. Available: http://www.epa.gov/endo/pubs/edspoverview/primer.htm

EDSP (Endocrine Disruptor Screening Program). 2006. Assays Under Consideration [website]. Endocrine Disruptor Screening Program, U.S. Environmental Protection Agency. Updated March 29, 2006. Available: http://www.epa.gov/scipoly/oscpendo/pubs/assayvalidation/consider.htm

EDSTAC (Endocrine Disruptor Screening and Testing Advisory Committee). 1998. Endocrine Disruptor Screening and Testing Advisory Committee (EDSTAC) Final Report. U.S. Environmental Protection Agency. August. http://www.epa.gov/scipoly/oscpendo/edspoverview/finalrpt.htm

European Commission. 2006. Environment fact sheet: REACH—A new chemicals policy for the EU. February. Available: http://ec.europa.eu/environment/chemicals/reach/fact_sheet.pdf

European Commission. 2002. Study on the Scientific Evaluation of 12 Substances in the Context of Endocrine Disrupter Priority List of Actions. Report No. UC 6052. November. Buckinghamshire, UK: European Commission. Available: http://ec.europa.eu/environment/endocrine/documents/wrc_report.pdf

Faustman E.M. and G.S. Omenn. 2001. Risk Assessment. In *Casarett and Doull's Toxicology: The Basic Science of Poisons*. Edited by C.D. Klaassen. New York: The McGraw-Hill Companies, Inc.

Fawell, J.K., D. Sheahan, H.A. James, M. Hurst, and S. Scott. 2001. Oestrogens and oestrogenic activity in raw and treated water in Severn Trent water. *Water Research*, 35(5):1240-1244.

Fisher, J.S. 2004. Are all EDC effects mediated via steroid hormone receptors? *Toxicology*, 205(1-2):33-41.

Ghijsen, R.T., and W. Hoogenboezem. 2000. Endocrine disrupting compounds in the Rhine and Meuse basin: occurrence in surface, process and drinking water. Amsterdam: Association of River Waterworks (RIWA).

Greer, M.A., G. Goodman, R.C. Pleus, and S.E. Greer. 2002. Health effects assessment for environmental perchlorate contamination: the dose response for inhibition of thyroidal radioiodine uptake in humans. *Environmental Health Perspectives*, 110(9):927-937.

GWRC (Global Water Research Coalition). 2003. Endocrine Disrupting Compounds: An Overview of Sources and Biological Methods for Measuring EDCs. September. Global Water Research Coalition (GWRC). 16 p.

GWRC (Global Water Research Coalition). 2003. Workshop on Methodologies for Chemical and Biological Analysis of EDC in Water Systems. Workshop Report. Draft Version. June. Global Water Research Coalition (GWRC). Karlsruhe, Germany. 62 p.

Harries, J.E., D.A. Sheahan, S. Jobling, P. Matthiessen, P. Neall, E.J. Routledge, R. Rycroft, J.P. Sumpter, and T. Tylor. 1996. A survey of estrogenic activity in United Kingdom inland waters. *Environmental Toxicology and Chemistry*, 15(11):1993-2002.

Harries, J.E., S. Jobling, P. Matthiessen, D.A. Sheahan, and J.P. Sumpter. 1995. Effects of Trace Organics on Fish - Phase 2. Report to the Department of the Environment. Marlow, UK: Foundation for Water Research.

Hogue, C. 2007. The future of U.S. chemical regulation. *Chemical & Engineering News* 85(2): 34-38. Available: http://pubs.acs.org/cen/government/85/8502regulation.html.

Holland, P.T. 2003. Analysis of endocrine active substances in food and the environment. *Pure & Applied Chemistry*, 75(11):1843-1857.

Hu, J., S. Cheng, T. Aizawa, Y. Terao, and S. Kunikane. 2003. Products of aqueous chlorination of 17β-estradiol and their estrogenic activities. *Environmental Science & Technology*, 37(24):5665-5670.

Hu, J.-Y., T. Aizawa, and S. Ookubo. 2002a. Products of aqueous chlorination of bisphenol A and their estrogenic activity. *Environmental Science & Technology*, 36(9):1980-1987.

Hu, J.-Y., and G.-H. Xie. 2002. Products of aqueous chlorination of 4-nonylphenol and their estrogenic activity. *Environmental Toxicology and Chemistry*, 21(10):2034-2039.

Huber, M.M., T.A. Ternes, and U. Vongunten. 2004. Removal of estrogenic activity and formation of oxidation products during ozonation of 17α-ethinylestradiol. *Environmental Science & Technology*, 38(19):5177-5186.

Huggett, D.B., C.M. Foran, B.W. Brooks, J. Weston, B. Peterson, K.E. Marsh, T.W. La Point, and D. Schlenk. 2003. Comparison of *in vitro* and *in vivo* bioassays for estrogenicity in effluent from North American municipal wastewater facilities. *Toxicological Sciences*, 72(1):77-83.

IEH (Institute for Environment and Health). 2005. Chemicals Purported to be Endocrine Disruptors: a Compilation of Published Lists (Web Report W20). March. Medical

Research Council (MRC), Institute for Environment and Health (IEH). Leicester, United Kingdom. http://www.silsoe.cranfield.ac.uk/ieh/pdf/w20.pdf.

Immunochemistry Technologies. 2005. Assay Development Services. What Is an Immunoassay or ELISA? [website]. Bloomington, MN: Immunochemistry Technologies, LLC. Available: http://www.immunochemistry.com/AssayDevWhatIsAnELISA.htm

Johnson, B.L., H.E. Hicks, W. Cibulas, O. Faroon, A.E. Ashizawa, C.T. De Rosa, V.J. Cogliano, and M. Clark.1996. Public Health Implications of Exposure to Polychlorinated Biphenyls (PCBs). December. Agency for Toxic Substances and Disease Registry (ATSDR), Atlanta, GA, and U.S. Environmental Protection Agency (U.S. EPA), Washington, DC. Available: http://www.atsdr.cdc.gov/DT/pcb007.html

Kavlock, R.J., G.P. Daston, C. Derosa, P. Fenner-Crisp, L.E. Gray, S. Kaattari, G. Lucier, M. Luster, M.J. Mac, C. Maczka, R. Miller, J. Moore, R. Rolland, G. Scott, D.M. Sheehan, T. Sinks, and H.A. Tilson. 1996. Research needs for the risk assessment of health and environmental effects of endocrine disruptors: A report of the U.S. EPA-sponsored workshop. *Environmental Health Perspectives*, 104(Supplement 4):715-740.

Körner, W., V. Hanf, W. Schuller, C. Kempter, J. Metzger, and H. Hagenmaier. 1999. Development of a sensitive E-screen assay for quantitative analysis of estrogenic activity in municipal sewage plant effluents. *The Science of the Total Environment*, 225(1-2):33-48.

Langley, G., T. Evans, S.T. Holgate, and A. Jones. 2007. Replacing animal experiments: choices, chances and challenges. *BioEssays*, 29(9):918-926.

Legler, J., M. Dennekampa, A.D. Vethaakc, A. Brouwera, J.H. Koemana, B. Van Der Burg, and A.J. Murka. 2002. Detection of estrogenic activity in sediment-associated compounds using *in vitro* reporter gene assays. *Science of the Total Environment*, 293(1-3):69-83.

Lenz, K., V. Beck, and M. Fuerhacker. 2004. Behaviour of bisphenol A (BPA), 4-nonylphenol (4-NP) and 4-nonylphenol ethoxylates (4-NP1EO, 4-NP2EO) in oxidative water treatment processes. *Water Science & Technology*, 50(5):141-147.

Leusch, F.D.L., H.F. Chapman, W. Körner, S.R. Gooneratne, and L.A. Tremblay. 2005. Efficacy of an advanced sewage treatment plant in southeast Queensland, Australia, to remove estrogenic chemicals. *Environmental Science & Technology*, 39(15):5781-5786.

Leusch, F.D., H.F. Chapman, M.R. Van Den Heuvel, B.L. Tan, S.R. Gooneratne, and L.A. Tremblay. 2006a. Bioassay-derived androgenic and estrogenic activity in municipal sewage in Australia and New Zealand. *Ecotoxicology and Environmental Safety*, 65(3):403-411.

Leusch, F.D., M.R. Van Den Heuvel, H.F. Chapman, S.R. Gooneratne, A.M. Eriksson, and L.A. Tremblay. 2006b. Development of methods for extraction and *in vitro* quantification of estrogenic and androgenic activity of wastewater samples. *Comparative Biochemistry and Physiology. Part C: Toxicology & Pharmacology*, 143(1):117-126.

Matsuoka, S., M. Kikuchi, S. Kimura, Y. Kurokawa, and S. Kawai. Determination of Estrogenic Substances in the Water of Muko River Using *in vitro* Assays, and the Degradation of Natural Estrogens by Aquatic Bacteria. 2005. *Journal of Health Science*, 51:178-184.

Meulenberg, E.P. 2006. Relevance of the Assessment of Thyroidal Activity in the (Water) Environment: A Deskresearch. Nieuwegein, The Netherlands: Association of River Waterworks - Rhine Water Works (RIWA).

Meulenberg, E.P., and G. Marchesini. 2006. Thyroid Hormone-like Activity: Biosensor Screening of Surface Water. Nieuwegein, The Netherlands: Association of River Waterworks - Rhine Water Works (RIWA).

Murk, A.J., J. Legler, M.M.H. Van Lipzig, J.H.N. Meerman, A.C. Belfroid, A. Spenkelink, B. Van Der Burg, G.B.J. Rijs, and D. Vethaak. 2002. Detection of estrogenic potency in wastewater and surface water with three *in vitro* bioassays. *Environmental Toxicology and Chemistry*, 21(1):16-23.

Musset, L. 2007. Personal communication. November 9. Paris, France: OECD Environment Directorate; Environment, Health, and Safety Division.

Nelson, M.A., E. Papastavros, M. Dodlinger, and D.S. Hage. 2007. Environmental analysis by on-line immunoextraction and reversed-phase liquid chromatography: optimization of the immunoextraction/RPLC interface. *Journal of Agricultural and Food Chemistry*, 55(10):3788-3797.

Oh, S.-M., K. Park, and K.-H. Chung. 2006. Combination of *in vitro* bioassays encompassing different mechanisms to determine the endocrine-disrupting effects of river water. *Science of the Total Environment*, 354(2-3):252-264.

NRC (National Research Council). 2005. Health implications of perchlorate ingestion. Washington, DC: National Research Council (NRC) of the National Academy of Sciences (NAS), National Academies Press. Available: http://www.nap.edu/catalog/11202.html.

NTP (National Toxicology Program). 2001. National Toxicology Program's Report of the Endocrine Disruptors Low-Dose Peer Review. Research Triangle Park, North Carolina: National Toxicology Program. Available: http://ntp-server.niehs.nih.gov/ntp/htdocs/liason/LowDosePeerFinalRpt.pdf.

Oh, S.-M., K. Park, and K.-H. Chung. 2006. Combination of *in vitro* bioassays encompassing different mechanisms to determine the endocrine-disrupting effects of river water. *Science of the Total Environment*, 354(2-3):252-264.

Quanrud, D.M., K. Quast, O. Conroy, M.M. Karpiscak, C.P. Gerba, K.E. Lansey, W.P. Ela, and R.G. Arnold. 2004. Estrogenic activity and volume fraction of waste water origin in monitoring wells along the Santa Cruz River, Arizona. *Ground Water Monitoring & Remediation*, 24(2):86-93.

Routledge, E.J. 2003. Identifying the causative agents: The use of combined chemical and biological strategies in monitoring programs. *Pure and Applied Chemistry*, 75(11-12): 2461-2466.

Schmidt, B.M.W., D. Gerdes, M. Feuring, E. Falkenstein, M. Christ, and M. Wehling. 2000. Rapid, nongenomic steroid actions: a new age? *Frontiers in Neuroendocrinology*, 21(1):57-94.

Schwarzenbach RP, Escher BI, Fenner K, Hofstetter TB, Johnson CA, von Gunten U, and Wehrli B. 2006. The challenge of micropollutants in aquatic systems. *Science,* 313 (5790):1072-1077.

Shim, W.B., Z.Y. Yang, J.Y. Kim, J.G. Choi, J.H. Je, S.J. Kang, A.Y. Kolosova, S.A. Eremin, and D.H. Chung. 2006. Immunochromatography using colloidal gold-antibody probe for the detection of atrazine in water samples. *Journal of Agricultural and Food Chemistry*, 54(26):9728-9734.

Snyder, S.; Vanderford, B.; Pearson, R.; Quiñones, O.; Yoon, Y. 2003. Analytical methods used to measure endocrine disrupting compounds in water. *Practice Periodical of Hazardous, Toxic, and Radioactive Waste Management* 7(4): 224-234.

Snyder, S.A., D.L. Villeneuve, E.M. Snyder, and J.P. Giesy. 2001. Identification and quantification of estrogen receptor agonists in wastewater effluents. *Environmental Science & Technology*, 35(18):3620-3625.

Snyder, S.A., E.M. Snyder, D.L. Villeneuve, K. Kannan, S.A. Villalobos, A. Blankenship, and J.P. Giesy. 2000. Instrumental and bioanalytical measures of endocrine disruptors in water. In *Analysis of Environmental Endocrine Disruptors: ACS Symposium Series 747*. Edited by L. H. KeithT. L. Jones-Lepp and L. L. Needham. Washington, DC: American Chemical Society.

Soto, A.M., J.M. Calabro, N.V. Prechtl, A.Y. Yau, E.F. Orlando, A. Daxenberger, A.S. Kolok, L.J. Guillette, Jr., B. Le Bizec, I.G. Lange, and C. Sonnenschein. 2004. Androgenic and estrogenic activity in water bodies receiving cattle feedlot effluent in eastern Nebraska, USA. *Environmental Health Perspectives*, 112(3):346-352.

Svenson, A., A.-S. Allard, and M. Ek. 2003. Removal of estrogenicity in Swedish municipal sewage treatment plants. *Water Research*, 37(18):4433-4443.

Timm, G. 2007. Personal communication. E-mail message on September 29, 2007. Washington, DC: U.S. Environmental Protection Agency, Office of Science Coordination and Policy.

U.S. EPA (United States Environmental Protection Agency). 2008. Endocrine Disruptor Screening Program: Statement of Policy; Notice. Endocrine Disruptor Screening Program: Priority-Setting Workshop; Notice. United States Environmental Protection Agency. *Federal Register* 63 (248), 71542-71568. Available: http://www.epa.gov/endo/pubs/122898frnotice.pdf.

U.S. EPA (United States Environmental Protection Agency). 2007. Draft list of initial pesticide active ingredients and pesticide inerts to be considered for screening under the Federal Food, Drug, and Cosmetic Act. *Federal Register*, 72(116):33486-33503. http://www.epa.gov/safewater/contaminants/index.html

U.S. EPA (United States Environmental Protection Agency). 2005. Endocrine Disruptor Screening Program; Chemical Selection Approach for Initial Round of Screening. September 27. U.S. Environmental Protection Agency (USEPA). *Federal Register* 70(186): 56449-56465.

U.S. EPA (United States Environmental Protection Agency). 1998. Endocrine Disruptor Screening and Testing Advisory Committee (EDSTAC) Final Report. U.S. Environmental Protection Agency. August. http://www.epa.gov/scipoly/oscpendo/edspoverview/finalrpt.htm

WHO. 2002. *Global Assessment of the State-of-the-Science of Endocrine Disruptors*. WHO/PCS/EDC/02.2. Edited by T Damstra, S Barlow, A Bergman, R Kavlock, and G Van Der Kraak: World Health Organization (WHO). http://www.who.int/ipcs/publications/new_issues/endocrine_disruptors/en/print.html

Zacharewski, T.R.;Fertuck, K.C.; Fielden, M.R.; Matthews, J.B. 2002. Mechanistically-based assays for the identification and characterization of endocrine disruptors. In: *Comprehensive Toxicology XIV: Cellular and Molecular Toxicology* (Volume Editors: J.P. Vanden Heuvel, W.F. Greenlee, G.H. Perdew, W. Mattes; Editors-in-Chief: I.G. Sipes, C.A. McQueen, A.J. Gandolfi). Elsevier Science. Amsterdam, The Netherlands. p.559-582.

Zacharewski, T. 1998. Identification and assessment of endocrine disruptors: limitations of *in vivo* and *in vitro* assays. *Environmental Health Perspectives*, 106(Suppl 2):577-582.

Zacharewski, T. 1997. *In vitro* bioassays for assessing estrogenic substances. *Environmental Science & Technology*, 31(3):613-623.

# CHAPTER 4
# OCCURRENCE

## INTRODUCTION

In the early 1990s, an increasing number of scientific studies began reporting the presence of organic compounds at trace concentrations in surface and groundwater under the impact of wastewater discharge. These compounds span a wide range of categories such as pharmaceutically active compounds and personal care products (PPCPs), household chemicals, endocrine disrupting compounds like steroid hormones, and others, which were released into the water through human activities or human metabolic processes. While the occurrence of these compounds in the aquatic environment represented a new finding, their presence was not new and their detection was driven mainly by the availability of more sensitive analytical instruments that allowed scientists to detect concentrations of compounds in water at the low parts-per-trillion (ppt) or nanogram-per-liter (ng/L) levels. Details of the various compounds identified in these studies were recently summarized in several review articles (Daughton and Ternes 1999; Heberer 2002; Snyder, Westerhoff et al. 2003). Analytical methods employed to quantify these trace organic compounds are summarized in Chapters 2 and 3. This chapter discusses the most frequently detected EDCs and PPCPs in source and drinking waters. As stated earlier, it is debatable which compounds should be considered as EDCs. Compounds that are not currently considered EDCs may be determined to have endocrine disruptive effects after further screening efforts are completed. With that in mind, this chapter is focusing on compounds that have been shown to have estrogenic and androgenic effects. For PPCPs, those compounds have been compiled that are most frequently reported in the current literature. Moreover, lists of the most widely used household chemicals and other trace organics have been presented as a demonstration of potential compounds of concern that may be detected in the future.

Drinking water sources that receive wastewater discharge or that are augmented in part through wastewater recycling can contain wastewater-derived trace organic compounds at concentrations that might cause potential adverse effects to aquatic life or human health. A study published by the U.S. EPA in 1980 estimated that about 15 million people in the United States were served by surface supplies containing at least 10 percent wastewater at low flow conditions and 4 million people use municipal supplies that contain 100 percent wastewater during low flow conditions (Swayne, Boone et al. 1980). With increasing water demand, dwindling water resources in many communities, and increasing interest in planned indirect potable reuse applications over the last 25 years, the proportion of wastewater impacted drinking water supplies in the U.S. has likely risen significantly.

The sources of EDCs and PPCPs in the environment are multifarious but are for the most part related to human activities. The most important sources for release of these compounds into surface and groundwater are discharges from municipal wastewater treatment plants, industrial manufacturing processes, leaky sewers, combined sewage overflows (CSOs), onsite wastewater systems (OWS), and confined animal feeding operations (CAFOs) (Drewes and Shore 2001; Kim and Carlson 2006). Therefore, the occurrence level of EDCs and PPCPs in drinking water sources primarily depends upon the degree of wastewater impact in the watershed. However, the degree of wastewater impact for the majority of occurrence studies is frequently either not reported or not known. Additional factors affecting occurrence are usage pattern for different compounds as well as prescription practices, which can vary with region, per-capita water consumption, which can result in different levels of dilution, and substitution and phase-out programs for specific chemicals, which denotes that the occurrence pattern of currently detected compounds is not static.

For source water conditions which are heavily impacted by wastewater discharge, the occurrence level of EDCs and PPCPs in treated wastewater effluents can be considered a "worst case" scenario for the potential presence of these compounds in drinking water sources. Therefore, this chapter will initially summarize findings regarding occurrence of compounds of interest in secondary and tertiary treated municipal wastewater effluents. A subsequent section will present current information regarding the occurrence of EDCs and PPCPs in drinking water sources, followed by a section addressing occurrence in raw and finished drinking water.

## OCCURRENCE OF EDCS AND PPCPS IN TREATED WASTEWATER EFFLUENTS

In order to compile information on the occurrence of trace organic compounds of interest in secondary or tertiary treated wastewater effluents, over 1,000 references were screened reporting occurrence in studies across the globe. This comprehensive review considered only articles that were peer-reviewed and reported both analytical methods employed and detailed experimental conditions. From these references over 100 papers published in 15 separate journals from 70 authors were identified meeting the review criteria of this study. Based upon this survey, the team identified 239 unique wastewater-derived organic micropollutants in domestic secondary effluents. These compounds were classified into 24 categories (i.e., antibacterial, antibiotic, etc.) as illustrated in Figure 4.1. It is noteworthy, that the number of compounds detected in each category does not necessarily reflect a real occurrence distribution but is biased through the selection of compounds targeted in studies currently published in the peer-reviewed literature. Based upon the findings of this survey, pharmaceutical residues, antibiotics, steroid hormones, and fragrances are the most commonly reported trace organic compounds currently reported to occur in secondary and tertiary treated municipal effluents.

The majority of studies on EDCs and PPCPs occurrence in treated wastewater published in the peer-reviewed literature today have been conducted in Europe and North America followed by Asia, South America and Australia. The following sections will mainly focus on occurrence of EDCs and PPCPs in Europe and North America.

The detection frequencies of select pharmaceutical residues and steroid hormones in secondary/tertiary treated wastewater effluents are presented in Figures 4.2 and 4.3. The targeted pharmaceutical residues exhibited a detection frequency of 70 percent in Europe for all compounds selected, whereas a lower detection frequency was observed in effluents samples collected in North America. In contrast, the detection frequency of steroid hormones exceeding 60 percent was very similar between Europe and North America. Of the four estrogens, 17α-ethynylestradiol exhibited the lowest detection frequency.

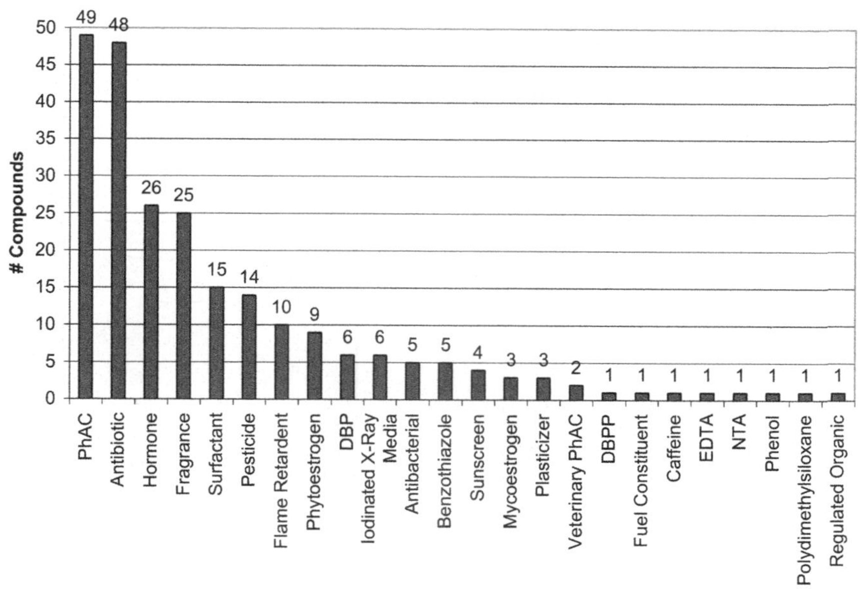


**Figure 4.1 Classes of compounds reported to occur in secondary/tertiary treated municipal wastewater effluents worldwide**

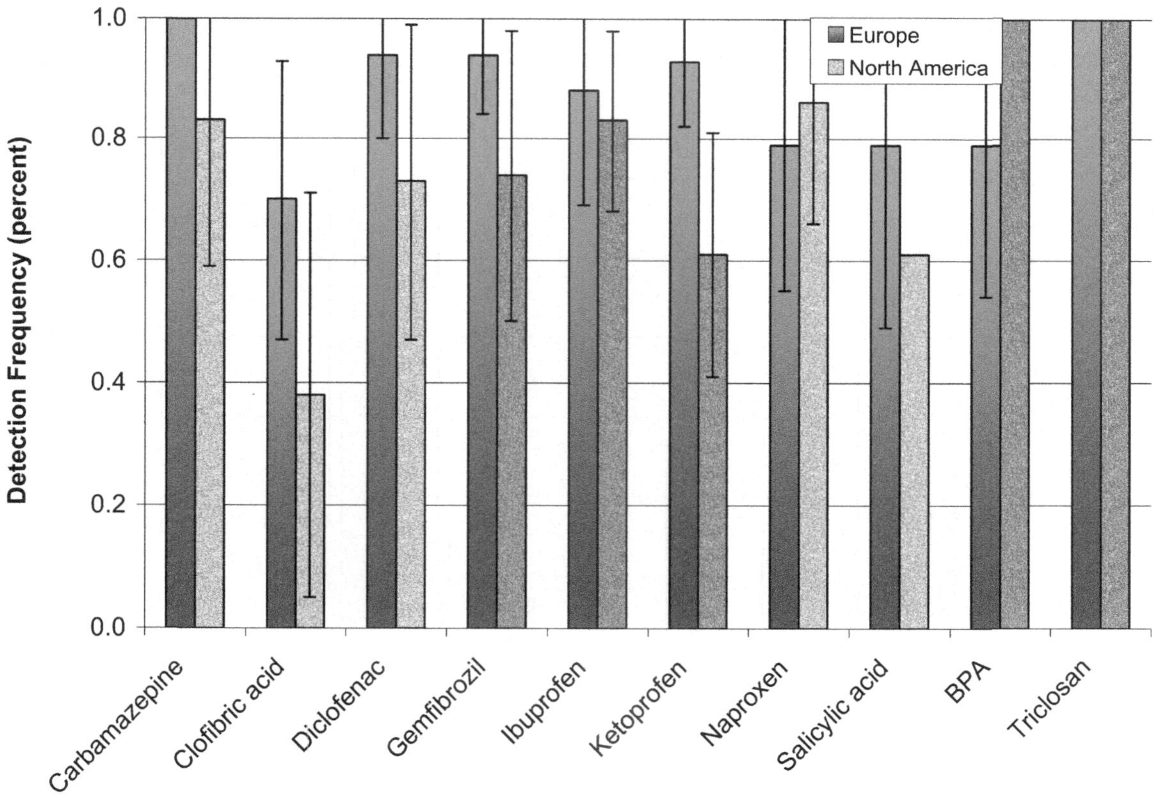


**Figure 4.2 Detection frequencies of select pharmaceutical residues in secondary/tertiary effluents in Europe and North America (error bars represents 1σ standard deviation)**

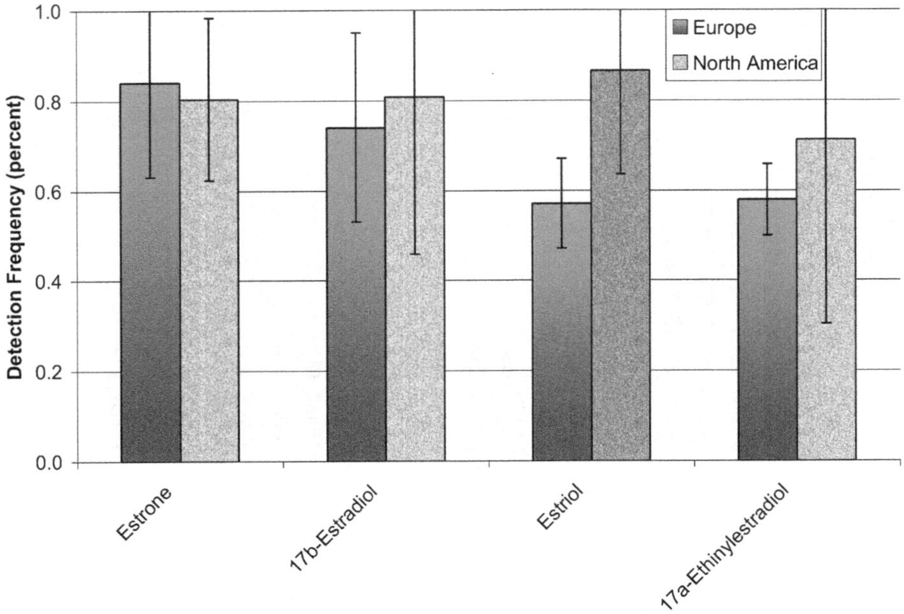


**Figure 4.3 Detection frequencies of select steroid hormones in secondary/tertiary effluents in Europe and North America (error bars represents 1σ standard deviation)**

Average concentrations, minimum and maximum concentrations of select compounds representing pharmaceutical residues, steroid hormones, and household chemicals reported to occur in secondary/tertiary treated effluents both in Europe and North America are presented in Figures 4.4 to 4.6.

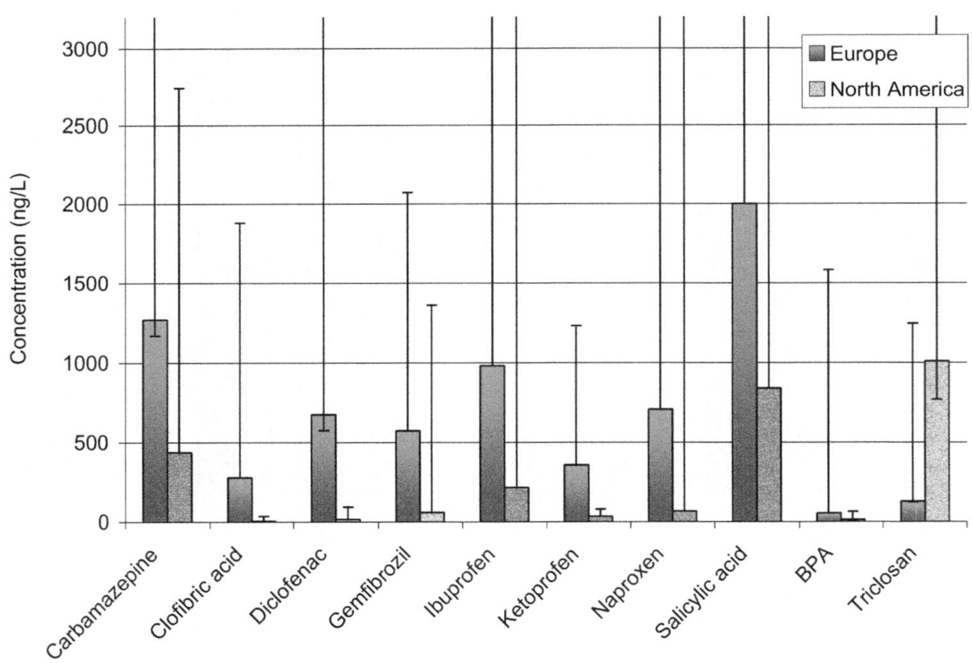


**Figure 4.4 Occurrence of select pharmaceutical residues in secondary/tertiary effluents in Europe and North America (error bars represents maximum and minimum concentrations reported)**

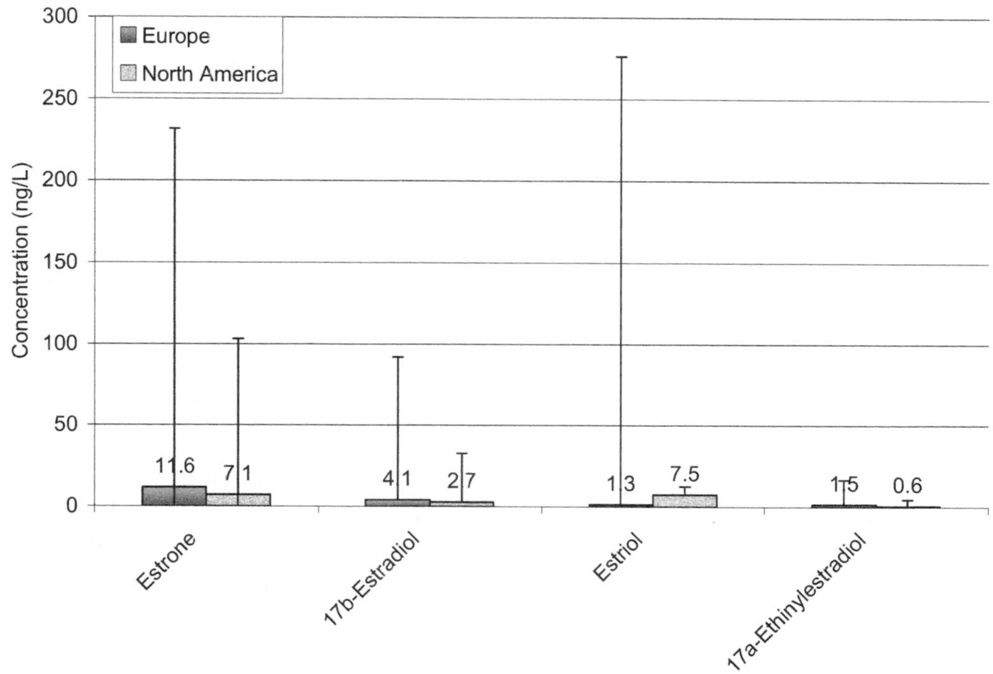


**Figure 4.5 Concentration of select steroid hormones in secondary/tertiary effluents in Europe and North America (error bars represents maximum and minimum concentrations reported)**

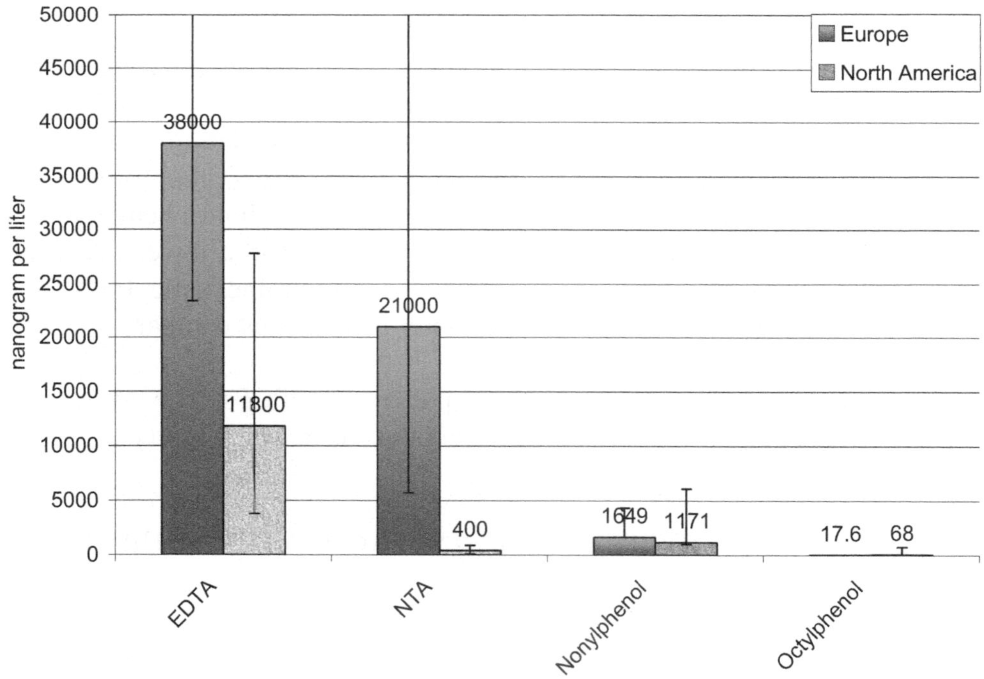


**Figure 4.6 Concentrations of household chemicals in secondary/tertiary effluents in Europe and North America (error bars represents maximum and minimum concentrations reported)**

The steroid hormones (Figure 4.5) exhibited similar concentrations in treated wastewater effluents in Europe and North America with median concentrations of less than 12 ng/L. Estriol concentrations quantified in treated wastewater effluents were slightly higher in North America as compared to Europe. The similar occurrence of steroids in Europe and North America is likely due to a general lower occurrence level of hormones in treated wastewater as compared to PPCPs and a rather efficient removal during conventional wastewater treatment. It is noteworthy that the level of occurrence of pharmaceutical residues and household chemicals was in general higher in European studies as compared to concentrations reported in studies conducted in North America. One exception from this pattern is triclosan, an antimicrobial that occurred approximately at 10-fold higher concentrations in effluents sampled in North America.

The occurrence of these target compounds can vary from region to region due to various factors as stated earlier, but two reasons are likely the cause for the difference in occurrence pattern: prescription and consumption practice and per-capita water consumption. The prescription practice for pharmaceuticals as well as the usage pattern for certain PPCPs can differ between regions and certainly deviates between different countries. For example, clofibric acid, a breakdown product of a blood lipid regulator, is widely administered in middle Europe but rarely used in the U.S. As a consequence, clofibric acid concentrations in treated wastewater in Europe exhibited approximately 50 times higher concentrations. A similar ratio was observed for the anti-inflammatory drug diclofenac that occurred at approximately 50 times higher concentrations in Europe. Unique prescription practices might also be reflected in the occurrence pattern of gemfibrozil, ketoprofen, and naproxen, which all exhibited significant higher concentration in European wastewater effluents. Triclosan is a key ingredient in antimicrobial solutions that are widely used in North America but less popular in Europe resulting in 10 times higher effluent concentrations in North America.

Another important factor to consider in the occurrence pattern of PPCPs in treated wastewater is the strength of the wastewater produced. The per-capita water consumption is usually a factor of 2 to 3 lower in Europe as compared to the U.S. resulting in higher concentrated wastewater in Europe (American Water Works Association 2007, Bundesverband der Gas- und Wasserwirtschaft 2007). For pharmaceutical drugs (e.g., ibuprofen, carbamazepine or salicylic acid (the breakdown product of Aspirin™)) or household chemicals (e.g., EDTA) that are popular in Europe and North America this would result in a lower occurrence level in North America. Indeed, concentrations of ibuprofen, carbamazepine, salicylic acid, and EDTA are by a factor of 2 to 4 higher in European wastewater effluents as compared to concentrations observed in North American effluents.

These findings reveal that the occurrence pattern of PPCPs in wastewater effluent is region or country specific and occurrence studies conducted in one country are not necessarily applicable to other regions of the world.

## OCCURRENCE OF EDCS AND PPCPS IN DRINKING WATER SOURCES

In addition to evaluating the occurrence of target compounds in treated wastewater effluents, the comprehensive literature review was expanded to assess occurrence of select EDCs and PPCPs in drinking water sources. Drinking water sources considered for this survey represent either surface water or groundwater. Table B.1 lists information from peer-reviewed studies regarding the class and subcategory of compounds, water source type (i.e., surface water, groundwater), country, the number of waters (i.e., differing lakes and/or rivers), sites (i.e., locations along a river) and samples (i.e., temporal samples), the estimated level of wastewater impact, the detection frequency and the range and median concentrations observed in the study. The table is organized into four compound classes: pharmaceuticals, steroid hormones, personal care products, and other chemicals. Because various analytical methods with different detection limits were employed in the quantification of the compounds of interest and due to the fact that many compounds occur in source waters at concentrations close to or below the detection limit, exact median concentrations with standard deviations for all compounds of interest cannot be calculated.

As discussed above, the occurrence of organic wastewater compounds in drinking water sources is country specific in regards to water consumption usage patterns, and prescription, chemical usage and wastewater disinfection practices. In addition to country specificity, the degree of wastewater impact of a surface water is unknown for the majority of studies, which likely explains the rather wide range of concentrations reported for individual compounds, thus making comparisons of concentrations between studies and regions of the world more difficult. However, several trends in the occurrence pattern of individual compounds can be highlighted. Organic wastewater compounds found in wastewater effluents are also detected in receiving waters, where, in general, median concentrations are significantly lower than observed in treated wastewater effluents. The detection frequencies of selected organic wastewater compounds in U.S., Germany, South Korea, and the United Kingdom surface waters are illustrated in Figure 4.7. The selected studies used to compile Figure 4.7 and their sampling site descriptions are shown in Table 4.1. Studies were selected based on the diversity and number of surface waters that were sampled for a given country. For the most part, most studies sampled surface waters that were known or suspected to be impacted by wastewater. Note when appropriate, detection frequencies of a given compound were averaged across studies for a given country and error bars represent the standard deviations. The detection frequency highly depends upon the reported minimum detection limit of a given analytical method for the particular study. The reported detection limits of individual compounds are shown in Table 4.2 for the studies used to compile Figure 4.7. The detection limits for a given compound across studies varied by 1.6 to 500 fold. For all the compounds but one compound (caffeine) the fold was greater than 10 times, which made comparing detection frequencies difficult. However, Figure 4.7 does illustrate that carbamazepine consistently had a high detection frequency (>88 percent) across three countries, U.S., Germany, South Korea, which is not surprising since carbamazepine is known to be relatively persistent during wastewater treatment and in the environment. The occurrence of carbamazepine indicates that this antiepileptic drug is widely used among the three countries. In addition, caffeine, a widely used stimulant, was detected frequently (>70 percent) in the U.S., Germany, South Korea, where reported detection limits (10-16 ng/L) were comparable among studies. Triclosan, an antimicrobial used in hand soaps, was frequently detected (62 percent) in U.S waters, however triclosan was not observed in South Korea waters even at a low reporting limit of 1 ng/L. This exemplifies chemical usage patterns can differ among countries.

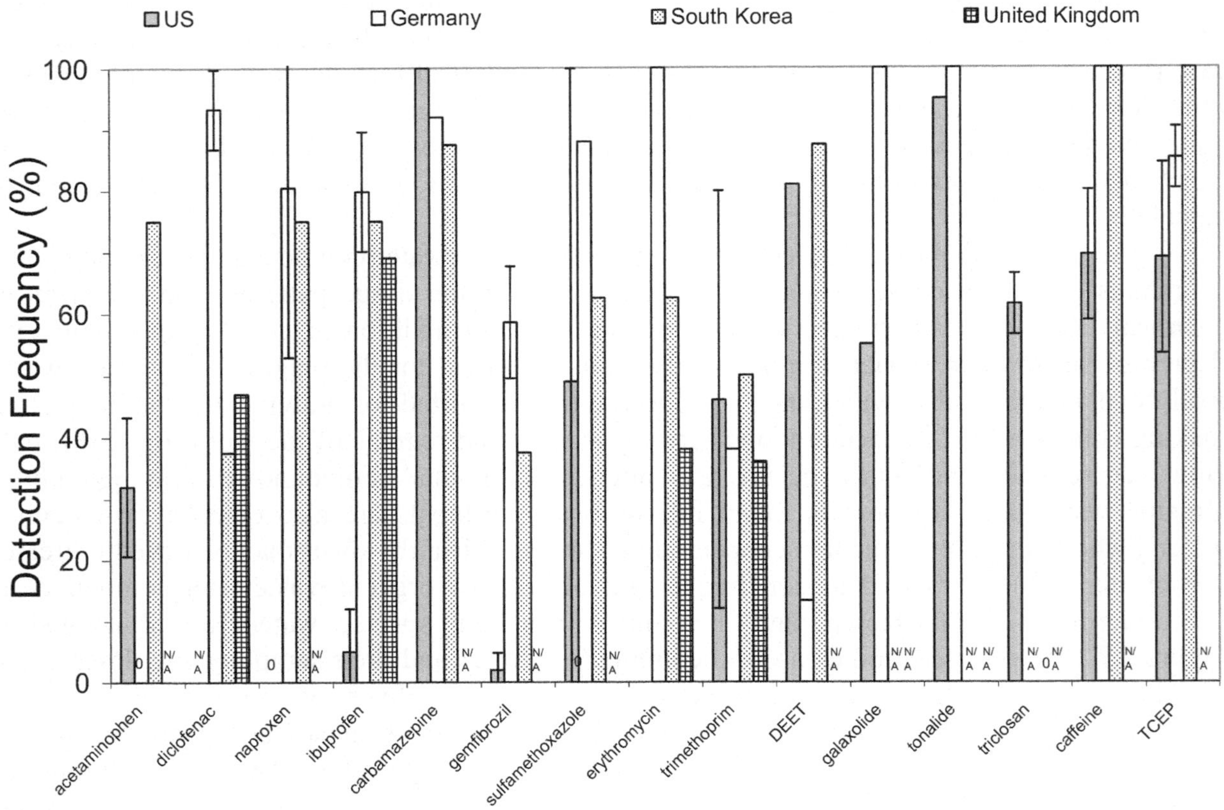


**Figure 4.7 Detection frequency of selected organic wastewater compounds in U.S., Germany, South Korea, and United Kingdom surface waters (N/A = not analyzed)**

**Table 4.1**
**Selected studies for occurrence data in surface waters**

| Study | # | Country | Description of Sampling Sites |
|---|---|---|---|
| (Kolpin, Furlong et al. 2002) | 1 | U.S. | Up to 104 stream sites were sampled nationwide, which were considered susceptible to contamination from human, industrial and agricultural wastewater. |
| (Glassmeyer, Furlong et al. 2005) | 2 | U.S. | The study focusd on 10 WWTPs nationwide. The sample sets consisted of one upstream and two downstream samples. The rivers ranged between low (~10%) to high (~90%) WW impacted surface waters. The detection frequency data only considers downstream data. |
| (Ternes 1998) | 3 | Germany | Up to 22 rivers and streams were sampled. |
| (Heberer, Schmidt-Baeumler et al. 1998; Heberer, Gramer et al. 1999) | 4,5 | Germany | 27 total samples were collected upstream and downstream of WWTP effluents from 12 different canals, rivers, and lakes of the city of Berlin. |
| (Wiegel, Aulinger et al. 2004) | 6 | Germany | Up to 25 sites were sampled from one river and 8 of its tributaries |
| (Fries and Puettmann 2001) | 7 | Germany | 11 total samples were collected from 6 rivers. Four different river sites were sampled twice representing two different times of the year. |
| (Fries and Puettmann 2003) | 8 | Germany | 9 total samples were collected from four sites of one river. Two sites were sampled at least at three different sampling times. |
| (Ternes, Bonerz et al. 2001) | 9 | Germany | Samples were collected from 11 German rivers and streams. All sites were at least 1 km, but generally between 5 and 10 km downstream of the WWTP effluent. |
| (Kim, Cho et al. 2007) | 10 | South Korea | 8 total samples were collected from three major rivers that were impacted by WWTP effluents. Samples were taken upstream and downstream though it is unknown whether upstream sites were also WW impacted. |
| (Ashton, Hilton et al. 2004) | 11 | U.K. | The study focusd on five WWTPs. One sample was taken 1 km upstream and 1 km downstream of WWTP effluent. The detection frequency data only considers downstream data. |

**Table 4.2**
**Reported detection limits (ng/L) of PPCPs in aforementioned studies**

| Study #*: | 1 | 2 | 3 | 4 | 5 | 6 | 7 | 8 | 9 | 10 | 11 |
|---|---|---|---|---|---|---|---|---|---|---|---|
| Country: | U.S. | U.S. | G | G | G | G | G | G | G | S.K. | U.K. |
| | RL | RL | LOD | - | LOD | LOQ | LOD | LOD | LOQ | RL | LOD |
| Acetaminophen | 84 | 36 | 150 | - | - | - | - | - | - | 1 | - |
| Diclofenac | - | - | 10 | na | - | 1 | - | - | - | 1 | 20 |
| Naproxen | - | na | 10 | - | - | 1 | - | - | - | 1 | - |
| Ibuprofen | 18 | 42 | 10 | na | - | 2 | - | - | - | 1 | 20 |
| Carbamazepine | - | 11 | 30 | - | - | - | - | - | - | 1 | - |
| Gemfibrozil | 15 | 13 | 10 | - | - | 2 | - | - | - | 1 | - |
| Sulfamethoxazole | 50 | 64 | - | - | - | na | - | - | - | 1 | 50 |
| Erythromycin | 50 | - | - | - | - | 30 | - | - | - | 1 | 10 |
| Trimethoprim | 30 | 13 | - | - | - | 30 | - | - | - | 1 | 10 |
| DEET | - | 500 | - | - | - | 20 | - | - | - | 1 | - |
| Galaxolide | - | 500 | - | - | 20 | - | - | - | - | - | - |
| Tonalide | - | 500 | - | - | 20 | - | - | - | - | - | - |
| Triclosan | 50 | 1000 | - | - | - | - | - | - | - | 1 | - |
| Caffeine | 14 | 16 | - | - | - | - | - | - | 10 | 10 | - |
| TCEP | 40 | 500 | - | - | - | - | 1 | 1 | - | 10 | - |

*Study number corresponds to the studies listed in Table 4.1
- Compound was not analyzed for the given study
Abbreviations: U.S.- United States, G – Germany, S.K.- South Korea, U.K.- United Kingdom, RL – reporting limit, LOD – limit of detection, LOQ – limit of quantification, na – compound analyzed but detection limit not available

Steroid concentrations in surface water are generally lower than that of other compounds. In Europe and the U.S. the median steroid concentrations have been, for the most part, less than 0.5 ng/L for 17α-ethynylestradiol, 17α-estradiol, 17β-estradiol, and estrone (Table B.1). However, it is noteworthy to point out the comprehensive USGS nationwide study (Kolpin, Furlong et al. 2002) reported higher detectable concentrations than the other studies, which could be due to interference from the analysis of unfiltered water by GC/MS (Ericson, Laenge et al. 2002; Kolpin, Furlong et al. 2002).

The percentage of treated wastewater impacting the receiving water is critical for assessing the degree of contamination. Although the exact degree of impact is mostly unknown for the majority of studies, the concentration level reported is usually significantly higher closer to the point of discharge. Dilution with non-impacted source water can frequently reduce concentrations close to or below the detection limit. Glassmeyer et al. (2005) examined ten locations across the United States where water samples were collected upstream and two successive points downstream from a wastewater treatment plant. The occurrence of selected organic wastewater compounds from two locations from the Glassmeyer et al. (2005) study that represent high (93 percent) and low (13 percent) wastewater impacted surface waters are illustrated in Figure 4.8. As expected the concentrations are markedly reduced downstream due to dilution for the low impacted surface water. Therefore, dilution can be more important during low-flow conditions versus high-flow conditions for a constant wastewater input into the surface water. More positive detections have been observed during low-flow conditions as compared to high-flow conditions (Barber, Furlong et al. 2003; Kolpin, Skopec et al. 2004).

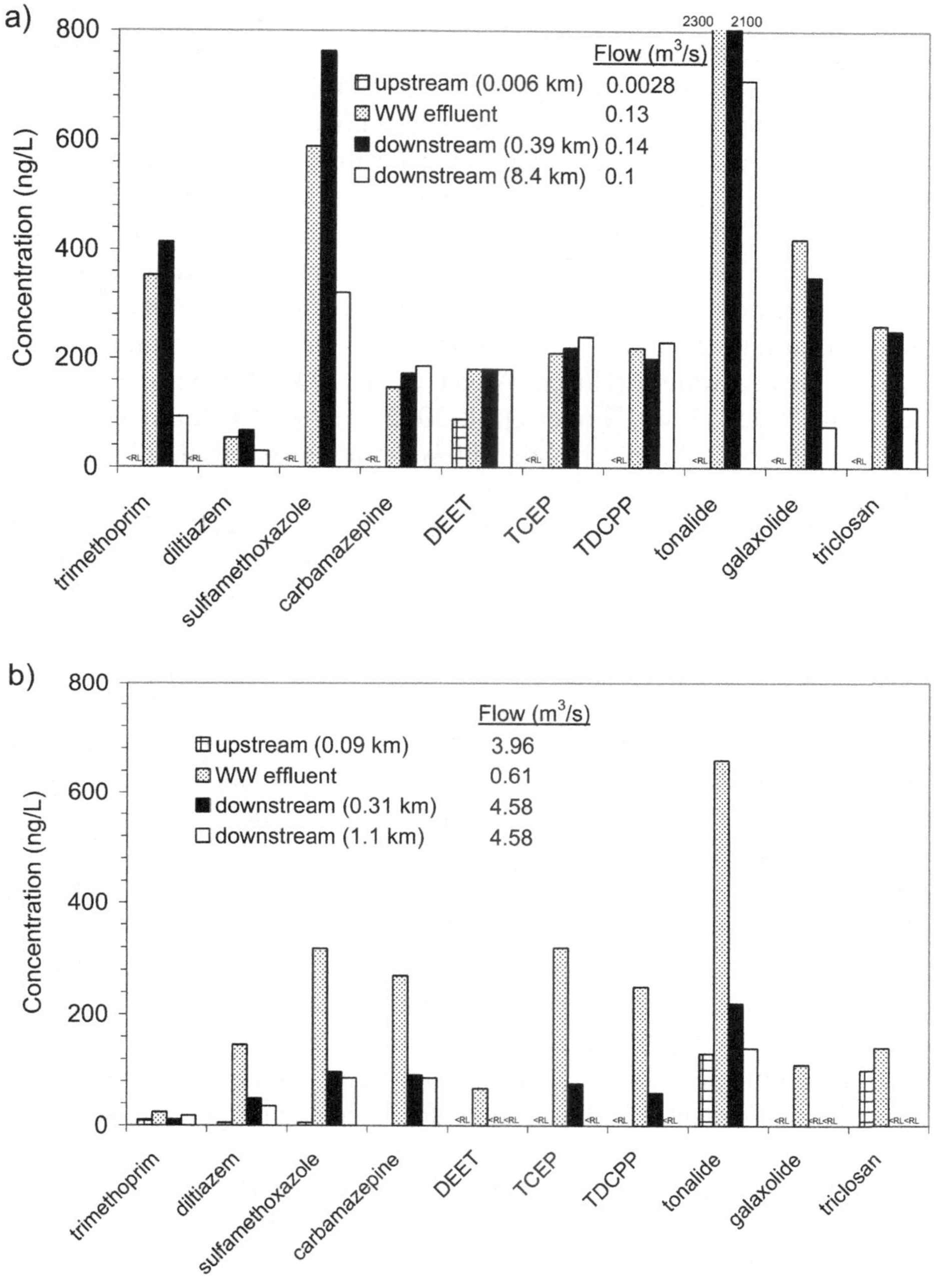


**Figure 4.8 Concentrations of selected organic wastewater compounds from two locations from the Glassmeyer et al. (2005) study that represent a) high (93%) and b) low (13%) wastewater impacted surface waters**

In addition to dilution, concentrations in surface waters can be potentially reduced due to phase partitioning and abiotic and biotic transformation processes. Trace organic wastewater compounds, such as fragrances galaxolide and tonalide, that have high octanol-water partition coefficients can adsorb onto river-bed sediments (Rimkus 1999). Therefore, the reduction in tonalide and galaxolide concentrations downstream (8.4 km) of the effluent of the high-impacted surface water (Figure 4.8a) could be due to partitioning to the river-bottom sediment. But for the most part, most pharmaceuticals have low sorption potential and volatility, thus sorption and volatilization removal mechanisms are negligible. Loffler et al. (2005) observed that sorption plays a minor role in removal for selected pharmaceutical compounds, i.e., carbamazepine, diazepam, clofibric acid, however the sorption can be important for formed residues, such as residues from the biodegradation of acetaminophen. Abiotic transformation processes include hydrolysis and photolysis. Pharmaceuticals are usually designed to be taken orally, thus these compounds are resistant to hydrolysis. Photolysis can occur either by direct absorption of light or indirectly by photosensitizers, such as nitrate and humic acids. Several pharmaceuticals, such as diclofenac, triclosan, sulfamethoxazole and propranolol have been found to be amenable to photolytic decay in surface waters (Buser, Poiger et al. 1998; Lindstrom, Buerge et al. 2002; Singer, Mueller et al. 2002; Tixier, Singer et al. 2002; Andreozzi, Raffaele et al. 2003; Tixier, Singer et al. 2003; Boreen, Arnold et al. 2004). Andreozzi et al. (2003) observed faster photodegradation for diclofenac and sulfamethoxazole in the presence of nitrate ions and faster photodegradation for sulfamethoxazole in the presence of humic acids. However, photolysis can be hindered by certain dissolved humic acids and turbidity, which can absorb solar radiation (Gao and Zepp 1998) and reduce the rate of photodegradation of other organic species present. Andreozzi et al. (2003) observed humic acids inhibited diclofenac reduction. Even carbamazepine has been found to be amenable to solar degradation, especially in the presence of nitrate, albeit at a slower rate as compared to the compounds discussed above (Andreozzi, Marotta et al. 2002; Andreozzi, Raffaele et al. 2003). Also, the degree of photodegradation is strongly influenced by solar intensity, which varies with latitude and season (Buser, Poiger et al. 1998; Lindstrom, Buerge et al. 2002; Tixier, Singer et al. 2002). Other organic wastewater compounds, such as ibuprofen, gemfibrozil, caffeine, and naproxen, can be transformed via biotransformation processes within the surface water (Buser, Poiger et al. 1999; Winkler, Lawrence et al. 2001; Buerge, Poiger et al. 2003; Fono, Kolodziej et al. 2006). Biotransformation mechanisms become more important for hydraulic travel times on the order of weeks. Fono and Sedlak (2006) examined the reduction of three pharmaceuticals in a river where wastewater effluent accounted for nearly the entire flow of the river over a travel time of approximately 2 weeks (Figure 4.9). Complementary bench-scale microcosm studies suggested biotransformation as the major removal mechanism. However, there are selected compounds that are less amenable to removal by surface water partitioning and transformation processes. Results presented in Figure 4.8a illustrate that carbamazepine, DEET, TCEP, and TDCPP concentrations are relatively constant 8.4 km downstream of the WW effluent, though these compound concentrations could be potentially reduced during longer travel times.

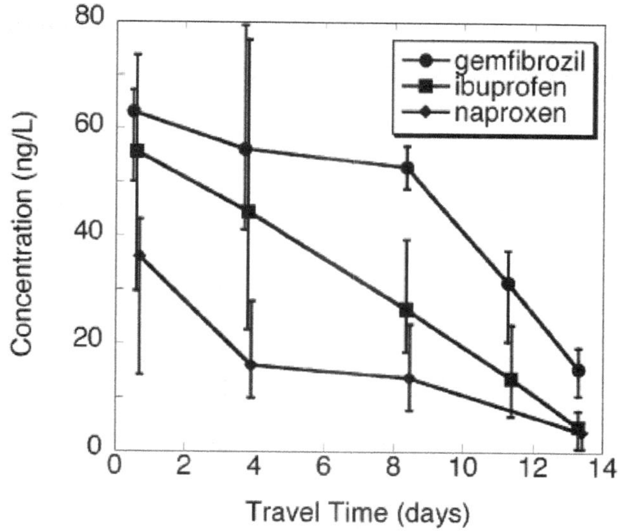


**Figure 4.9 Concentrations of three pharmaceuticals in a river where wastewater effluent accounted for nearly the entire flow of the river. Figure extracted from Fono and Sedlak (2006)**

## OCCURRENCE OF EDCS AND PPCPS IN RAW AND FINISHED DRINKING WATER

Few studies have documented occurrence of EDCs and PPCPs in drinking-water systems. These studies are listed in Table B.2. For every compound reported, this table provides insight into the treatment train employed as well as the raw and finished water concentration observed. Facilities considered in this survey exhibit raw water concentrations, which are in the low ppt-concentration range and within a range of concentrations typical for impaired source water (see Table B.2). The occurrence of EDCs and PPCPs in drinking water sources highly depends upon the degree of wastewater impact upon the source water. Stackelberg et al. (2004) evaluated the occurrence of PPCPs within a drinking water treatment (DWT) plant within a highly urbanized area where more than 50 WWTP discharge effluent to the two streams that provide source water for the 235 million L/d DWT plant. The process train includes the following processes in order: coagulation/flocculation/sedimentation ($FeCl_3$ = 20-45 mg/L), disinfection (NaClO; contact time of 200-300 minutes), GAC filtration (after 2 months operating time; contact time of 1.5 to 3 min), and 2$^{nd}$ disinfection (chlorine residual ~ 1.2 mg/L). The average concentrations before and after the respective treatment processes are presented in Figure 4.10. Based upon findings of this study, clarification had little impact on the removal of PPCPs, whereas chlorination and GAC filtration were more effective removal mechanisms depending on the individual compound. In the finished water all compounds were reduced to various degrees, however some compounds were detected in the finished water indicating incomplete degradation or removal. For example, carbamazepine and DEET were detected in every finished water sample and cotinine and tonalide were detected in 75 percent and 50 percent, respectively, of finished-water samples. Ternes et al. (2002) examined the occurrence of a few pharmaceuticals within a full-scale drinking water treatment plant (Figure 4.11), which includes the following processes in order: preozonation (ozone dose: 0.7-1.0 mg/L; contact time of 3 min), coagulation/ flocculation/sedimentation ($FeCl_3$), disinfection (ozone dose: 1.0-1.5 mg/L; contact time of 10 min), and GAC filtration. Preozonation reduced carbamazepine and diclofenac near and below the detection limit, respectively, and carbamazepine was not detected after ozone disinfection. Clofibric acid was not detected only after GAC filtration. Further insight in the efficacy of unit processes employed during water treatment regarding the removal of compounds of concern is provided in Chapter 5.

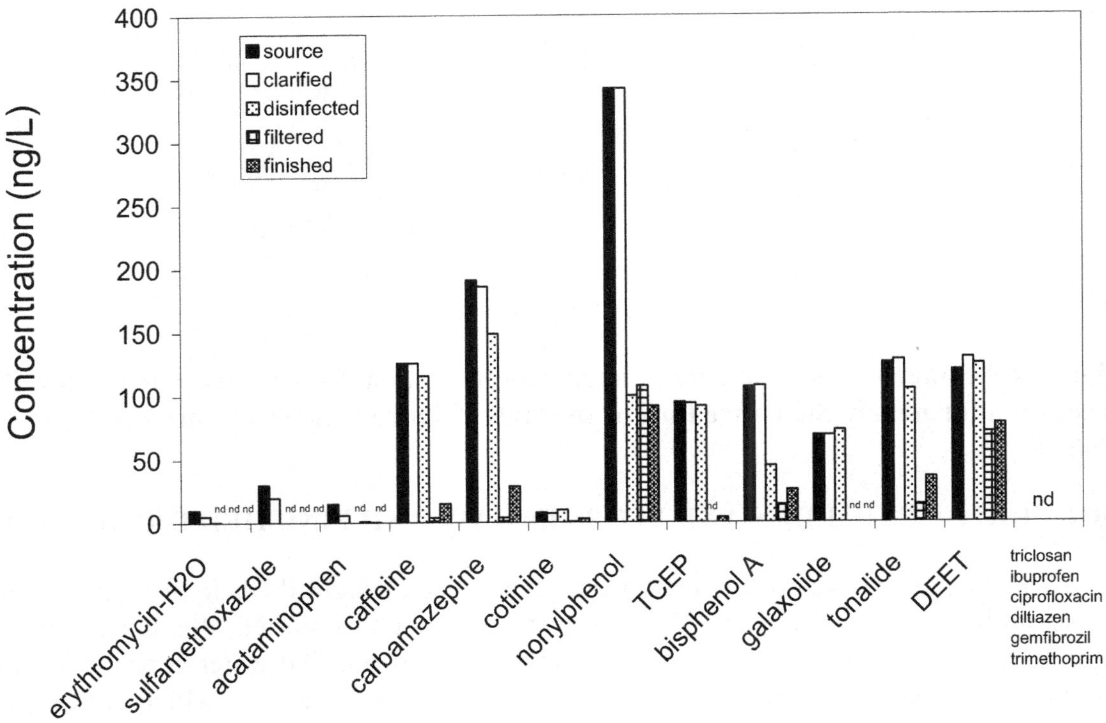


**Figure 4.10 Average concentrations before, during, and after full-scale drinking water treatment. Data is from Stackelberg et al. (2004)**

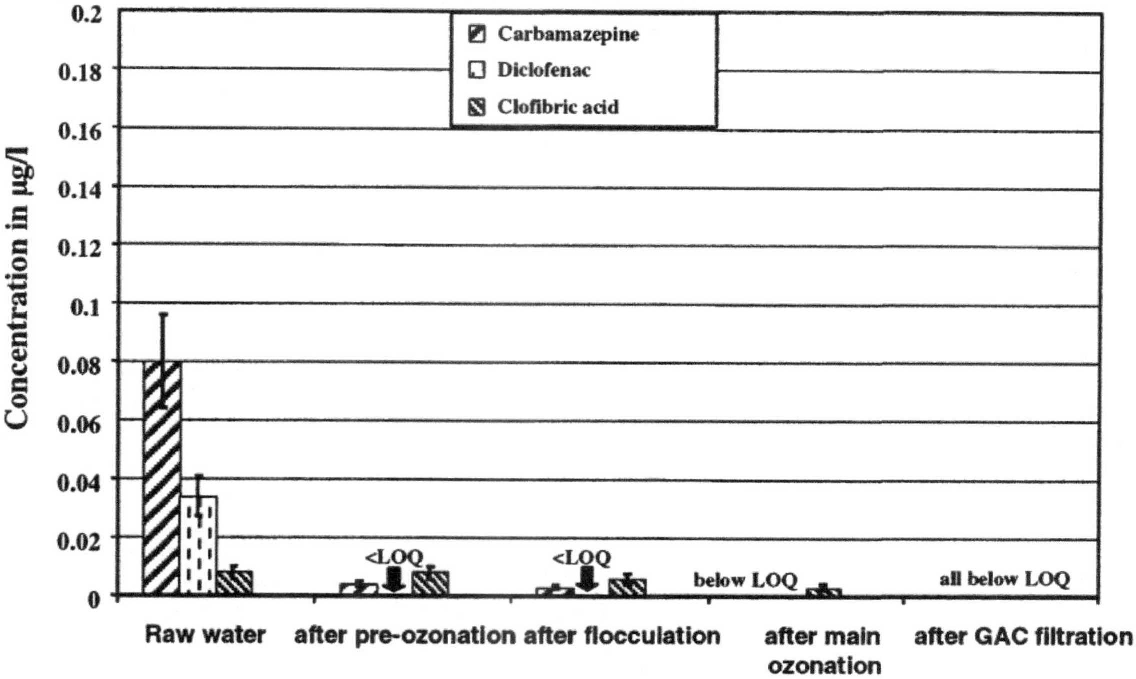


**Figure 4.11 Concentrations of three pharmaceuticals in a full-scale drinking water treatment plant. Figure extracted from Ternes et al. (2002)**

## CONCLUSIONS

The most important sources for release of EDCs and PPCPs into surface and groundwater are discharges from municipal wastewater treatment plants, industrial manufacturing processes, leaky sewers, combined sewage overflows (CSPOs), onsite wastewater systems (OWS), and combined animal feeding operations (CAFOs). Pharmaceutical residues, antibiotics, steroid hormones, and fragrances are the most commonly reported trace organic compounds currently reported to occur in secondary and tertiary treated municipal effluents as well as in surface water and groundwater receiving these wastewater discharges. Most studies considered in this report show occurrence of certain compounds in source water receiving various degrees of wastewater discharge, which likely explains the rather wide range of concentrations reported for individual compounds. Although, the exact degree of impact is mostly unknown, the concentration level reported is usually significantly higher the closer the sampling occurred at the point of discharge. The highest concentrations are usually observed in recycled water used for planned indirect potable reuse applications. Dilution with non impacted source water and photolysis and biodegradation processes can frequently reduce concentrations close to or below the detection limits. In general, median concentrations reported for the compounds of interest in receiving streams are, as expected, significantly lower than observed in treated wastewater effluents. Due to differences in prescription practice and per-capita water consumption, certain pharmaceutical residues exhibit lower concentrations in U.S. source waters as compared to Europe. Steroid hormones usually occur in source waters in Europe and North America at similar concentrations frequently close or below the detection limits employed in the studies. Since detection of these compounds is mainly driven by the availability of sensitive analytical instruments, which will allow scientists to detect compounds at increasingly lower concentrations, occurrence of compounds targeted in this study as well as compounds currently not identified in various types of water will likely increase in the future.

## REFERENCES


American Water Works Association. 2007. Water Use Statistics [Online]. Available: <http://www.drinktap.org/consumerdnn/Home/WaterInformation/Conservation/WaterUseStatistics/tabid/85/Default.aspx>. [cited August 31, 2007]

Bundesverband der Gas- und Wasserwirtschaft. 2007. Wasserfakten im Ueberblick [Online]. Available: <http://www.bgw.de/files/wasserfakten-im-ueberblick-juli-2007.pdf>. [cited August 31, 2007]

Andreozzi, R., R. Marotta, G. Pinto, and A. Pollio. 2002. Carbamazepine in water: persistence in the environment, ozonation treatment and preliminary assessment on algal toxicity. *Water Research*, 36(11):2869-2877.

Andreozzi, R., M. Raffaele, and P. Nicklas. 2003. Pharmaceuticals in STP effluents and their solar photodegradation in aquatic environment. *Chemosphere*, 50(10):1319-1330.

Ashton, D., M. Hilton, and K. Thomas. 2004. Investigating the environmental transport of human pharmaceuticals to streams in the United Kingdom. *Science of the Total Environment*, 333:167-184.

Barber, L.B., E.T. Furlong, S.H. Keefe, G.K. Brown, and J.D. Cahill. 2003. Natural and contaminant organic compounds in the Boulder Creek watershed, Colorado during high-flow and low-flow conditions, 2000. In *Comprehensive Water Quality of the Boulder Creek Watershed, Colorado, During High-Flow and Low-Flow Conditions, 2000*. Edited by Denver, Colorado: USGS.

Boreen, A.L., W.A. Arnold, and K. Mcneill. 2004. Photochemical fate of sulfa drugs in the aquatic environment: sulfa drugs containing five-membered heterocyclic gorups. *Environmental Science & Technology*, 38(14):3933-3940.

Buerge, I.J., T. Poiger, M.D. Mueller, and H.-R. Buser. 2003. Caffeine, an Anthropogenic Marker for Wastewater Contamination of Surface Waters. *Environmental Science and Technology*, 37(4):691-700.

Bundesverband der Gas- und Wasserwirtschaft. 2007. Wasserfakten im Ueberblick [Online]. Available: <http://www.bgw.de/files/wasserfakten-im-ueberblick-juli-2007.pdf>. [cited August 31, 2007]

Buser, H.R., T. Poiger, and M.D. Muller. 1998. Occurrence and fate of the pharmaceutical drug diclofenac in surface waters: Rapid photodegradation in a lake. *Environmental Science & Technology*, 32(22):3449-3456.

Buser, H.R., T. Poiger, and M.D. Muller. 1999. Occurrence and environmental behavior of the chiral pharmaceutical drug ibuprofen in surface waters and in wastewater. *Environmental Science & Technology*, 33(15):2529-2535.

Daughton, C.G., and T.A. Ternes. 1999. Pharmaceuticals and personal care products in the environment: agents of subtle change? *Environmental Health Perspectives*, 107(6):907-938.

Drewes, J.E., and L.S. Shore. 2001. Concerns about pharmaceuticals in water reuse, groundwater recharge, and animal waste. *ACS Symposium Series*, 791:206-228.

Ericson, J.F., R. Laenge, and D.E. Sullivan. 2002. Comment on "Pharmaceuticals, hormones, and other organic wastewater contaminants in U.S. streams, 1999-2000. A national reconnaissance". *Environmental Science & Technology*, 38(18):4005-4006.

Fono, L.J., E.P. Kolodziej, and D.L. Sedlak. 2006. Attenuation of wastewater-derived contaminants in an effluent-dominated river. *Environmental Science & Technology*, 40:7257-7262.

Fries, E., and W. Puettmann. 2001. Occurrence of organophosphate esters in surface water and ground water in Germany. *Journal of Environmental Monitoring*, 3(6):621-626.

Fries, E., and W. Puettmann. 2003. Monitoring of the three organophosphate esters TBP, TCEP and TBEP in river water and ground water (Oder, Germany). *Journal of Environmental Monitoring*, 5(2):346-352.

Gao, H., and R.G. Zepp. 1998. Factors influencing photoreactions of dissolved organic matter in a coastal river of the Southeastern United States. *Environmental Science & Technology*, 32(19):3632-3638.

Glassmeyer, S.T., E.T. Furlong, D.W. Kolpin, J.D. Cahill, S.D. Zaugg, S.L. Werner, M.T. Meyer, and D.D. Kryak. 2005. Transport of chemical and microbial compounds from known wastewater discharges: potential for use as indicators of human fecal contamination. *Environmental Science & Technology*, 39:5157-5169.

Heberer, T., K. Schmidt-Baeumler, and H.-J. Stan. 1998. Occurrence and distribution of organic contaminants in the aquatic system in Berlin, Germany. Part 1. Drug residues and other polar contaminants in Berlin surface and groundwater. *Acta Hydrochimica Et Hydrobiologica*, 26(5):272-278.

Heberer, T., S. Gramer, and H.-J. Stan. 1999. Occurrence and distribution of organic contaminants in the aquatic system in Berlin, Germany. Part 3. Determination of synthetic musks in Berlin surface water applying solid-phase microextraction (SPME) and gas chromatography-mass spectrometry (GC-MS). *Acta Hydrochimica Et Hydrobiologica*, 27(3):150-156.

Heberer, T. 2002. Occurrence, fate, and removal of pharmaceutical residues in the aquatic environment: a review of recent research data. *Toxicology Letters*, 131(1-2):5-17.

Kim, S.-C., and K. Carlson. 2006. Occurrence of ionophore antibiotics in water and sediments of a mixed-landscape watershed. *Water Research*, 40:2549-2560.

Kim, S.D., J. Cho, I.S. Kim, B.J. Vanderford, and S.A. Snyder. 2007. Occurrence and removal of pharmaceuticals and endocrine disruptors in South Korean surface, drinking, and waste waters. *Water Research*, 41(5):1013-1021.

Kolpin, D.W., E.T. Furlong, M.T. Meyer, E.M. Thurman, S.D. Zaugg, L.B. Barber, and H.T. Buxton. 2002a. Pharmaceuticals, hormones, and other organic waste contaminants in U.S. streams, 1999-2000: a national reconnaissance. *Environmental Science & Technology*, 36(6):1202-1211.

Kolpin, D.W., E.T. Furlong, M.T. Meyer, E.M. Thurman, S.D. Zaugg, L.B. Barber, and H.T. Buxton. 2002b. Response to comment on "Pharmaceuticals, hormones, and other organic wastewater contaminants in U.S. streams, 1999-2000: A national reconnaissance". *Environmental Science & Technology*, 36(18):4007-4008.

Kolpin, D.W., M. Skopec, M.T. Meyer, E.T. Furlong, and S.D. Zaugg. 2004. Urban contribution of pharmaceuticals and other organic wastewater contaminants to streams during differing flow conditions. *Science Of The Total Environment*, 328(1-3):119-130.

Lindstrom, A., I.J. Buerge, T. Poiger, P.A. Bergqvist, M.D. Muller, and H.R. Buser. 2002. Occurrence and environmental behavior of the bactericide triclosan and its methyl derivative in surface waters and in wastewater. *Environmental Science & Technology*, 36(11):2322-2329.

Loffler, D., J. Rombke, M. Meller, and T.A. Ternes. 2005. Environmental fate of pharmaceuticals in water/sediment systems. *Environmental Science & Technology*, 39(14):5209-5218.

Rimkus, G.G. 1999. Polycyclic musk fragrances in the aquatic environment. *Toxicology Letters*, 111(1-2):37-56.

Singer, H., S. Mueller, C. Tixier, and L. Pillonel. 2002. Triclosan: occurrence and fate of a widely used biocide in the aquatic environment: field measurements in wastewater treatment plants, surface waters, and lake sediments. *Environmental Science & Technology*, 36(23):4998-5004.

Snyder, S.A., P. Westerhoff, Y. Yoon, and D.L. Sedlak. 2003. Pharmaceuticals, personal care products, and endocrine disruptors in water: implications for the water industry. *Environmental Engineering Science*, 20(5):449-469.

Stackelberg, P.E., E.T. Furlong, M.T. Meyer, S.D. Zaugg, A.K. Henderson, and D.B. Reissman. 2004. Persistence of pharmaceutical compounds and other organic wastewater contaminants in a conventional drinking-water-treatment plant. *Science Of The Total Environment*, 329(1-3):99-113.

Swayne, M.D., G.H. Boone, D. Bauer, and J.S. Lee. 1980. Wastewater in Receiving Waters at Water Supply Abstraction Points. Final rept. Aug 77-Jan 79. Edited by Cincinnati: Sponsor: Municipal Environmental Research Lab.

Ternes, T., M. Bonerz, and T. Schmidt. 2001. Determination of neutral pharmaceuticals in wastewater and rivers by liquid chromatography-electrospray tandem mass spectrometry. *Journal of Chromatography A*, 938(1-2):175-185.

Ternes, T.A. 1998. Occurrence of drugs in German sewage treatment plants and rivers. *Water Research*, 32(11):3245-3260.

Ternes, T.A., M. Meisenheimer, D. McDowell, F. Sacher, H.-J. Brauch, B. Haist-Gulde, G. Preuss, U. Wilme, and N. Zulei-Seibert. 2002. Removal of pharmaceuticals during drinking water treatment. *Environmental Science & Technology*, 36(17):3855-3863.

Tixier, C., H.P. Singer, S. Canonica, and S.R. Mueller. 2002. Phototransformation of triclosan in surface waters: a relevant elimination process for this widely used biocide-laboratory studies, field measurements, and modeling. *Environmental Science & Technology*, 36(16):3482-3489.

Tixier, C., H.P. Singer, S. Oellers, and S.R. Mueller. 2003. Occurrence and fate of carbamazepine, clofibric acid, diclofenac, ibuprofen, ketoprofen, and naproxen in surface waters. *Environmental Science & Technology*, 37(6):1061-1068.

Wiegel, S., A. Aulinger, R. Brockmeyer, H. Harms, J. Löffler, H. Reincke, R. Schmidt, B. Stachel, W. Von Tümpling, and A. Wanke. 2004. Pharmaceuticals in the river Elbe and its tributaries. *Chemosphere*, 57(2):107-126.

Winkler, M., J.R. Lawrence, and T.R. Neu. 2001. Selective degradation of ibuprofen and clofibric acid in two model river biofilm systems. *Water Research*, 35(13):3197-3205.

# CHAPTER 5
# WATER TREATMENT

## INTRODUCTION

The purpose of this chapter is to identify the efficiency of individual unit operations comprising an overall drinking water train to remove EDCs and PPCPs from impacted source water. The chapter is structured in key operations employed in conventional and advanced drinking water treatment applications, such as coagulation/flocculation, disinfection processes, granular and powdered activated carbon, advanced oxidation processes, low-pressure and high-pressure membranes, and natural treatment systems such as riverbank filtration.

## COAGULATION/FLOCCULATION

Information regarding the removal efficiency of select EDCs and PPCPs during coagulation/flocculation is summarized in Table C.1. Coagulation herein is defined as the process by which chemicals are added to water to cause destabilization of colloidal particles, allowing aggregation through flocculation, followed by sedimentation and filtration. There are limited studies reporting the removal of EDC and PPCPs via conventional drinking water treatment processes. A few studies reported that pharmaceuticals (i.e., sulfamethazine, trimethoprim), steroids (i.e., estrone), personal care products (i.e., DEET) and other compounds (i.e., chlorinated flame retardants such as TCEP) are poorly removed by ferric or alum coagulation (Adams, Wang et al. 2002; Ternes, Meisenheimer et al. 2002; Ballard and MacKay 2005; Westerhoff, Yoon et al. 2005; Vieno, Tuhkanen et al. 2006). For trace organic compounds selected, Adams et al. (2002) and Westerhoff et al. (2005) reported similar removals between alum and ferric coagulation. Westerhoff et al. (2005) also observed aliphatic compounds, such as TCEP, had the lowest removals and neutral compounds with higher log $K_{OW}$ (> 3) exhibited better removal. Since most compounds are relatively polar (log $K_{OW}$ < 3), this suggests only a few EDCs and PPCPs (i.e., nonylphenol, bisphenol A, EE2) could be associated with organic phases of particles and could be removed during chemical precipitation. With few exceptions coagulation and flocculation is inappropriate to remove EDCs and PPCPs.

## DISINFECTION

Information regarding the removal efficiency of select EDCs and PPCPs during water disinfection processes is summarized in Table C.2. Removal efficiencies are reported for chlorine dioxide, chlorine, ozone, and UV irradiation.

In order to meet federal and state drinking water regulations, drinking water treatment systems must employ disinfection. Since disinfection processes are oxidation processes, these unit operations have the potential to transform EDCs and PPCPs. In drinking water treatment systems, chlorine, chlorine dioxide, ozone, and UV are frequently used as disinfectants. Chlorine, chlorine dioxide and ozone oxidations of EDCs and PPCPs are selective for certain chemical structures and functional groups. Among these three oxidants, ozone is the most reactive. Snyder et al. (2003) proposed some similarities regarding the reactivity of the before mentioned oxidants with organic compounds:

- Dissociated acidic compounds are more reactive than protonated forms, but nondissociated bases are more reactive when not protonated.
- General order of reactivity from highest to lowest for aromatic or aliphatic compounds: thiols > amines > hydroxyl > carboxyl
- Aromatic compounds are more reactive than aliphatic compounds.

Chlorine can be used to remove EDCs and PPCPs to concentrations below the detection limit (Adams, Wang et al. 2002; Gallard, Leclercq et al. 2004; Lee, Kamata et al. 2004; Boyd, Zhang et al. 2005; Dodd, Shah et al. 2005; Westerhoff, Yoon et al. 2005; Bedner and MacCrehan 2006). Westerhoff et al. (2005) observed removal exceeding 95 percent of the following compounds: acetaminophen, diclofenac, estrone, 17β-estradiol, 17α-ethynylstradiol, naproxen, sulfamethoxazole, triclosan, erythromycin-$H_2O$, carbamazepine, gemfibrozil, and trimethoprim. Pinkston and Sedlak (Pinkston and Sedlak 2004) also observed significant reactivity of chorine with gemfibrozil, naproxen, acetaminophen, atenolol, metoprolol, and propranolol. Westerhoff et al. (2005) observed that compounds with primary or secondary amines (i.e., diclofenac, sulfamethoxazole, trimethoprim) and compounds with phenols (i.e., estrone, 17β-estradiol, 17α-ethynylstradiol, and acetaminophen) or substituted phenols (i.e., triclosan) were very reactive with chlorine. Alum et al. (2004) and Lee et al. (2004) also reported bisphenol A and nonylphenol, phenolic compounds, were very reactive with chlorine. However, chlorine substituted aromatic rings (i.e., diazepam) were less reactive with chlorination because the chlorine atom is electron-withdrawing (Westerhoff, Yoon et al. 2005). Westerhoff et al. (2005) also observed chlorine was inefficient at removing ibuprofen, DEET, iopromide, and TCEP. These compounds have either electron-withdrawing groups or lack conjugated carbon bonds. Pinkston and Sedlak (2004) also observed that ibuprofen and ketoprofen were not removed after a reaction time of 5 days (excess chlorine). This is expected since these compounds have an electron-withdrawing functional group on the aromatic ring. It is expected that certain EDCs and PPCPs will be significantly transformed under the conditions typically encountered in many chlorine disinfection systems.

Chlorine dioxide is generally a stronger oxidant than free chlorine. Huber et al. (2005) observed appreciable removals of sulfamethazine, sulfamethoxazole, estrone, 17β-estradiol, 17α-ethynylstradiol, roxithromycin, erythromycin-$H_2O$, and diclofenac by chorine dioxide. However, caffeine, clofibric acid, gemfibrozil, ketoprofen, naproxen, iopromide were recalcitrant to chlorine dioxide oxidation. Compared to chlorine, the rate constants for the reactions of chlorine dioxide with 17α-ethynylstradiol and naproxen are higher than chlorine rate constants (Huber, Korhonen et al. 2005). However, for aliphatic primary and secondary amine functional groups it is known that these groups react faster with chlorine than chlorine dioxide (Huber, Korhonen et al. 2005). Chlorine and chlorine dioxide react primarily with electron functional groups like amines and phenols. Huber et al. (2005) also compared the ozone and chlorine dioxide rate constants for 17α-ethynylstradiol, diclofenac, sulfamethoxazole, roxithromycin, naproxen, and observed the rate constants were always higher for ozone. Both of these oxidants react primarily with electron functional groups like amines and phenols. However, ozone also attacks carbon-carbon double bonds and activated benzene rings. Chlorine dioxide is effective at oxidizing certain compound classes such as the subcategories of sulfonamide and antibiotics, and estrogens. It is expected that certain EDCs and PPCPs will be significantly transformed under the conditions typically encountered in many chlorine dioxide disinfection systems.

Ozonation is a strong oxidant and very effective in the transformation of EDCs and PPCPs (Adams, Wang et al. 2002; Andreozzi, Marotta et al. 2002; Ternes, Meisenheimer et al. 2002; Huber, Canonica et al. 2003; Alum, Yoon et al. 2004; Huber, Göbel et al. 2005; McDowell, Huber et al. 2005; Westerhoff, Yoon et al. 2005). Ozone is a selective electrophile and reacts with electron functional groups like amines and phenols, carbon-carbon double bonds, and activated benzene rings. McDowell et al. (2005) observed that ozone reacts rapidly with the double bond in carbamazepine. Huber et al. (2005) reported many pharmaceuticals (i.e., sulfamethoxazole, roxithroymcin, diclofenac, and naproxen) and steroids (i.e., estrone, 17β-estradiol, 17α-ethynylstradiol) were oxidized by more than 90-99 percent for ozone doses $\geq 2$ mg/L in wastewater effluents. However, Huber et al. (2005) observed X-ray contrast media (i.e., iopromide, deteriorate) were only partly oxidized. Westerhoff et al. (2005) reported that ozone oxidized 20 EDCs, PPCPs and other compounds (Table C.2) by more than 80 percent except for iopromide, TCEP and ibuprofen. TCEP is an aliphatic compound with polar (chlorine) functional groups and ibuprofen has an electron-withdrawing functional group on the aromatic ring. Ozone will have higher compound removals compared to chlorine. For nucleophilic sites, ozone is a better oxidant. Also, during ozonation, hydroxyl radicals will form, which are powerful oxidants that react nonselectively with most organic compounds. Ozonation is a very effective treatment process to oxidize EDCs and PPCPs.

Ultraviolet irradiation is a very efficient disinfection process. UV can also be used for treatment of EDC and PPCPs with chromophores that adsorb at UV wavelengths. However, typical UV doses required for disinfection (i.e., <5-30 mJ/cm$^2$) are orders of magnitude lower than those used for destructive treatment of EDC and PPCPs. Even at higher doses of 3,000 mJ/cm$^2$ only 50 to 80 percent of the pharmaceuticals (i.e., trimethoprim, sulfamethazine) could be removed (Adams, Wang et al. 2002). Therefore, trimethoprim, sulfamethazine will have low susceptibility to typical UV dosages. Rosenfeldt and Linden (2004) also reported poor removal (<22 percent) of 17β-estradiol, 17α-ethynylstradiol, and bisphenol A at UV doses of 3,000 mJ/cm$^2$. Thus, UV at disinfection dosages is inappropriate to remove EDCs and PPCPs.

## GRANULAR OR POWDERED ACTIVATED CARBON (GAC/PAC)

Activated carbon can be used to remove EDCs and PPCPs completely or partially from drinking water (Fuerhacker, Dürauer et al. 2001; Tanghe and Verstraete 2001; Adams, Wang et al. 2002; Ternes, Meisenheimer et al. 2002; Yoon, Westerhoff et al. 2003; Vieno, Tuhkanen et al. 2005; Westerhoff, Yoon et al. 2005). Activated carbon can be used in the following two types: powdered activated carbon (PAC) and granular activated carbon (GAC). PAC is usually added in contact basins (1-5 hours) prior to coagulation at dosages of 5-50 mg/L. At typical plant PAC dosages (see Table C.3) PAC can effectively remove pharmaceuticals (i.e., trimethoprim), steroids (i.e., 17β-estradiol), personal care products (i.e., triclosan), and other compounds (i.e., bisphenol A) (Adams, Wang et al. 2002; Yoon, Koyanagi et al. 2003; Westerhoff, Yoon et al. 2005). Note these experiments were performed at the bench-scale and Bruce et al. (2002) showed that bench-scale batch contact times can represent full-scale basin contact times quite well. GAC systems operate as filters with contact times less than 30 minutes. GAC systems remain in operation for months to years, and unlike PAC systems where the solid-phase concentration is in equilibrium with the effluent liquid concentrations, the solid-phase concentration in GAC systems equilibrate with influent liquid concentrations.

The performance of activated carbon adsorption depends upon the properties of the activated carbon sorbent (i.e., surface area, pore size distribution, surface charge, oxygen content) and on the properties of the solute (i.e., shape, size, charge, and hydrophobicity). Activated carbon with the point of zero charge near the ambient pH will have less ionization and potentially have a greater affinity for hydrophobic compounds (Yoon, Westerhoff et al. 2003). Nevskaia et al. (2001) observed that inorganic impurities (i.e., carbonate) on the surface of the activated carbon can affect nonylphenol adsorption. They observed the introduction of acidic oxygen surface groups on the activated carbon surface enhanced nonylphenol adsorption. Hydrophobic interaction is the dominant mechanism of removal of organic compounds; therefore activated carbon efficiently removes most nonpolar organic compounds (log $K_{OW} > 2$). Yoon et al. (2003) observed the order of removal of three estrogenic compounds corresponded with the hydrophobicity (i.e., log $K_{OW}$ values) of the compounds. Additionally, Westerhoff et al. (2005) observed that the compound hydrophobicity expressed as the log $K_{OW}$ value predicted reasonably well the removal of 22 EDCs and PPCPs.

The influent concentrations, DOC concentrations and contact time can also affect the performance of activated carbon adsorption. Influent concentrations can affect the performance of GAC columns, but for PAC applications the removal of micropollutants at the parts-per-billion and parts-per-trillion levels has been shown to be independent of initial concentration (Knappe, Matsui et al. 1998). NOM in water will compete for adsorption sites and decrease the activated carbon capacity for micropollutants. Adams et al. (2002) observed a decrease in removal of antibiotics (i.e., trimethoprim) in the presence of NOM compared to the absence of NOM. In addition, Tanghe and Verstraete (2001) observed the sorption capacity of nonylphenol was adversely affected in the presence of dissolved humic acid.

Activated carbon adsorption can readily remove EDCs and PPCPs except for a limited number of polar compounds, such as the iodinated contrast media and Sulfamethoxazole that are insufficiently removed by activated carbon.

## ADVANCED OXIDATION PROCESSES (AOP)

Information regarding the removal efficiency of select EDCs and PPCPs during advanced oxidation processes (AOP) is presented in Table C.4. AOPs can form hydroxyl radicals which can nonselectively attack and transform EDCs and PPCPs. AOPs include UV/hydrogen peroxide, ozone/hydrogen peroxide, and UV/ozone. Zwiener and Frimmel (2000) observed removals exceeding 90 percent for diclofenac, ibuprofen, and clofibric acid at ozone and hydrogen peroxide doses greater than 3.7 and 1.4 mg/L, respectively ($O_3$:$H_2O_2$ = 2.5 mg/mg; 10 minute contact time). For a UV/$H_2O_2$ process, Rosenfeldt and Linden (2004) observed removal exceeding 90 percent removals of bisphenol A, 17β-estradiol, and 17α-ethynylstradiol at a UV dose of 1,000 mJ/cm$^2$ and a hydrogen peroxide dose of 15 mg/L. Huber et al. (2003) observed that the ozone/hydrogen peroxide AOP process considerably increased the removal of ibuprofen from 40 to 80 percent. Westerhoff et al. (2005) observed that an addition of a small amounts of $H_2O_2$ (0.025 mg of $H_2O_2$/mg of $O_3$) prior to ozonation generally improved the extent of EDC and PPCP oxidation by 5-15 percent as compared to ozone alone. In this study testosterone exhibited more than 20 percent higher oxidation in the presence of hydrogen peroxide. AOPs are very effective treatment processed for oxidizing EDCs and PPCPs, however compared to ozone, AOPs provide only small increase in removal efficiency.

## LOW-PRESSURE MEMBRANES (MICROFILTRATION/ULTRAFILTRATION)

Low-pressure membranes represent either microfiltration (MF) or ultrafiltration (UF) membranes. Common pore sizes of microfiltration (MF) and ultrafiltration (UF) membranes are well above several thousand daltons. Since the molecular weight of the majority of EDCs and PPCPs range between 200 and 400 g mole$^{-1}$, steric exclusion of pharmaceutical residues in MF and UF membranes is not relevant. However, highly hydrophobic compounds can still adsorb onto the MF and UF membrane surface and might partition through the membrane into the filtrate. For example, Chang et al. (2003) conducted controlled experiments with hollow fiber MF membranes and reported a significant accumulation of estrone on the membrane surface. Yoon et al. (2007) performed ultrafiltration experiments using pharmaceuticals (i.e., gemfibrozil, ibuprofen), steroids (i.e., estrone), PCPs (i.e., DEET) and other compounds (i.e., TCEP). In general the researchers observed partial adsorption of compounds with log $K_{OW}$ larger than 2.8 (i.e., testosterone, triclosan).

## HIGH-PRESSURE MEMBRANES (NANOFILTRATION/REVERSE OSMOSIS)

High-pressure membranes are nanofiltration (NF) or reverse osmosis (RO) membranes. Information regarding the removal efficiency of select EDCs and PPCPs during high-pressure membrane treatment is presented in Table C.5. NF/RO membrane operations can physically remove EDCs and PPCPs and greater than 90 percent removal can be achieved by RO and tight NF membranes. Some of the mechanisms affecting the permeation of EDCs and PPCPs during RO and NF treatment are fairly well understood, such as (1) physical sieving of solutes larger than the molecular cutoff (MWCO) of a membrane. Other mechanisms of rejection, such as (2) electrostatic repulsion and (3) hydrophobic-hydrophobic interactions between membrane and solute, have been studied mainly at the bench-scale (Kiso, Nishimura et al. 2000; Kiso, Kon et al. 2001; Kiso, Sugiura et al. 2001; Ozaki and Li 2002; Ngheim, Schaefer et al. 2004). The following key solute parameters have been identified by Bellona et al. (2004) in a comprehensive literature review to primarily affect solute rejection during NF/RO: molecular weight, molecular size (i.e., length and width), acid dissociation constant (pKa), hydrophobicity /hydrophilicity (i.e., log $K_{ow}$), and diffusion coefficient. Bellona et al. (2004) also identified key membrane properties that affect rejection, which included MWCO, pore size, surface charge, hydrophobicity/hydrophilicity and membrane surface morphology.

Xu et al. (2005) and Nghiem et al. (2005) examined the rejection of pharmaceuticals due to electrostatic repulsion and they observed that the membrane surface charge of NF and RO membranes determines the rejection of ionic hydrophilic solutes, such acidic pharmaceuticals (i.e., naproxen, diclofenac, ibuprofen, ketoprofen, gemfibrozil) and dissociated antibiotics (i.e., sulfamethoxazole). Speciation of acidic pharmaceuticals as a function of pH may result in a dramatic change of rejection (Bellona and Drewes 2005; Nghiem, Schaefer et al. 2005; Xu, Drewes et al. 2005). Also, fouling can affect the rejection of charged compounds due to the foulant altering the membrane surface characteristics (i.e., surface charge). Xu et al. (2006) observed that the transport of ionic organic compounds was hindered as a result of electrostatic exclusion likely due to a more negative membrane surface charge caused by the foulant.

For neutral pharmaceutical compounds, intrinsic physicochemical properties (i.e., size, polarity, hydrophobicity) of the compounds can affect their rejection. Xu et al. (2005) performed bench-scale experiments and observed the solute size and MWCO of the membrane affected the rejection of hydrophilic and hydrophobic nonionic compounds, such as neutral pharmaceuticals (i.e., carbamazepine, primidone) and steroids (i.e., EE2). Steric exclusion is the major mechanism for retention of uncharged and nonadsorptive organic solutes. When electrostatic repulsion is absent, compounds with a sufficiently high dipole moment (polarity) will approach the membrane pores in a particular orientation, which can affect the separation process (Nghiem, Schaefer et al. 2005). The influence of molecular polarity is significant of compounds that are cylindrical in shape, such as sulfamethoxazole (Nghiem, Schaefer et al. 2005). Nghiem et al. (2005) reported the lower retention of neutral sulfamethoxazole as compared to neutral carbamazepine (NF-270, pH 3.5) was due to the lower dipole moment and more bulky structure of carbamazepine even though they approximately had similar molecular weights. Note the molecular polarity interaction mechanism remains steric in nature.

Yoon et al. (2006) and Nghiem et al. (2004) reported that the last rejection mechanism, hydrophobic-hydrophobic interactions, was responsible for the tendency of hydrophobic estrogens to adsorb onto hydrophobic nanofiltration membrane surfaces. Nghiem et al. (2004) observed steroid (i.e., 17β-estradiol, estrone, testosterone) rejection decreased rapidly initially, which was caused by the hormones adsorbing onto the membrane surface and subsequent increased permeate concentration due to diffusion across the skin layer. Ng and Elimelech (2004) also reported that 17β-estradiol rejection with time was much more severe when colloidal fouling occurred. They speculate that the cake layer formed due to fouling hinders the back diffusion of 17β-estradiol from the membrane surface back to the bulk solution. Consequently, buildup of 17β-estradiol at the membrane surface provided a larger concentration gradient for its diffusion across the RO membrane. However, partial rejection or evidence for steroid partitioning with time was not observed during sampling of full-scale RO installations (Drewes, Bellona et al. 2005), which is likely due to the formation of a fouling layer while RO membranes are employed on surface or recycled water. This fouling layer can act as a second barrier for hydrophobic compounds.

In general, RO and *tight* NF membranes are very effective treatment processes for removing EDCs and PPCPs from drinking water.

## RIVERBANK FILTRATION (RBF)

Riverbank filtration (RBF) is a natural process that has been used for public and industrial water supply in Europe for more than a century and for nearly half a century in the United States. The process in principle acting like a slow-sand filter has been shown to be effective in moderating the peak concentrations of various contaminants present in river or lake water (for lake bank filtration) and in many cases involving removal of contaminants to below detection levels. As water passes through the subsurface, several processes during RBF remove or dilute contaminants. These processes include sorption, biodegradation, dilution, mineralization, precipitation, ion exchange, and filtration. Sorption and biodegradation are often the most important removal processes in RBF. The performance of riverbank filtration systems depends upon well type and pumping rates, travel time of surface water between the river and the pumping wells, source water quality, site hydrogeologic conditions, biogeochemical reactions in sediments and aquifer, and the quality of background groundwater. Studies from Europe and the United States indicate significant removal of dissolved organic carbon, nitrate, pesticides, pathogens, and turbidity, and recent studies also indicate efficient removal of select EDCs and PPCPs during RBF. Information regarding the removal efficiency of select EDCs and PPCPs during riverbank filtration is presented in Table C.6. Due to the small concentration of organic trace compounds (ng/L range) usually present in impaired surface waters, it is unlikely that energy from micropollutant metabolism is sufficient for biomass maintenance and growth. At trace levels, chemicals are therefore potentially transformed by cometabolism (co-utilization). Several boundary conditions have been identified in a recent study (Drewes et al. 2008) that need to be fulfilled in order to establish biotransformation of EDCs and PPCPs, such as predominant redox conditions, availability of biodegradable dissolved organic carbon (BDOC) as primary substrate, and the presence of an adapted community of microorganisms.

In subsurface systems, biodegradable organic carbon is often promoting the removal of river-borne pollutants. Heterotrophic denitrification is often limited due to a lack of organic carbon which originates in part from infiltrated river water and organic matter fixed in the sediments of the aquifer material (Grischek, Hiscock et al. 1998). It has also been reported that biodegradable organic matter plays a dominant role in the degradation of certain trace organic pollutants through a co-metabolic degradation pathway. During RBF, co-metabolic removal of trace pollutants has been previously observed for certain aromatic and aliphatic amines (Boernick, Eppinger et al. 2001; Paul, Boernick et al. 2001) and halogenated organics (Drewes and Jekel 1996). Recent studies by Rauch et al. (2004) suggest that quantity and composition of NOM affect the removal efficiency of organic micropollutants during soil infiltration in groundwater recharge systems. Hydrophilic and colloidal carbon fractions have been found to stimulate soil microbial activity (Rauch and Drewes 2004) and thereby promoting the metabolic or cometabolic removal of certain organic micropollutants (Rauch, Munoz et al. 2004; Drewes, Hoppe, et al. 2008).

Extensive studies were conducted to investigate the fate of pesticides during RBF. Findings from studies reported by Ray et al. (1998) and Verstraeten et al. (1999) indicated a 50 to 75 percent removal of atrazine during RBF, although the underlying removal mechanisms are not clear. Sontheimer (1991) conducted adsorption experiments with atrazine and observed poor removal and attributed this to low biodegradability of the pesticide and low adsorption capacity of the sediment. Subsurface systems, such as RBF or soil-aquifer treatment, demonstrated high removal efficiencies for EDCs and PPCPs (Drewes and Shore 2001; Heberer 2002; Montgomery-Brown, Drewes et al. 2003; Verstraeten, Heberer et al. 2003; Kreuzinger, Clara et al. 2004; Mansell, Drewes et al. 2004). Mansell et al. (2004) and Mansell and Drewes (2004) reported that steroidal hormones (such as 17β-estradiol, estriol, and testosterone) were efficiently removed after short travel times. Montgomery-Brown et al. (2003) extensively studied fate and transport of alkylphenolpolyethoxylates (APECs) and nonylphenol during SAT and findings of this study indicate a significant removal after travel times of approximately two weeks. Riverbank filtration, however, showed some limitations in removing organic contaminants. Certain trace pollutants have been regularly found in the product water of RBF systems, among them urotropin (aliphatic amine) and 1,5-naphthalindisulfonate (aromatic sulfonate) (Brauch, Sacher et al. 2000). Recent studies investigating the fate of PPCPs reported that antiepileptic drugs (e.g., carbamazepine, primidone), some blood-lipid regulator (e.g., clofibric acid), antibiotics (e.g., sulfamethoxazole), X-ray contrast media (e.g., iopromide), and fragrances (e.g., galaxolide and tonalide) were present in both river water and bank filtered water (Kuehn and Mueller 2000; Drewes, Fox et al. 2001; Drewes and Shore 2001; Drewes, Heberer et al. 2003; Kreuzinger, Clara et al. 2004; Schmidt, Lange et al. 2004). A partial reduction in concentration was only achieved under certain redox conditions and through dilution with local groundwater.

The interactions between hydrodynamics, redox conditions in the subsurface and microbial interactions are key considerations for the removal mechanisms of organic matter and a fundamental understanding of the removal mechanisms is still lacking.

## CONCLUSIONS

In order to treat impaired water sources to remove EDCs and PPCPs from water requires a sequence of diverse treatment processes that can tackle the wide range of physicochemical properties of EDCs and PPCPs. There is not a single treatment process that provides a 100 percent barrier to EDCs and PPCPs in drinking water.

Conventional drinking water treatment consisting of coagulation/flocculation followed by sedimentation and filtration commonly employed for surface water is not capable of removing EDCs and PPCPs occurring at the parts-per-trillion level. Removal of EDCs and PPCPs, however, can be expected during drinking water disinfection using chlorine, chlorine dioxide, and ozone. No removal is achieved by UV radiation applied at disinfection dosages. Especially, steroid hormones are well destructed during chlorine disinfection. Stronger oxidants, such as ozone or advanced oxidation processes (AOP), can remove a wide range of EDCs and PPCPs, but exhibit limited removal regarding chlorinated flame retardants (e.g, TCEP, TCPP) or the mosquito repellent DEET. Activated carbon can remove a wide range of hydrophobic EDCs and PPCPs including steroid hormones but has limitations regarding polar compounds, such as x-ray contrast agents. Low-pressure membranes, such as MF and UF, are ineffective in removing EDCs and PPCPs from water. High-pressure membranes, such as NF or RO, however, can remove a wide range of EDCs and PPCPs. Problematic for high-pressure membranes are low-molecular weight organics, such as N-nitrosamines or certain pharmaceuticals (i.e., acetaminophen, phenacetine). Natural processes, such as riverbank filtration (RBF) or soil-aquifer treatment (SAT), can be employed either as an additional treatment step for wastewater prior to discharge to the environment or as a pre-treatment to subsequent drinking water treatment. RBF and SAT both provide a significant barrier for a majority of EDCs and PPCPs, which the exception of recalcitrant compounds that are not amendable to a biological attack.

## REFERENCES


Adams, C., Y. Wang, K. Loftin, and M. Meyer. 2002. Removal of antibiotics from surface and distilled water in conventional water treatment processes. *Journal of Environmental Engineering*, 128(3):253-260.

Alum, A., Y. Yoon, P. Westerhoff, and M. Abbaszadegan. 2004. Oxidation of bisphenol A, 17.beta.-estradiol, and 17.alpha.-ethynyl estradiol and byproduct estrogenicity. *Environmental Toxicology*, 19(3):257-264.

Andreozzi, R., R. Marotta, G. Pinto, and A. Pollio. 2002. Carbamazepine in water: persistence in the environment, ozonation treatment and preliminary assessment on algal toxicity. *Water Research*, 36(11):2869-2877.

Ballard, B.D., and A.A. MacKay. 2005. Estimating the Removal of Anthropogenic Organic Chemicals from Raw Drinking Water by Coagulation Flocculation. *Journal of Environmental Engineering (Reston, VA, United States)*, 131(1):108-118.

Bedner, M., and W.A. Maccrehan. 2006. Reactions of the amine-containing drugs fluoxetine and metoprolol during chlorination and dechlorination processes used in wastewater treatment. *Chemosphere*, 65:2130-2137.

Bellona, C., J.E. Drewes, P. Xu, and G. Amy. 2004. Factors affecting the rejection of organic solutes during NF/RO treatment - A literature review. *Water Research*, 38:2795-2809.

Bellona, C., and J.E. Drewes. 2005. The role of membrane surface charge and solute physico-chemical properties in the rejection of organic acids by NF membranes. *Journal of Membrane Science*, 249:227-234.

Boernick, H., P. Eppinger, T. Grischek, and E. Worch. 2001. Simulation of biological degradation of aromatic amines in river bed sediments. *Water Research*, 35(3):619-624.

Boyd, G.R., S. Zhang, and D.A. Grimm. 2005. Naproxen removal from water by chlorination and biofilm processes. *Water Research*, 39:668-676.

Brauch, H.-J., F. Sacher, E. Denecke, and T. Tacke. 2000. Efficiency of bank filtration for the removal of hydrophilic organic micropollutants. *GWF, Wasser/Abwasser*, 141(4):226-234.

Bruce, D.P., P. Westerhoff, and A. Brawley-Chesworth. 2002. Removal of 2-nethylisoborneol and geosmin in surface water treatment plants in Arizona. *Journal of Water Supply: Research and Technology - AQUA*, 51(4):183-198.

Chang, S., T.D. Waite, A.I. Schafer, and A.G. Fane. 2003. Adsorption of the endocrine-active compound estrone on microfiltration hollow fiber membranes. *Environmental Science & Technology*, 37(14):3158-3163.

Dodd, M.C., A.D. Shah, U. Von Gunten, and C.-H. Huang. 2005. Interactions of fluoroquinolone antibacterial agents with aqueous chlorine: reaction kinetics, mechanisms, and transformation pathways. *Environmental Science & Technology*, 39(18):7065-7076.

Drewes, J.E., and M. Jekel. 1996. Simulation of groundwater recharge with advanced treated wastewater. *Water Science and Technology*, 33(10-11):409-418.

Drewes, J.E., P. Fox, and M. Jekel. 2001. Occurrence of iodinated X-ray contrast media in domestic effluents and their fate during indirect potable reuse. *Journal of Environmental Science and Health Part a- Toxic/Hazardous Substances & Environmental Engineering*, 36(9):1633-1645.

Drewes, J.E., and L.S. Shore. 2001. Concerns about pharmaceuticals in water reuse, groundwater recharge, and animal waste. *ACS Symposium Series*, 791:206-228.

Drewes, J.E., T. Heberer, T. Rauch, and K. Reddersen. 2003. Fate of pharmaceuticals during groundwater recharge. *Ground Water Monitoring & Remediation*, 23(3):64-72.

Drewes, J.E., C. Bellona, M. Oedekoven, P. Xu, T.-U. Kim, and G. Amy. 2005. Rejection of wastewater-derived micropollutants in high-pressure membrane applications leading to indirect potable reuse. *Environmental Progress*, 24(4):400-409.

Fuerhacker, M., A. Dürauer, and A. Jungbauer. 2001. Adsorption isotherms on 17ß-estradiol on granular activated carbon (GAC). *Chemosphere*, 44:1573-1579.

Gallard, H., A. Leclercq, and J. Croué. 2004. Chlorination of bisphenol A: kinetics and by-products formation. *Chemosphere.*, 56(5):465-473.

Grischek, T., K.M. Hiscock, T. Metschies, and P.F. Dennis. 1998. Factors affecting denitrification during infiltration of river water into a sand and gravel aquifer in Saxony, Germany. *Water Research*, 32(2):450-460.

Heberer, T. 2002. Occurrence, fate, and removal of pharmaceutical residues in the aquatic environment: a review of recent research data. *Toxicology Letters*, 131(1-2):5-17.

Huber, M.M., S. Canonica, G.-Y. Park, and U. Von Gunten. 2003. Oxidation of pharmaceuticals during ozonation and advanced oxidation processes. *Environmental Science & Technology*, 37(5):1016-1024.

Huber, M.M., S. Korhonen, T.A. Ternes, and U. Von Gunten. 2005a. Oxidation of pharmaceuticals during water treatment with chlorine dioxide. *Water Research Water Research*, 39:3607-3617.

Huber, M.M., A. Göbel, A. Joss, N. Hermann, D. Löffler, C.S. Mcardell, A. Reid, H. Siegrist, T.A. Ternes, and U. Von Gunten. 2005b. Oxidation of pharmaceuticals during ozonation of municipal wastewater effluents: a pilot study. *Environmental Science & Technology*, 39(11):4290-4299.

Kiso, Y., Y. Nishimura, T. Kitao, and K. Nishimura. 2000. Rejection properties of non-phenylic pesticides with nanofiltration membranes. *Journal of Membrane Science*, 171(2):229-237.

Kiso, Y., Y. Sugiura, T. Kitao, and K. Nishimura. 2001a. Effects of hydrophobicity and molecular size on rejection of aromatic pesticides with nanofiltration membranes. *Journal of Membrane Science*, 192(1-2):1-10.

Kiso, Y., T. Kon, T. Kitao, and K. Nishimura. 2001b. Rejection properties of alkyl phthalates with nanofiltration membranes. *Journal of Membrane Science*, 182(1-2):205-214.

Knappe, D.R.U., Y. Matsui, V.L. Snoeyink, P. Roche, M.J. Prados, and M.-M. Bourbigot. 1998. Predicting the capacity of powder activated carbon for trace organic compounds in natural waters. *Environmental Science & Technology*, 32(11):1694-1698.

Kreuzinger, N., M. Clara, B. Strenn, and B. Vogel. 2004. Investigation on the behaviour of selected pharmaceuticals in the groundwater after infiltration of treated wastewater. *Water Science and Technology*, 50(2):221-228.

Kuehn, W., and U. Mueller. 2000. Riverbank filtration. An overview. *Journal - American Water Works Association*, 92(12):60-69.

Lee, B.-C., M. Kamata, Y. Akatsuka, M. Takeda, K. Ohno, T. Kamei, and Y. Magara. 2004. Effects of chlorine on the decrease of estrogenic chemicals. *Water Research*, 38(3):733-739.

Mansell, J., J.E. Drewes, and T. Rauch. 2004. Removal mechanisms of endocrine disrupting compounds (steroids) during soil aquifer treatment. *Water Science and Technology*, 50(2):229-237.

Mansell, J., and J. Drewes. 2004. Fate of steroidal hormones during soil-aquifer treatment. *Ground Water Monitoring & Remediation*, 24(2):94-101.

McDowell, D.C., M.M. Huber, M. Wagner, U. Von Gunten, and T.A. Ternes. 2005. Ozonation of carbamazepine in drinking water: identification and kinetic study of major oxidation products. *Environmental Science & Technology*, 39(20):8014-8022.

Montgomery-Brown, J., J. Drewes, P. Fox, and M. Reinhard. 2003. Behavior of alkylphenol polyethoxylate metabolites during soil aquifer treatment. *Water Research*, 37:3672-3681.

Nevskaia, D.M., A. Sepulveda-Escribano, and A. Guerrero-Ruiz. 2001. Surface properties of activated carbons in relation to their ability to adsorb nonylphenol aqueous contaminant. *Physical Chemistry Chemical Physics*, 3:463-468.

Ng, H.Y., and M. Elimelech. 2004. Influence of colloidal fouling on rejection of trace organic contaminants by reverse osmosis. *Journal of Membrane Science*, 244:215-226.

Ngheim, L.D., A.I. Schaefer, and M. Elimelech. 2004. Removal of natural hormones by nanofiltration membranes: measurement, modeling, and mechanisms *Environmental Science & Technology*, 38:1888-1896.

Nghiem, L., A. Schaefer, and M. Elimelech. 2005a. Nanofiltration of hormone mimicking trace organic contaminants. *Separation Science and Technology*, 40(13):2633-2649.

Nghiem, L.D., A.I. Schaefer, and M. Elimelech. 2005b. Pharmaceutical retention mechanisms by nanofiltration membranes. *Environmental Science & Technology*, 39:7698-7705.

Ozaki, H., and H. Li. 2002. Rejection of organic compounds by ultra-low pressure reverse osmosis membrane. *Water Research*, 36:123-130.

Paul, S., H. Boernick, T. Grischek, and E. Worch. 2001. The relevance of aliphatic amines for drinking water production from river Elbe bank filtrate. *Vom Wasser*, 97:125-133.

Pinkston, K.E., and D.L. Sedlak. 2004. Transformation of aromatic ether- and amine-containing pharmaceuticals during chlorine disinfection. *Environmental Science & Technology*, 38(4019-4025).

Rauch, T., and J.E. Drewes. 2004. Assessing the removal potential of soil-aquifer treatment systems for bulk organic matter. *Water Science and Technology*, 50(2):245-253.

Rauch, T., J. Munoz, and J.E. Drewes. 2004. Metabolic removal of organic micropollutants in artificial groundwater recharge systems using water of impaired quality, In *WEFTEC*. Edited by New Orleans, Louisiana.

Ray, C., D.T.W. Soong, G.S. Roadcap, and D.K. Borah. 1998. Agricultural chemical: Impacts on riparian wells during floods. *Journal of the American Water Works Association*, 90(7):90-100.

Rosenfeldt, E., and K.G. Linden. 2004. Degradation of endocrine disrupting chemicals bisphenol A, ethinyl estradiol, and estradiol during UV photolysis and advanced oxidation processes. *Environmental Science & Technology*, 38(20):5476-5483.

Schmidt, C., F. Lange, and H.-J. Brauch. 2004. Assessing the impact of different redox conditions and residence times on the fate of organic micropollutants during riverbank filtration: 4th International Symposium on Pharmaceutically Active Compounds and Endocrine Disrupting Chemicals.

Snyder, S.A., P. Westerhoff, Y. Yoon, and D.L. Sedlak. 2003. Pharmaceuticals, personal care products, and endocrine disruptors in water: implications for the water industry. *Environmental Engineering Science*, 20(5):449-469.

Sontheimer, H. 1991. *Trinkwasser aus dem Rhein?* Sankt Augustin, Germany: Academia Verlag.

Tanghe, T., and W. Verstraete. 2001. Adsorption of nonylphenol onto granular activated carbon. *Water Air and Soil Pollution*, 131:61-72.

Ternes, T.A., M. Meisenheimer, D. McDowell, F. Sacher, H.-J. Brauch, B. Haist-Gulde, G. Preuss, U. Wilme, and N. Zulei-Seibert. 2002. Removal of pharmaceuticals during drinking water treatment. *Environmental Science & Technology*, 36(17):3855-3863.

Verstraeten, I.I., J.D. Carr, G.V. Steele, E.M. Thurman, K.C. Bastian, and D.F. Dormedy. 1999. Surface-water/groundwater interaction: Herbicide transport into municipal collector wells. *Journal of Environmental Quality*, 28(5):1396-1405.

Verstraeten, I.I., T. Heberer, J. Vogel, T.F. Speth, S. Zuehlke, and U. Duennbier. 2003. Occurrence of endocrine-disrupting and other wastewater compounds during water treatment with case studies from Lincoln, Nebraska and Berlin, Germany. *Practice Periodical of Hazardous,Toxic, and Radioactive Waste Management*, 7(4):253-263.

Vieno, N., T. Tuhkanen, and L. Kronberg. 2006. Removal of pharmaceuticals in drinking water treatment: effect of chemical coagulation. *Enivronmental Technology*, 27:183-192.

Vieno, N.M., T. Tuhkanen, and L. Kronberg. 2005. Seasonal variation in the occurrence of pharmaceuticals in effluents from a sewage treatment plant and in the recipient water. *Environmental Science & Technology*, 39(21):8220-8226.

Westerhoff, P., Y. Yoon, S. Snyder, and E. Wert. 2005. Fate of endocrine-disruptor, pharmaceutical, and personal care product chemicals during simulated drinking water treatment processes *Environmental Science & Technology*, 39(17):6649-6663.

Xu, P., J.E. Drewes, C. Bellona, G. Amy, T.-U. Kim, M. Adam, and T. Heberer. 2005. Rejection of emerging organic micropollutants in nanofiltration-reverse osmosis membrane applications. *Water Environment Research*, 77(1):40-48.

Xu, P., J.E. Drewes, T.-U. Kim, C. Bellona, and G. Amy. 2006. Effect of membrane fouling on transport of organic contaminants in NF/RO membrane applications. *Journal of Membrane Science*, 279:165-175.

Yoon, B.O., S. Koyanagi, T. Asano, M. Hara, and A. Higuchi. 2003a. Removal of endocrine disruptors by selective sorption method using polydimethylsiloxane membranes. *Journal of Membrane Science*, 213(1-2):137-144.

Yoon, Y., P. Westerhoff, S.A. Snyder, and M. Esparza. 2003b. HPLC-fluorescence detection and adsorption of bisphenol A, 17β-estradiol, and 17α-ethynyl estradiol on powdered activated carbon. *Water Research*, 37(14):3530-3537.

Yoon, Y., P. Westerhoff, S.A. Snyder, and E.C. Wert. 2006. Nanofiltration and ultrafiltration of endocrine disrupting compounds,pharmaceuticals and personal care products. *Journal of Membrane Science*, 270:88-100.

Yoon, Y., P. Westerhoff, S.A. Snyder, E.C. Wert, and J. Yoon. 2007. Removal of endocrine disrupting compounds and pharmaceuticals by nanofiltration and ultrafiltration membranes. *Desalination*, 202(1-3):16-23.

Zwiener, C., and F.H. Frimmel. 2000. Oxidative treatment of pharmaceuticals in water. *Water Research*, 34(6):1881-1885.

# CHAPTER 6
# HUMAN HEALTH

## OVERVIEW

In this chapter, research on and available evidence for potential effects of EDCs/PPCPs in drinking water on human health are breifly reviewed. The focus of the review for both EDCs and PPCPs is on potential for effects resulting from concentrations that do occur or that could reasonably be expected to occur in drinking water. Data gaps such as lack of occurrence and toxicity data that prevent or limit human health risk assessment are identified.

Human health risk assessment for chemicals requires data describing both exposure and toxicity potential. Limited research has been conducted on the potential for long-term exposure to EDCs and PPCPs at trace concentrations found in drinking water to impact human health. Even for pharmaceuticals, for which the toxicity database is generally larger than for EDCs or personal care products, toxicity testing usually focuses on much higher therapeutic doses, and potential effects in non-target populations (e.g., pregnant women, infants, and children) receiving unintentional exposure are not known. Risk assessments conducted to date have not reported that the trace concentrations of pharmaceuticals detected in drinking water pose an unacceptable health risk to consumers, and likewise, there currently is no convincing evidence that EDCs at levels occurring in drinking water have produced unacceptable health risk to humans (Snyder et al. 2005). However, more research is needed to investigate other compounds to see if this trend holds true. Furthermore, reliable data on occurrence of EDCs/PPCPs in drinking water are sparse, and human exposure to these chemicals through other routes (e.g., diet, medical treatment) must be considered. Drinking water will likely be a minor contributor to total EDC/PPCP exposure for most people. Risk assessment is complicated by the likelihood that people are exposed to multiple EDCs and PPCPs at very low levels in drinking water and other environmental matrices (e.g., air, ingested food), and the implications of these combined exposures are unknown.

## HUMAN HEALTH RISK ASSESSMENT

Human health risks due to chemical exposure result from an interaction between the exposure itself and the potential for effects. The mere detection of EDCs or PPCPs in drinking water does not imply that these contaminants will cause effects in people who consume the water. The intensity and timing of exposure, pharmacokinetics and metabolism, and toxicity of the chemical are all factors that determine whether adverse effects will occur.

### Intensity of Exposure

The dose of a chemical received must be sufficiently large to cause an effect. Additionally, there is the frequency of exposure. In risk assessment, the cautious assumption is that there is daily exposure. For some chemicals, there is a threshold below which no response (or no adverse response) is expected. For others (e.g., certain carcinogens), one cannot assume that a threshold exists, and exposure to even a single molecule of a substance is assumed to have a minute but definite chance of causing an adverse response (Faustman and Omenn 2001).

## Timing of Exposure

Certain life stages or life processes tend to be particularly sensitive to chemical insult. Human subpopulations including pregnant women, infants, children, the elderly, and those with compromised health (e.g., thyroid deficient, immunodeficient) and life cycle stages such as reproduction and development tend to be particularly vulnerable. (Faustman and Omenn 2001, Gilbert 2004).

## Pharmacokinetics and Metabolism

Occurrence in this context is the presence of a chemical in drinking water. Exposure assessment involves evaluating more than just occurrence. In order for long-term, low-dose chemical exposure to cause an adverse effect, the chemical and/or its toxic metabolites must reach sensitive parts of the body at concentrations that are sufficiently great and for a period of time long enough to induce a response that results in a negative impact. The body possesses protective mechanisms, and a chemical occurring in drinking water must overwhelm these defenses to cause an adverse effect. The following processes are involved in evaluation of exposure and can affect the ability of a chemical in drinking water to cause injury to a consumer (Eaton and Klaassen 2001).

- ***Absorption.*** A contaminant must be absorbed into or interact with the body to cause an effect, and the body has protective barriers that can sometimes reduce or prevent absorption of a chemical. For example, skin can serve as a protective barrier against uptake of some contaminants from water during bathing. The low pH of the stomach can degrade or destroy some contaminants such that they are not absorbed to a significant extent into the bloodstream.
- ***Metabolism and excretion.*** Excretion refers to processes by which substances (e.g., chemical contaminants, natural hormones) are cleared from the body. Depending on their properties, chemicals may be eliminated, for example, in breath, urine, feces, semen, mucus secretions, sweat, tears, hair, and fingernails. Some chemicals may be transferred across the placenta into a developing fetus or excreted into breast milk. Metabolism generally increases excretion and thus decreases toxicity of chemicals. In some cases, metabolites are more toxic than the parent chemical or metabolic activation is required for a chemical to exert toxic effects.
- ***Bioaccumulation.*** The concentrations of bioaccumulative chemicals collect inside the body with successive or continuous exposure over time because mechanisms to excrete the chemicals like metabolism or excretion are slow or are overwhelmed. Bioaccumulative contaminants are commonly found in fat, but some associate with other tissues in the body such as bones or blood cells. Some chemicals are sequestered within the body so that they do not contact sensitive receptors or tissues and thus do not cause an effect, or at least not an immediate effect. For example, bioaccumulative chemicals that are stored in fat might cause little or no harm until the fat stores are mobilized for use by the body, releasing the associated contaminants.

- ***Contact with sensitive parts of the body.*** As explained above, a chemical must reach sensitive parts of the body (e.g., receptors, cells, tissues, organs) at concentrations great enough and for a sufficient period of time to overwhelm bodily defense mechanisms that prevent an adverse effect. There are numerous examples of defense mechanisms that can reduce or prevent toxicologic effects: DNA repair mechanisms might reverse DNA damage before subsequent effects can occur, and homeostatic feedback loops in the endocrine system can sometimes compensate for short-term effects of certain EDCs on hormone levels by adjusting them back into their normal range before adverse effects occur.

**Toxicity**

Toxicity is the degree of danger posed by a substance to a living organism (RAIS 2007). The mechanism of action, type and severity of effects, and potency are all factors in the evaluation of the toxic potential of a chemical.

- ***Mechanism of action.*** The mechanism of action is the collection of events beginning with the initial chemical interaction with the body and ultimately resulting in a toxicologic effect. Different subpopulations can vary in their sensitivity to chemicals that act through a particular mechanism of action. For example, people suffering from iodide deficiency are more likely to be adversely affected by exposure to perchlorate, an EDC that inhibits the uptake of iodide into the thyroid.
- ***Type and severity of effects.*** Acutely toxic chemicals might cause death at sufficiently great doses over short periods of time. However, concern in this report is focused on low-level exposure to EDCs and pharmaceuticals that might cause chronic toxicity following long-term exposure. Some toxicologic effects are reversible (e.g., transient alteration of enzyme activities), while others are irreversible (e.g., permanent developmental defects). Chronic toxicologic effects of chemicals in general range in severity from relatively minor (e.g., slight changes in appearance) to disabling or life-threatening (e.g., sterility, cancer).
- ***Potency.*** Potency refers to the ability of a chemical to produce an effect relative to other chemicals that act in the same manner or produce similar effects. For example, the natural hormone 17ß-estradiol and the pharmaceutical ethynylestradiol are among the most potent estrogenic chemicals, meaning that they exert a more powerful estrogen-like effect than other estrogenic EDCs.

**Exposure**

Chapter 4 presents occurrence data for selected EDCs in drinking water. Few monitoring efforts have been directed at trace levels of EDCs in drinking water, and for the majority of EDCs, the database is insufficient to conduct an adequate assessment of exposure through drinking water. However, some chemicals that have been purported to be EDCs are currently regulated on the basis of other potential toxicologic effects (some of which, e.g., reproductive problems, might be mediated through the endocrine system), and extensive monitoring data are available for these chemicals in drinking water. Examples include atrazine, benzo(a)pyrene, hexachlorobenzene, lindane, methoxychlor, and others (U.S. EPA 2007). Whether the detection limits currently in use are sufficiently low to be protective against possible endocrine-mediated effects remains to be seen. Though people clearly are exposed to low levels of known or potential EDCs in drinking water, other routes of exposure (e.g., diet) generally are expected to contribute more to the total EDC exposure for most people. Although certain EDCs (phytoestrogens, endocrine-active pharmaceuticals) are intentionally ingested for a therapeutic effect, exposure to these chemicals through drinking water is not desirable because it is involuntary and cannot be easily controlled to avoid exposure of nontarget groups that might be harmed.

**Potential Health Effects**

The susceptibility of humans to endocrine disrupting effects from exposure to EDCs became evident when it was discovered in 1971 that female children of women who took diethylstilbestrol (DES) during pregnancy had an increased incidence of a rare vaginal cancer (McLachlan 2006). DES is a synthetic estrogenic medication once prescribed to pregnant women to prevent miscarriages or premature deliveries. In these cases DES was administered at therapeutic doses, which are much greater than the doses of EDCs received through typical environmental exposures.

Industrial chemicals, organochlorine pesticides, and naturally-occurring chemicals considered to be EDCs have been implicated as potential causes for a suite of human health effects that might be mediated through the endocrine system: malformations of the reproductive tract (including cryptorchidism and hypospadias in males); precocious puberty; cancers of the breast, ovary, and prostate; low sperm counts and other indicators of poor semen quality; endometriosis; premature menopause (WHO 2002); changes in secondary sex characteristics (e.g., breast development in pre-pubertal boys) (Henley et al. 2007); and neurobehavioral and developmental deficits (including impaired cognitive ability) (Johnson et al. 1996, WHO 2002). However, cause-effect relationships between low-level environmental exposures to specific EDCs and human health effects have not been established, and in some cases, the existence of the effects has been disputed (WHO 2002). According to a World Health Organization report (WHO 2002), "Analysis of human data by itself, while generating concerns, has so far failed to provide firm evidence of direct causal associations between low-level (i.e., levels measured in the general population) exposure to EDCs and adverse health outcomes." This same report describes criteria that can be used to evaluate cause-effect linkages between environmental exposures to EDCs and human health effects. Anderson (2005) also concluded, "Large studies have not indicated any association with a list of effects that have sometimes been attributed to environmental exposure to EDCs: low sperm counts, premature puberty in girls, testicular cancer in young men, and breast cancer in some women."

However, this does not imply that such effects are impossible or that further study will not provide convincing evidence in the future. The following excerpt from the WHO (2002) report summarizes the findings related to human health effects of environmental exposure to EDCs.

Although it is clear that certain environmental chemicals can interfere with normal hormonal processes, there is weak evidence that human health has been adversely affected by exposure to endocrine-active chemicals. However, there is sufficient evidence to conclude that adverse endocrine-mediated effects have occurred in some wildlife species. Laboratory studies support these conclusions.

Generally, studies examining EDC-induced effects in humans have yielded inconsistent and inconclusive results, which is responsible for the overall data being classified as "weak." This classification is not meant to downplay the potential effects of EDCs; rather, it highlights the need for more rigorous studies. This document has identified a number of inherent challenges and confounding factors that contribute to the difficulties in understanding the risks that EDCs pose to human health. The only evidence showing that humans are susceptible to EDCs is currently provided by studies of high exposure levels. Our understanding of the effects of chronic, low levels of EDCs is much more obscure. In particular, the relationship between early-life exposures to EDCs in humans and functioning in adult life is poorly understood. This is a concern because laboratory animal studies have indicated that early life stages may be especially sensitive to the effects of EDCs. Only recently have human epidemiological studies been conducted with the necessary rigor to sufficiently address potential cause-and-effect relationships in regards to EDC exposures.

Assessments of the potential for endocrine disruption in humans due to exposure to EDCs in drinking water, groundwater, surface water, reuse water, and wastewater generally have not indicated a risk (Aherne and Briggs 1989; Bursch et al 2004; Harries et al. 1995, 1996; NRC 2005; WHO 2002). Studies evaluating the health effects of using treated wastewater for groundwater recharge in Los Angeles County showed, after nearly 30 years of recharge, no association between exposure to reuse water and increased cancer rates, mortality, infectious disease, or adverse birth outcomes including effects on prenatal development, infant mortality, and birth defects (U.S. EPA 2004 Guidelines for Water). Anderson (2005) noted that no studies have effectively linked low concentrations of EDCs in wastewater to adverse effects in people. Aherne and Briggs (1989) concluded, "Norethisterone and ethinyloestradiol concentrations in sewage effluent, reservoirs, rivers, and potable water have been estimated at less than 20 ng/L, a value unlikely to present a significant risk to human health."

Following an in-depth review of the U.S. EPA's proposed reference dose for the EDC perchlorate, NRC (2005) reported:

- The available epidemiologic studies are not consistent with the hypothesis that perchlorate exposure at the doses investigated causes any of the following health effects: congenital hypothyroidism, significant changes in thyroid function in newborns, or hypothyroidism and other thyroid disorders in adults. There are not enough epidemiologic data to determine whether there is an association between perchlorate exposure and adverse developmental effects on the nervous systems of children or between perchlorate exposure and thyroid cancer. The study committee believes thyroid cancer to be an unlikely outcome of perchlorate exposure in humans.
- Clinical studies in humans provide more useful data. In a study by Greer et al. (2002), groups of healthy men and women were given perchlorate in doses of 0.007 to 0.5 mg/kg of body weight per day for 14 days.1 The study identified a no-observed effect level for inhibition of iodide uptake by the thyroid at 0.007 mg/kg per day. Furthermore, there were no significant changes in thyroid hormone concentrations at any dose. These findings are supported by the results of other similar studies, one of which was of 6-months duration. In addition, a 1984 study of long-term treatment of patients with hyperthyroidism found that moderately high doses of perchlorate did not cause hypothyroidism, even when administered after the patients' blood levels of thyroid hormones had returned to normal.

According to (Bursch et al. 2004), "A consortium of Austrian scientists (ARCEM) carried out a multidisciplinary environmental study on Austrian surface and ground waters including chemical monitoring, bioindication, risk assessment and risk management for selected endocrine disrupters: 17ß-estradiol, estriol, estrone, 17α-ethinylestradiol, 4-nonylphenol, 4-nonylphenol ethoxylates (4-NP1EO, 4-NP2EO) and their degradation products, octylphenol, octylphenol ethoxylates (OP1EO, OP2EO) as well as bisphenol A." For humans, they considered potential exposure via either drinking water (ground water) or fish consumption. They concluded that the levels of exposure to compounds under study were less than those considered to result in human health risks.

A report by Harries et al. (1995) describes the results of research conducted from 1992-1995 that assessed whether estrogenic effects could be detected in male fish held in cages in rivers at locations downstream of WWTP discharges, at locations where raw water is collected for drinking water treatment plants, or in raw water storage reservoirs. Caged male rainbow trout were held in the river or raw water storage reservoir for 3-6 weeks before they were tested for evidence of vitellogenin production (Harries et al. 1995, Harries et al. 1996). Vitellogenin is an egg yolk precursor protein produced by female fish in response to elevated blood estrogen levels in preparation for egg production. Vitellogenin usually is not produced in significant amounts in sexually immature fish or in male fish due to their lower blood estrogen levels, but vitellogenin production can be induced in immature or male fish exposed to exogenous (externally acquired) estrogenic chemicals. For this reason, vitellogenin induction in male fish in particular has been used as an indicator of the presence of estrogenic chemicals in water.

According to the summary of the report by Harries et al. (1995), "Of the 15 raw water storage reservoirs that were surveyed, it was shown that none produced a vitellogenin response in caged fish. On the basis that biologically significant amounts of substances, which are responsible for the oestrogenic response (in fish), were not detected in all 15 raw water storage reservoirs most likely to be affected, there is no evidence of a risk to drinking water supplies, and for the present there is no need for further research." It appears that the results of the same reservoir study were later published elsewhere (Harries et al.1996). According to the later report (Harries et al. 1996), trout were placed in 15 raw water storage reservoirs in southeast England during the summer of 1993 for an exposure period of 6 weeks. These reservoirs received water mainly from river abstractions downstream of municipal WWTP discharges, but also from surface runoff and groundwater wells. No biologically significant increases in plasma vitellogenin were observed in fish caged in any of the reservoirs. The authors concluded, "This appears to lessen, if not eliminate, the role of drinking water as a potential source of environmental estrogens to humans." However, because vitellogenin production occurs in egg-laying animals, the relevance of vitellogenin induction in fish exposed to estrogens in water to humans who consume drinking water obtained from the same source is not entirely clear. Routes of exposure and uptake also differ between fish and humans.

## PPCPS

### Exposure

Chapter 4 presents occurrence data for selected PPCPs. In comparison with the database for EDC occurrence in drinking water, even fewer data are available for trace levels of PPCPs in drinking water. The occurrence data that are available generally consist of individual samples or small sets of samples that represent only a snapshot of occurrence. Consequently, the currently available database is insufficient to allow an adequate assessment of exposure to PPCPs through drinking water. However, in comparison with most EDCs and personal care products, pharmaceuticals commonly are better characterized with regard to pharmacokinetics and metabolism, which are also important factors in exposure characterization.

### Potential Health Effects

Pharmaceuticals have undoubtedly improved and lengthened human life. The increased production and use of antibiotics in the 1940s allowed ready treatment of previously terrifying infectious diseases such as rheumatic fever, syphilis, pneumonia and tuberculosis, and the ability to successfully complete previously dangerous medical operations (Bud 2005) It is likely that pharmaceuticals have been entering the environment for as long as people have been using them. However, only recently, with the advent of exquisitely sensitive analytical instrumentation and methods for detection of trace organics in water, have we been capable of detecting the presence of minute concentrations of pharmaceuticals in wastewater, groundwater, surface water, and drinking water. The modes of action and potential toxicity of therapeutic doses of pharmaceuticals are often well-characterized when compared with what is known about the same for other waterborne contaminants. However, little is known about the potential effects of unintended, long-term, chronic exposure to the low levels of pharmaceuticals found, or likely to be found, in drinking water. This is particularly true for non-target populations for whom pharmaceuticals are not intended.

According to Daughton and Ternes (1999), based on information about the activity and potential toxicity of pharmaceuticals, possible health effects of long-term exposure to pharmaceuticals could include endocrine disruption, induction of antibiotic resistance in human pathogens, genotoxicity, carcinogenicity, allergic reactions, and reproductive and/or developmental effects. Some potential effects of pharmaceuticals are difficult to predict. For example, some pharmaceuticals (e.g., antibiotics) are known to cause allergic reactions in humans, but predicting the likelihood of allergic effects is difficult because people sensitized to a chemical can respond at extremely low levels of exposure. Additionally, the possibility of microbial resistance induced by antibiotics in the environment is a subject of some controversy. Microbial resistance to antibiotics has been noted in surface water and sewage effluent (Harwood et al. 2001, Sayah et al. 2005, Hamelin et al. 2007). However, (Ayscough, J.Fawell et al. 2000) concluded that the most likely cause for antibiotic resistance in organisms in the environment is not the induction of resistance caused by exposure to low levels of antibiotics in the environment, but rather the excretion of resistant organisms by humans and animals receiving antibiotic treatment or transfer of plasmids from excreted resistant organisms to non-resistant organisms in the environment.

Drugs are designed to be biologically active. While beneficial for one population group, the same biological activity can be hazardous to other population groups or at certain times—for example, during pregnancy or when taken with other medications. Additionally, published "side effect" profiles provide evidence that unintended effects are known to occur at therapeutic doses, and animal toxicity studies at therapeutic or higher doses often show an increased incidence in undesired effects such as cancer or reproductive effects. Drug allergies and idiosyncratic responses are also known to occur. In some cases, studies conducted after marketing a drug reveal unexpected effects over the course of use (e.g., associations with developmental or neurological effects or cancer). The U.S. Food and Drug Administration (FDA) monitors such responses through its Adverse Event Reporting System (AERS) program (FDA 2007). The most commonly affected systems are the nervous, blood, cardiovascular, and respiratory systems. Because drugs are often not studied in children, information about potential risks of exposure to this group may be limited to evidence from animal studies.

Prior to marketing a pharmaceutical in the United States, the manufacturer is required to collect data to establish its safety. Studies are conducted to assess the drug's behavior, toxicological effects in animals, and safety and efficacy in humans. A typical suite of animal toxicology studies includes acute studies (exposure of one day or less to evaluate effects that happen very rapidly after an exposure), subchronic studies (up to three months in duration), chronic studies (repeated longer-term exposure, usually up to two years, to assess the potential for cancer, organ problems, nervous system impairment, reproductive effects, etc.), studies specific to effects on reproduction and offspring development (with administration just prior to, during, and immediately after pregnancy), and genotoxicity studies (to assess whether the agent causes genetic mutations). Studies generally use therapeutic or higher doses that are delivered for a relatively short period of time. Data on the hazards associated with low-level exposure to PPCPs or EDCs are largely limited to aquatic studies with fish or other aquatic species and often involve mixtures of chemicals. Continual, life-long exposure to trace levels is a largely unexplored domain of toxicology.

According to a review entitled *"Pharmaceuticals and Personal Care Products in the Water Cycle"* (Puijker and Mons, 2004). "Evaluations of human health risk have been made before by Richardson and Bowron (1985), Christensen (1998), Mons *et al.* (2000), Webb (2001) and Schulman et al. (2002). They all concluded that no appreciable risk for humans exists at low levels measured in drinking water...Mons et al. (2003) calculated that lifetime consumption of drinking water with these concentrations of pharmaceuticals would result in a maximum consumption of 5% of one (1) daily therapeutic dose."

Several authors (Schulman et al. 2002, Schwab et al. 2005, Webb et al. 2003, Webb 2001) assessed the potential for adverse health effects from exposure to pharmaceuticals in drinking water by comparing exposures one would get from drinking the water—assuming exposure to concentrations that have been measured or concentrations that have been estimated assuming dilution of effluents— to the therapeutic dose, divided by uncertainty or safety factors to extrapolate from these doses to safe levels for populations "that could include sensitive individuals". International regulatory agencies and scientific expert committees (e.g., the European Medicines Agency (EMEA) and the Joint FAO/WHO Expert Committee on Food Additives (JECFA)) also have established ADIs for veterinary pharmaceuticals based on minimum therapeutic doses divided by uncertainty factors (EMEA 2000, JEFCA 1994, JEFCA 1998). While therapeutic doses can be useful benchmarks, they have their limitations. The dose-response relationships of pharmaceuticals in their target organisms are generally well-documented, but uncertainty exists about the potential effects of these pharmaceuticals on non-target population groups (such as the fetus, infants, or the elderly) and about the effect of such factors as differences in dosing and timing of exposure or exposure to multiple chemicals at the same time (Jones et al. 2004). Moreover, the therapeutic effects of some drugs, such as chemotherapeutic agents, are toxic (purposefully), carcinogenic, or mutagenic. Thus therapeutic doses must be considered carefully. Nevertheless, some have used theraupetic doses as points of departure and Table 6.1 lists ADIs published by some of these entities for human and veterinary pharmaceuticals.

Other authors (Christensen 1998, Schulman et al. 2002) evaluated human health risks of pharmaceuticals in drinking water based on no-effect levels or on specific information on the chemical's toxicity and concluded that no appreciable risk for humans exists at the low levels measured in drinking water. Recently Versteegh et al. (2003) calculated provisional "no-effect levels" for a number of pharmaceuticals detected in Dutch drinking water, based on ADI or maximum residue levels (MRL) for veterinary pharmaceuticals in milk, and these investigators drew similar conclusions.

Dolan et al. (2005) proposed thresholds of toxicologic concern (TTCs) for exposure to pharmaceutical residues in food and other media. The TTC approach assumes that any intake below a published and documented toxicologic threshold does not pose a health concern. They suggest TTCs of 1 μg/day for compounds that are likely to be carcinogenic, 10 μg/day for "relatively unstudied compounds" with evidence of pharmacological activity at relatively low doses but no evidence of carcinogenicity, and 100 μg/day for relatively unstudied compounds that have no a priori evidence of unusual toxicity or potency and no evidence of carcinogenicity. These values correspond to ADIs of 0.014 μg/kg-day, 0.14 μg/kg-day, and 1.4 μg/kg-day, respectively, assuming exposure to a 70 kg adult.

**Table 6.1**
**Selected acceptable daily intakes (ADIs) published for human and veterinary pharmaceuticals**

| Substance | ADI (μg/kg-d) | Basis |
|---|---|---|
| **Schwab et al. 2005** | | |
| Acetaminophen | 340 | Therapeutic effect |
| Albuterol | 2.8 | Therapeutic effect |
| Cimetidine | 29 | Therapeutic effect |
| Ciprofloxacin | 1.6 | Sensitivity of human intestinal microflora |
| Codeine | 2 | Therapeutic effect |
| Dehydronifedipine | 100 | Animal study NOEL |
| Digoxigenin | 0.07 | Therapeutic effect |
| Digoxin | 0.07 | Therapeutic effect |
| Diltiazem | 14 | Therapeutic effect |
| Doxycycline | 30 | Sensitivity of human intestinal microflora |
| Enalaprilat | 70 | Therapeutic effect |
| Erythromycin-H2O | 40 | Therapeutic effect |
| Fluoxetine | 2.9 | Therapeutic effect |
| Gemfibrozil | 55 | Therapeutic effect |
| Ibuprofen | 110 | Therapeutic effect |
| Lincomycin | 25 | Sensitivity of human intestinal microflora |
| Metformin | 62 | Therapeutic effect |
| Norfloxacin | 190 | GI effect at lowest therapeutic dose |
| Oxytetracycline | 30 | Sensitivity of human intestinal microflora |
| Paroxetine | 2.9 | Therapeutic effect |
| Ranitidine | 11 | Therapeutic effect |
| Sulfamethoxazole | 130 | Animal study NOEL |
| Sulfathiazole | 50 | Changes in thyroid tissue in animals |
| Tetracycline | 30 | Sensitivity of human intestinal microflora |
| Trimethoprim | 4.2 | Sensitivity of human intestinal microflora |
| Warfarin sodium | 0.16 | Therapeutic effect |
| **Schulman et al. 2002** | | |
| Acetyl salicylic acid | 14 | Therapeutic effect |
| Clofibric acid | 240 | Therapeutic effect |

(continued)

Table 6.1 (Continued)

| Substance | ADI (μg/kg-d) | Basis |
|---|---|---|
| Cyclophosphamide | 0.014 | Animal study tumor incidence |
| Indomethacin | 24 | Subtherapeutic effect |
| **Webb et al. 2003** | | |
| Acetyl salicylic acid | 8.3 | EMEA or JECFA values |
| Benzylpenicillin | 0.5 | EMEA or JECFA values |
| Carazolol | 0.1 | EMEA or JECFA values |
| Chloramphenicol | No safe level | EMEA or JECFA values |
| Clenbuterol | 0.0042 | EMEA or JECFA values |
| Doxycycline | 3 | EMEA or JECFA values |
| Erythromycin | 5 | EMEA or JECFA values |
| 17ß-Oestradiol | 0.05 | EMEA or JECFA values |
| Ketoprofen | 5 | EMEA or JECFA values |
| Naficillin, oxacillin, cloxacillin, dicloxacillin | 4.4 (group) | EMEA or JECFA values |
| Sulfamethazine | 50 | EMEA or JECFA values |
| Tetracycline, oxytetracycline and chlorotetracycline | 3 (group) | EMEA or JECFA values |

Notes: ADI—Acceptable Daily Intake; EMEA—European Medicines Agency; JECFA—Joint FAO/WHO Expert Committee on Food Additives; NOEL—No observed effect level

Table 6.2 lists pharmaceuticals that have been detected in source water and raw and finished drinking water (see Chapter 4 for more details), their maximum detected concentrations, and their lowest therapeutic doses (i.e., the lowest recommended dosage level indicated on the package labeling, assumed to be the lowest exposure level at which the chemical produces the desired pharmacologic effect). The table also shows water concentrations for each pharmaceutical that would correlate to the lowest therapeutic dose, if one assumes that a person drinks two liters of water at this concentration every day. To provide an additional margin of safety, these therapeutic dose concentrations were divided by a factor of 1,000. As shown, in each case, the highest detected concentration of a pharmaceutical in source water or raw or finished drinking water is well below the concentration based on the lowest therapeutic dose divided by 1,000. Overall, the maximum detected concentrations in drinking water are a factor of 5 to 12,000 lower than the levels calculated as "safe" using this method and are far below unadjusted therapeutic dose levels. Additionally, assuming that people are exposed to the highest detected concentration is conservative because people do not generally consume untreated surface water and all of the maximum detected concentrations presented in this table are surface water or groundwater concentrations. Table 6.3 shows how therapeutic concentrations compare to maximum concentrations measured in finished drinking water.

**Table 6.2**

**Comparison of maximum detected pharmaceutical concentrations in drinking water sources (surface or groundwater) and water concentrations equivalent to the therapeutic dose/1,000**

| Compound | Compound Group | Maximum Conc. (ng/L) | Median Conc. (ng/L) | Country | Reference | Minimum Therapeutic Dose (mg/d) | DWEL Therapeutic Dose/ 1000 (ng/L) [h] | Margin of Safety |
|---|---|---|---|---|---|---|---|---|
| Acetaminophen | Analgesic | 10,000 | <9 | US | Kolpin et al., 2002 | 160 b | 560,000 | 56 |
| Atenolol | β-blocker | 241 | 17 | Italy | Calamari et al., 2003 | 25 c | 13,000 | 54 |
| Bezafibrate | Lipid regulator | 3,100 | 350 | Germany | Ternes 1998a | 400 c | 200,000 | 65 |
| Carbamazepine | Antiepileptic drugs | 7,100 | 400 | Germany | Wiegel et al. 2004 | 10 d | 35,000 | 5 |
| Ciprofloxacin | Antibiotic | 30 | 20 | US | Kolpin et al., 2002 | 100 e | 350,000 | 12,000 |
| Clarithromycin | Antibiotic | 260 | <20 | Germany | Hirsch et al., 1999 | 500 c | 248,500 | 960 |
| Clofibric acid | Lipid regulator | 7,300 | - | Germany | Heberer et al., 1997 | 2,000 c | 1,000,000 | 140 |
| Diatrizoate | Contrast media | 4,000 | 4,000 | Germany | Putschew et al., 2000 | NA | NA | NA |
| Diazepam | Muscle Relaxant: | 33 | <10 | Germany | Ternes et al., 2001 | 2 c | 1,015 | 31 |
| Diclofenac | Analgesic | 1,200 | 150 | Germany | Ternes 1998a | 100 c | 49,000 | 41 |
| Diltiazem | Anti-hypertensive | 106 | <12 | US | Kolpin et al., 2004 | 120 c | 59,500 | 560 |
| Erythromycin-H2O | Antibiotic | 1,700 | <50 | US | Kolpin et al., 2002 | 300 e | 1,050,000 | 620 |
| Gemfibrozil | Lipid regulator | 1,550 | 228 | Spain | Farré et al., 2001 | 1,200 c | 595,000 | 380 |
| Ibuprofen | Analgesic | 5,850 | <280 | US | Loraine and Pettigrove 2006 | 50 f | 350,000 | 60 |
| Iopromide | Contrast media | 1,600 | 1,600 | Germany | Putschew et al., 2000 | NA | NA | NA |
| Ketoprofen | Analgesic | 300 | <28 | Spain | Farré et al., 2001 | 75 c | 39,000 | 130 |
| Lincomycin | Antibiotic | 730 | <50 | US | Kolpin et al., 2002 | 100 e | 350,000 | 480 |
| Metoprolol | β-blocker | 2,200 | 45 | Germany | Ternes 1998a | 25 c | 13,000 | 6 |
| Naproxen | Analgesic | 2,000 | 67 | Spain | Farré et al., 2001 | 125 c | 455,000 | 230 |
| Primidone | Antiepileptic drugs | 635 | - | Germany | Heberer et al., 2002 | 50 c | 24,850 | 39 |
| Propranolol | β-blocker | 590 | 12 | Germany | Ternes 1998a | 80 c | 39,000 | 66 |
| Roxithromycin | Antibiotic | 560 | <20 | Germany | Hirsch et al., 1999 | 600 c | 300,000 | 540 |
| Sulfamethoxazole | Antibiotic | 1,900 | <50 | US | Kolpin et al., 2002 | 400 g | 2,800,000 | 1,500 |
| Trimethoprim | Antibiotic | 710 | <30 | US | Kolpin et al., 2002 | 80 e | 280,000 | 390 |

Conc.- concentration

a. AwwaRF #3085: Toxicological Relevance of Pharmaceuticals and Endocrine Disruptors in Drinking Water
b. Dose recommended for a child, age 2-3 years; body weight of 10 kg assumed
c. Dose recommended for an adult; body weight of 70 kg assumed
d. Dose recommended for a child, age less than 6 years; body weight of 10 kg assumed
e. Dose recommended for a “child” or “pediatric” ; body weight of 10 kg assumed
f. Dose recommended for a child, age 6-12 months; body weight of 5 kg assumed
g. Dose recommended for a child, age greater than 2 months; body weight of 2 kg assumed
h. Calculated by first estimating the minimum therapeutic dose in mg/kg-day, based on the subject’s assumed body weight, then multiplying the dose by an assumed adult body weight of 70 kg, dividing by an assumed drinking water ingestion rate of 2 L/day, and multiplying by a conversion factor of 1,000,000 (ng/mg)

**Table 6.3**
**Comparison of maximum detected pharmaceutical concentrations in finished drinking water* and water concentrations equivalent to the therapeutic dose/1,000**

| Compound | Compound Group | Maximum Finished Concentration (ng/L) | Median Finished Concentration (ng/L) | Corresponding Raw Water Concentration (ng/L) | Reference | Minimum Therapeutic Dose (mg/d) | DWEL Therapeutic Dose/ 1000 (ng/L) § | Margin of Safety |
|---|---|---|---|---|---|---|---|---|
| Bezafibrate | Lipid regulator | 100 | 100 | 170 | Heberer et al., 2001b | 400 † | 200,000 | 2,000 |
| Carbamazepine | Antiepileptic | 250 | 120 | 60-400 | Stackelberg et al., 2004 | 10 ‡ | 35,000 | 140 |
| Clofibric acid | Lipid regulator | 70 | 70 | 50 | Heberer et al., 2001b | 2,000 † | 1,000,000 | 14,000 |
| Diatrizoate | Contrast media | NA | 4,000 | 2,000 | Putschew et al., 2001 | NA | NA | NA |
| Diclofenac | Analgesic | 930 | <280 | NA | Loraine and Pettigrove 2006 | 100 † | 49,000 | 53 |
| Iopromide | Contrast media | <50 | <50 | 1,600 | Putschew et al., 2000 | NA | NA | NA |
| Ketoprofen | Analgesic | <5 | <5 | 9.5 | Vieno et al., 2005 | 75 † | 39,000 | 7,800 |
| Naproxen | Analgesic | <5 | <5 | 7.25 | Vieno et al., 2005 | 125 † | 455,000 | 91,000 |
| Primidone | Antiepileptic | 15 | 15 | 105 | Heberer et al., 2001b | 50 † | 24,850 | 1,700 |

*AwwaRF #3085: Toxicological Relevance of Pharmaceuticals and Endocrine Disruptors in Drinking Water
† Dose recommended for an adult; body weight of 70 kg assumed
‡ Dose recommended for a child, age less than 6 years; body weight of 10 kg assumed
§ Calculated by first estimating the minimum therapeutic dose in mg/kg-day, based on the subject's assumed body weight, then multiplying the dose by an assumed adult body weight of 70 kg, dividing by an assumed drinking water ingestion rate of 2 L/day, and multiplying by a conversion factor of 1,000,000 (ng/mg)

## REGULATIONS

EDCs in drinking water are not regulated, and there generally are no guidelines describing safe or acceptable levels of EDCs in drinking water. One exception is perchlorate, an EDC that occurs in drinking water in the United States. Perchlorate was listed on the U.S. EPA's second Drinking Water Contaminant Candidate List (CCL 2) (U.S. EPA 2006 Drinking water CCL). The U.S. EPA has developed a reference dose for perchlorate based on its effects on the thyroid, and the resulting drinking water equivalent level (DWEL) is 24.5 μg/L. However, the DWEL is not a drinking water standard but an interim value that can be used to develop a standard. There is no federal maximum contaminant level (MCL) for perchlorate, and it appears unlikely that a federal standard will be set within the next several years. California Department of Health Services set a public health goal (PHG) of 6 μg/L for perchlorate in drinking water and has proposed an MCL of 6 μg/L (Ca DHS 2006 Perchlorate). Massachusetts Department of Environmental Protection promulgated a drinking water standard of 2 ppb (2 μg/L) for perchlorate; this is the first drinking water standard for perchlorate in the United States (Massachusetts DEP 2006). While other potential EDCs are regulated in drinking water, they are not regulated specifically on the basis of their potential to cause endocrine disruption. For example, the current federal maximum contaminant level (MCL) for atrazine is based on its potential to cause cardiovascular system or reproductive problems (specifically decreased body weight gain) (U.S. EPA. 2007. Drinking Water Contaminants). Although reproductive problems could be caused by endocrine disruption, they also can occur through other mechanisms.

The U.S. EPA developed the EDSP in response to a Congressional mandate in section Section 408(p) of the Federal Food, Drug, and Cosmetic Act (FFDCA). The statute grants U.S. EPA discretionary authority to "provide for the testing of any other substance that may have an effect that is cumulative to an effect of a pesticide chemical if the Administrator determines that a substantial population may be exposed to such a substance" (U.S. EPA 2007. Draft list). Furthermore, Section 1457 of the Safe Drinking Water Act (SDWA) endows U.S. EPA with discretionary authority to provide for testing (under FFDCA Section 408(p)) "of any other substances that may be found in sources of drinking water if the Administrator determines that a substantial population may be exposed to such substance" (U.S. EPA 2007. Draft list). However, the EDSP remains incomplete. Initial screening of a list of pesticide ingredients is expected to be required in 2008, but the more definitive Tier 2 tests are still under development. The Endocrine Disruptor Screening and Testing Advisory Committee (EDSTAC) recommended that U.S. EPA should include in the EDSP representative mixtures of the most commonly occurring disinfection byproducts for screening and possible testing for endocrine disruption (EDSTAC 1998 Endocrine Disruptor). Later, U.S. EPA tabled the evaluation of mixtures for endocrine activity until methods can be developed and validated for individual chemicals (U.S. EPA 2007. Draft list). Federal regulations based on endocrine effects for EDCs in drinking water are not expected within the next several years.

Pharmaceuticals in drinking water also are not regulated, with the exception of fluoride added intentionally to potable water for benefits to dental health. The California Department of Health Services Draft Groundwater Recharge Reuse Regulations (Ca DHS 2003) include monitoring requirements for specified EDCs and pharmaceuticals but do not establish standards. To the best of the authors' knowledge, no regulatory limits have been established in the United States for EDCs or pharmaceuticals in municipal WWTP effluents. However, the FDA requires environmental risk assessments for pharmaceuticals submitted for new drug approvals if the predicted environmental concentration exceeds 1 μg/L (FDA 1998, 2000; Velagaleti et al. 2002).

## PROPOSED APPROACH FOR IDENTIFYING LEVELS OF CONCERN FOR PHARMACEUTICALS AND EDCS IN DRINKING WATER

Although no specific regulatory guidance exists to mandate the manner in which human health risks of exposure to PPCPs or EDCs in drinking water are assessed, standard methods exist for determining exposure levels to environmental contaminants that are not likely to be associated with adverse health effects (WHO 1994, U.S. EPA 2000 Methodology, ATSDR 2007 Minimal). These exposure levels, often termed acceptable daily intakes (ADIs)[4], are typically defined by reviewing animal toxicology data and, for effects other than carcinogenicity, identifying a point of departure (POD) upon which to base the ADI. The POD typically is the highest dose at which an effect is *not* seen (the *no* observed adverse effect level, or NOAEL) or the lowest dose at which an effect *is* seen (the *lowest* observed adverse effect level, or LOAEL). Below this dose, there is no evidence of a statistically or biologically significant increase in adverse effects, although some changes may occur that are not considered adverse (*e.g.*, changes in certain enzyme levels). The point of departure is then divided by uncertainty factors (UFs) to derive an ADI considered to be protective to broader population groups, including sensitive populations:

$$\text{ADI}\,(\text{mg/kg}-\text{d}) = \frac{\text{NOAEL or LOAEL}\,(\text{mg/kg}-\text{d})}{\text{UFs}}$$

An ADI is commonly defined as the amount of a chemical to which a person can be exposed on a daily basis over an extended period of time (usually a lifetime) without suffering a deleterious effect (U.S. EPA 1993 Reference dose). ADIs are often presented in terms of dose per day (*e.g.*, mg/d) or dose per unit of body weight per day (*e.g.*, mg/kg-d). Generally, several UFs are applied, individually ranging in value from 3 to 10, with each factor representing a specific area of uncertainty in the available data. For example, if the POD is based on an animal study, a factor of 10 may be applied to account for possible differences in responsiveness between animals and humans. A second factor of 10 may account for variation in susceptibility among humans. Other factors may account for database deficiencies, such as when no or minimal information exists on reproductive effects or longer-term exposures. When high-quality toxicity data are available, combined uncertainty factors typically range from 30 to 1,000.

For contaminants present in drinking water, ADIs can be converted to drinking water equivalent levels (DWELs) by making an assumption about the volume of water a person drinks in a day. As shown below, an ADI in mg/kg-d can be converted to a DWEL in nanograms per liter (ng/L; equivalent to a part per trillion, or ppt) by multiplying it by an assumed body weight (*e.g.*, 70 kg, the U.S. EPA default body weight of an adult male) and a metric unit conversion factor, and dividing by an average daily drinking water ingestion rate (*e.g.*, 2 liters per day) (U.S. EPA 2006 Setting standards):

---

[4] ADIs are variously termed reference doses (RfDs), minimal risk levels (MRLs), and tolerable daily intakes (TDIs) by U.S. EPA, the Center for Disease Control's Agency for Toxic Substances and Disease Registry (ATSDR), and the World Health Organization (WHO), respectively.

$$\text{DWEL (ng/L)} = \frac{\text{ADI (mg/kg} - \text{d)} \times 70\text{ kg} \times 1{,}000{,}000\text{ ng/mg}}{2\text{ L/d}}$$

Often, when setting action levels for contaminants in surface water systems, regulatory agencies incorporate an additional factor into the above equation to account for the possibility that on any given day, a person could be exposed to the substance through some source other than drinking water ingestion. This factor, called a relative source contribution (or RSC), typically ranges in value from 20 to 80% based on knowledge about likely sources of exposure to the chemical (U.S. EPA 2000 Methodology; Ca DHS 2006 Drinking water; U.S. EPA 2006 Setting standards). Thus, the resulting DWEL will be 20 to 80% of the value based solely on ingestion of drinking water. For pesticides under the Food Quality Protection Agent (FQPA), U.S. EPA uses a "risk cup" approach to account for aggregate exposure to residues of pesticides through all potential routes of exposure (including food, drinking water, and residential use) and exposure to all other pesticides with a common mechanism of toxicity (U.S. EPA 2000 Report on FQPA). Under this approach, the allowed percentage of pesticide in drinking water may be less than 20% of the RfD if significant additional sources are expected.

Some pharmaceuticals have been shown to be carcinogenic in animal bioassays conducted as part of the drug development process. Likewise, certain EDCs are carcinogens or have been implicated in hormonally-mediated cancers (WHO 2002). Carcinogenicity could be a concern for these chemicals, particularly at low chronic exposure levels. For example, gemfibrozil, a drug taken to reduce blood lipid levels, was shown in a two-year cancer bioassay in rats to produce liver tumors, although it is not clear that the required metabolic step that produces reactive chemical metabolites (peroxisome proliferation) is seen in humans (Fitzgerald et al. 1981). For chemicals that show positive evidence of carcinogenicity in high dose animal studies, extrapolation models can be used to predict the tumorigenic response at lower doses. Resulting "slope factors" (SFs) predict the lifetime excess risk of cancer to a person per 1 mg/kg-day of exposure to the chemical of interest.

DWELs for carcinogens can be estimated assuming that a *de minimis* lifetime excess cancer risk of one additional cancer per one million lifetime exposures ($10^{-6}$) is acceptable, and that a 70-kg person consumes two liters of water daily (365 days per year) for 30 years over a 70-year lifetime (equivalent to 25,550 days):

$$\text{DWEL (ng/L)} = \frac{10^{-6} \times 70\text{ kg} \times 25{,}550\text{ d} \times 1{,}000{,}000\text{ ng/mg}}{\text{SF (mg/kg} - \text{d)}^{-1} \times 2\text{ L/d} \times 30\text{ yr} \times 365\text{ d/yr}}$$

For each chemical of interest, the lesser of the two DWELs (for carcinogenic and noncarcinogenic effects) can then be compared to occurrence data for the same chemical in drinking water. While a wide range of EDCs and PPCPs have been detected in effluent, surface, and drinking water, the number of samples overall is generally is insufficient to characterize the spatial or temporal variation for most of the unregulated chemicals. For the majority of these compounds, data are insufficient to adequately characterize exposure, and therefore risk, to aquatic organisms or humans. However, preliminary, screening-level risk assessments using conservative assumptions about rates and levels of exposure can be undertaken for these chemicals until more information on the duration and timing of exposure can be gathered. These assessments provide perspective regarding the contaminants that are likely to pose the greatest risks and the likely maximum levels of these risks. For example, for many EDCs and PPCPs, drinking water occurrence and concentration data are not available for comparison to a DWEL. Because a significant number of EDCs and PPCPs found in source waters for drinking water abstraction arise primarily from municipal WWTP effluent, concentration data for WWTP effluents can be used to provide a conservative estimate of exposure for comparison to DWELs. Exposure estimates based on concentrations in WWTP effluents are conservative because discharged contaminants are subject to dilution and removal processes that effectively clear them from water, including biological and physical degradation, volatilization, and partitioning to sediments[5]. Drinking water treatment processes also can further reduce concentrations of many EDCs and PPCPs.

The approach described above is more closely associated with the US EPA approach and guidance on human health risk assessment. This is the process that was used in AwwaRF 3085. Other methods were used as well for assessing carcinogenic risk. The interested reader is suggested to review that report for more information on the risk assessment process for PPCPs and EDCs.

5 Contaminants that bind to sediment might become unavailable, at least temporarily, but sediments also can serve as "sinks" from which contaminants are later reintroduced into water. Some contaminants in sediment are degraded there by biological or physical processes. Even when binding to sediment is practically irreversible, some contaminants can become available again due to sediment re-suspension during flooding, dredging, and other types of disturbance. Burrowing benthic organisms can re-suspend sediment or release contaminants from sediment as they digest it.

## REFERENCES


Aherne, G.W., and R. Briggs. 1989. The relevance of the presence of certain synthetic steroids in the aquatic environment. *Journal of Pharmacy and Pharmacology*, 41:735-736.

Anderson, Paul D. 2005. Technical Brief: Endocrine Disrupting Compounds and Implications for Wastewater Treatment. Water Environment Research Foundation (WERF). Alexandria, Virginia, USA. 36 p.

ATSDR (Agency for Toxic Substances and Disease Registry). 2007. Minimal Risk Levels (MRLs) for Hazardous Substances. [Online] Available : <www.atsdr.cdc.gov/mrls.html. [cited June 1, 2007]

Ayscough N.J., J. Fawell, G. Franklin, and W. Young. 2000. *Review of Human Pharmaceuticals in the Environment*. Research and Development Technical Report P390. Environment Agency R&D Dissemination Centre. Wilts, UK.

Bud R. 2005. *The History of Antibiotics*. The Medicine Chest. The Wellcome Trust. London, England.

Bursch, W., M. Fuerhacker, M. Gemeiner, B. Grillitsch, A. Jungbauer, N. Kreuzinger, E. Moestl, S. Scharf, E. Schmid, S. Skutan, and I. Walter. 2004. Endocrine disrupters in the aquatic environment: the Austrian approach - ARCEM. *Water Science & Technology*, 50(5):293-300.

Ca DHS (California Department of Health Services). 2006. *Drinking Water Notification Levels and Response Levels: An Overview*. California Department of Health Services (Ca DHS), Drinking Water Program. June 28. [Online] Available: www.dhs.ca.gov/ps/ddwem/chemicals/AL/notificationoverview.pdf

Ca DHS (California Department of Health Services). 2006. Perchlorate in California Drinking Water: Overview and Links. December 1. California Department of Health Services (Ca DHS). http://www.dhs.ca.gov/ps/ddwem/chemicals/perchl/perchlindex.htm.

Ca DHS (California Department of Health Services). 2003. DHS' Groundwater Recharge Reuse Regulations. July 2003 Draft. Title 22, California Code of Regulations. Division 4. Environmental Health. Chapter 3. Recycling Criteria. Last update 21 July 2003. California Department of Health Services (Ca DHS). Sacramento, California. http://www.dhs.ca.gov/ps/ddwem/publications/waterrecycling/recycledraftreg-07-21-03.htm.

Christensen, F.M. 1998. Pharmaceuticals in the environment- a human risk? *Regulatory Toxicology and Pharmacology*, 28(3):212-221.

Daughton, C.G., and T.A. Ternes. 1999. Pharmaceuticals and personal care products in the environment: agents of subtle change? *Environmental Health Perspectives*, 107(Supplement 6):907-938.

Dolan D.G., B.D. Naumann, E.V. Sargent, A. Maier, and M. Dourson. 2005. Application of the threshold of toxicological concern concept to pharmaceutical manufacturing operations. *Regul Toxicol Pharmacol.*, 43(1):1-9. Erratum in: *Regul. Toxicol. Pharmacol.* 2006 Mar;44(2):189.

Eaton D.L. and C.D. Klaassen. 2001. Risk Assessment. In *Casarett and Doull's Toxicology: The Basic Science of Poisons*. Edited by C.D. Klaassen. New York: The McGraw-Hill Companies, Inc.

EDSTAC (Endocrine Disruptor Screening and Testing Advisory Committee). 1998. Endocrine Disruptor Screening and Testing Advisory Committee (EDSTAC) Final Report. U.S. Environmental Protection Agency. August. http://www.epa.gov/scipoly/oscpendo/edspoverview/finalrpt.htm

EMEA (European Medicines Agency). 2000. Clenbuterol Summary Report. Committee for Veterinary Medical Products. EMEA/MRL/723/99-FINAL. February. Available online at: http://www.emea.eu.int/pdfs/vet/mrls/072399en.pdf.

Faustman E.M. and G.S. Omenn. 2001. Risk Assessment. In *Casarett and Doull's Toxicology: The Basic Science of Poisons*. Edited by C.D. Klaassen. New York: The McGraw-Hill Companies, Inc.

Fawell, J.K., D. Sheahan, H.A. James, M. Hurst, and S. Scott. 2001. Oestrogens and oestrogenic activity in raw and treated water in Severn Trent water. *Water Research*, 35(5):1240-1244.

FDA (United States Food and Drug Administration). 2007. Food and Drug Information Medwatch: The FDA Safety Information and Adverse Event Reporting Program. U.S. Food and Drug Administration. Revised July 3, 2007. Available: https://www.accessdata.fda.gov/scripts/medwatch/medwatch_online.cfm [cited August 16, 2007]

FDA (United States Food and Drug Administration). 2000. Regulations in 21 CFR part 25 United States Food and Drug Administration. Available: http://www.fda.gov/cvm/Guidance/guide89.PDF

FDA (United States Food and Drug Administration). 1998. United States Food and Drug Administration. Available: http://www.fda.gov/Cder/guidance/1730fnl.pdf.

Fitzgerald J.E., J.L. Sanyer, J.L. Schardein, R.S. Lake, E.J. McGuire, and F.A. de la Iglesia. 1981. Carcinogen Bioassay and Mutagenicity Studies with the Hypolipidemic Agent Gemfibrozil. *J Natl Cancer Inst*, 67(5):1105-16.

Gilbert, S.G. 2004. *A Small Dose of Toxicology: The Health Effects of Common Chemicals*. New York: CRC Press.

Greer, M.A., G. Goodman, R.C. Pleus, and S.E. Greer. 2002. Health effects assessment for environmental perchlorate contamination: the dose response for inhibition of thyroidal radioiodine uptake in humans. *Environmental Health Perspectives*, 110(9):927-937.

Hamelin K., G. Bruant, A. El-Shaarawi, S. Hill, T.A. Edge, J. Fairbrother, J. Harel, C. Maynard, L. Masson, and R. Brousseau. 2007. Occurrence of virulence and antimicrobial resistance genes in Escherichia coli from different aquatic ecosystems within the St. Clair River and Detroit River areas. *Appl Environ Microbiol*, 73(2):477-84.

Harries, J.E., D.A. Sheahan, S. Jobling, P. Matthiessen, P. Neall, E.J. Routledge, R. Rycroft, J.P. Sumpter, and T. Tylor. 1996. A survey of estrogenic activity in United Kingdom inland waters. *Environmental Toxicology and Chemistry*, 15(11):1993-2002.

Harries, J.E., S. Jobling, P. Matthiessen, D.A. Sheahan, and J.P. Sumpter. 1995. Effects of Trace Organics on Fish - Phase 2. Report to the Department of the Environment. Marlow, UK: Foundation for Water Research.

Harwood V.J., M. Brownell, W. Perusek, and J.E. Whitlock. 2001. Vancomycin-resistant Enterococcus spp. isolated from wastewater and chicken feces in the United States. *Appl Environ Microbiol*, 67(10):4930-3.

Henley, D.V., N. Lipson, K.S. Korach, and C.A. Bloch. 2007. Prepubertal gynecomastia linked to lavender and tea tree oils. *New England Journal of Medicine*, 356(5):479-485.

JEFCA (Joint FAO/WHO Expert Committee on Food Additives). 1994. Toxicological evaluation of certain veterinary drug residues in food. Sulfadimidine. WHO Food Additives Series, No. 33. Food and Agriculture Organization of the United Nations and the World Health Organization (FAO/WHO), Geneva, Switzerland.

JEFCA (Joint FAO/WHO Expert Committee on Food Additives). 1998. Toxicological evaluation of certain veterinary drug residues in food. Tetracyclines: oxytetracycline, chlortetracycline, and tetracycline (addendum). WHO Food Additive Series 41. Food and Agriculture Organization of the United Nations and the World Health Organization (FAO/WHO), Geneva, Switzerland.

Johnson, B.L., H.E. Hicks, W. Cibulas, O. Faroon, A.E. Ashizawa, C.T. De Rosa, V.J. Cogliano, and M. Clark.1996. Public Health Implications of Exposure to Polychlorinated Biphenyls (PCBs). December. Agency for Toxic Substances and Disease Registry (ATSDR), Atlanta, GA, and U.S. Environmental Protection Agency (U.S. EPA), Washington, DC. Available: http://www.atsdr.cdc.gov/DT/pcb007.html

Jones O.A., N. Voulvoulis, and J.N. Lester. 2004. Potential Ecological and Human Health Risks Associated with the Presence Of Pharmaceutically Active Compounds in the Aquatic Environment. *Critical Reviews in Toxicology*, 34 (4):335-50.

Massachusetts DEP. 2006. Perchlorate Fact Sheet. July. Available: http://mass.gov/dep/toxics/pchlorqa.pdf.

McLachlan JA. 2006. Commentary: Prenatal exposure to diethylstilbestrol (DES): a continuing story. *International Journal of Epidemiology* 35 (4): 868-870. http://ije.oxfordjournals.org

Mons, M. N., J. Genderen, and A. M. Dijk-Looijaard. 2000. Inventory of the presence of pharmaceuticals in Dutch water. *KIWA Ondorzoek en Advies*, pp. 39.

NRC. 2005. Report in Brief. Health Implications of Perchlorate Ingestion. Washington, DC: National Research Council (NRC) of the National Academy of Sciences (NAS).

Puijker, L.; Mons, M. 2004. *Pharmaceuticals and Personal Care Products in the Water Cycle.* Report No. KWR 04.013. March 2004. Global Water Research Coalition (GWRC) and Kiwa Water Research. Nieuwegein, The Netherlands. 37 p.

RAIS (Risk Assessment Information System). 2007. Risk Assessment Information System (RAIS). Website sponsored by the U.S. Department of Energy, Office of Environmental Management, Oak Ridge Operations Office through a contract with Bechtel Jacobs Company LLC. Last updated March 24, 2006. Accessed September 20, 2007. http://rais.ornl.gov/homepage/glossary.shtml.

Richardson, M. L. and J. M. Bowron. 1985. The fate of pharmaceutical chemicals in the environment. *J Pharm Pharmacol.*, 37(1):1-12.

Sayah R.S., J.B. Kaneene, Y. Johnson, and R. Miller. 2005. Patterns of antimicrobial resistance observed in Escherichia coli isolates obtained from domestic- and wild-animal fecal samples, human septage, and surface water. *Appl Environ Microbiol*, 71(3):1394-404.

Schulman, L.J., E.V. Sargen, B.D. Naumann, E.C. Faria, D.G. Dolan, and J.P. Wargo. 2002. A human health risk assessment of pharmaceuticals in the aquatic environment. *Human and Ecological Risk Assessment*, 8(4):657-680.

Schwab, B.W., E.P. Hayes, J.M. Fiori, F.J. Mastrocco, N.M. Roden, D. Cragin, R.D. Meyerhoff, V.J. D'aco, and P.D. Anderson. 2005. Human pharmaceuticals in US surface waters: A human health risk assessment. *Regulatory Toxicology and Pharmacology*, 42(3):296-312.

Snyder, E.M., R.C. Pleus, and S.A. Snyder. 2005. Pharmaceuticals and EDCs in the US water industry - an update. *Journal of the American Water Works Association*, 97(11):32-36.

Snyder, S., B. Vanderford, R. Pearson, O. Quiñones, and Y. Yoon. 2003. Analytical methods used to measure endocrine disrupting compounds in water. *Practice Periodical of Hazardous, Toxic, and Radioactive Waste Management*, 7(4):224-234.

Stahlschmidt-Allner, P., B. Allner, J.R., Römbke, J., and T. Knacker. 1997. Endocrine disrupters in the aquatic environment. *Environmental Science & Pollution Research*, 4(3):155-162.

U.S. EPA (United States Environmental Protection Agency). 1993. *Reference Dose (RfD): Description and Use in Health Risk Assessments*. Integrated Risk Information System. March 15. [Online] Available: www.epa.gov/iris/rfd.htm

U.S. EPA (United States Environmental Protection Agency). 2000a. *Methodology for Deriving Ambient Water Quality Criteria for the Protection of Human Health*. Office of Water. EPA-822-B-00-004. [Online] Available: <www.epa.gov/waterscience/criteria/humanhealth/method/complete.pdf#search=%22Methodology%20for%20Deriving%20Ambient%20Water%20Quality%20Criteria%20for%20the%20Protection%20of%22 >

U.S. EPA (United States Environmental Protection Agency). 2000b. *Report on FQPA Tolerance Reassessment Progress and Interim Risk Management Decision: Mevinphos*. Office of Prevention, Pesticides, and Toxic Substances. EPA 738-R-00-014. September.

U.S. EPA (United States Environmental Protection Agency). 2006. Setting Standards for Safe Drinking Water. Office of Water. [Online] Available: www.epa.gov/ogwdw/standard/setting.html

U.S. EPA (United States Environmental Protection Agency). 2004. Guidelines for water reuse. September. Washington, DC: U.S. Environmental Protection Agency. EPA/625/R-04/108. http://www.epa.gov/nrmrl/pubs/625r04108/625r04108.htm.

U.S. EPA (United States Environmental Protection Agency). 2006. Drinking Water Contaminant Candidate List and Regulatory Determinations. United States Environmental Protection Agency (U.S. EPA). Last updated November 3, 2006. http://www.epa.gov/safewater/ccl/ccl2.html.

U.S. EPA (United States Environmental Protection Agency). 2007. Drinking Water Contaminants. U.S. Environmental Protection Agency (U.S. EPA). Last updated September 10, 2007.

U.S. EPA (United States Environmental Protection Agency). 2007. Draft list of initial pesticide active ingredients and pesticide inerts to be considered for screening under the Federal Food, Drug, and Cosmetic Act. *Federal Register*, 72(116):33486-33503. http://www.epa.gov/safewater/contaminants/index.html

Velagaleti, R., P.K. Burns, M. Gill, and J. Prothro. 2002. Impact of current good manufacturing practices and emission regulations and guidances on the discharge of pharmaceutical chemicals into the environment from manufacturing, use, and disposal. *Environmental Health Perspectives*, 110(3):213-220.

Versteegh J.F.M., A.A.M. Stolker, W. Niesing, and J.J.A. Muller. 2003. Geneesmiddelen In Drinkwater En Drinkwaterbronnen {Pharmaceuticals In Drinking Water And Drinking Water Sources. In Dutch}. RIVM-report nr. 703719004/2003 RIVM. Bilthoven, Netherlands.

Webb S., T. Ternes, M. Gibert, and K. Olejniczak. 2003. Indirect Human Exposure to Pharmaceuticals Via Drinking Water. *Toxicol Lett*, 142 (3):157-67.

Webb S.F. 2001. A Data-Based Perspective On The Environmental Risk Assessment Of Human Pharmaceuticals I. Collation Of Available Ecotoxicity Data. In *Pharmaceuticals in the Environment*, edited by K Kummerer. New York: Springer; pp. 175-201.

WHO (World Health Organization). 2002. Global Assessment of the State-of-the-Science of Endocrine Disruptors. International Program on Chemical Safety. Edited by Damstra, T.; Barlow, S.; Bergman, A.; Kavlock, R.; Van Der Kraak, G. World Health Organization. WHO/PCS/EDC/02.2. http://www.who.int/ipcs/publications/new_issues/endocrine_disruptors/en/print.html.

WHO (World Health Organization). 1994. *Environmental Health Criteria 170. Assessing Human Health Risks Of Chemicals: Derivation Of Guidance Values For Health-Based Exposure Limits*. World Health Organization, Geneva. [Online] Available: www.inchem.org/documents/ehc/ehc/ehc170.htm#SectionNumber:2.2

# CHAPTER 7
# COMMUNICATION

The Oxford English Dictionary defines a medicine as: A substance or preparation used in the treatment of illness; a drug; esp. one taken by mouth (OED 1989). It is the general understanding that medicines are intended to be biologically active—indeed, they are known to have powerful effects in some people. Through the use of exquisitely sensitive analytical instruments and methods, it has become apparent that pharmaceuticals and personal care products (PPCPs) and endocrine disrupting compounds (EDCs) are increasingly being detected in drinking water sources. What effect might these medicines have on healthy people if they are exposed inadvertently to a multitude of these compounds in drinking water, albeit at very low levels, day after day?

The issue of occurrence of EDCs and PPCPs in drinking water and potential health effects of exposure has gained increasing media attention, piquing public interest with reports on such issues as feminization of male animals and suggestions of increased risks of adverse effects on different organ systems or developmental effects in unborn children. Clearly, if these risks are significant, then reducing exposures to these compounds in water is of immediate importance. However, a rational evaluation of what these risks are requires examination of the likelihood of the adverse effects.

It is customary to look to our regulatory agencies to establish acceptable levels of exposure to chemicals found in the environment. However, regulatory agencies in the United States have not yet established risk-based guidelines for acceptable levels of most EDCs and PPCPs (other than perchlorate) in water. Consequently, drinking water utilities are left to respond to public concerns on their own. The purpose of this chapter is to provide an understanding of the current regulatory status of this issue with regard to protecting human health, and discuss what is known about the health risks at detected levels and how these risks compare to other types of common risks. This chapter also provides some strategies utility managers might consider in determining how to respond to public concerns about increasing detections of EDCs and PPCPs. It is hoped that this information can assist utility managers and public relations staff in effectively addressing concerns as they arise.

## REGULATORY REQUIREMENTS: WHAT LEGAL STANDARDS FOR EDCS AND PPCPS IN WATER ARE IN PLACE?

Under The Safe Drinking Water Act (SDWA), passed in 1974 and amended in 1986 and 1996, the U.S. EPA was given the authority to set drinking water standards to control the level of contaminants in the nation's drinking water. For contaminants identified as regulatory priorities, U.S. EPA sets non-enforceable Maximum Contaminant Level Goals (MCLGs). MCLGs represent the maximum level of a contaminant in drinking water at which no known or anticipated adverse effect on the health of drinking water consumers is expected to occur, allowing for an adequate margin of safety (U.S. EPA 2006). Based on these values and considering the availability of treatment techniques and their costs, U.S. EPA sets enforceable Maximum Contaminant Levels (MCLs), which are the maximum levels of contaminants allowed in water delivered to any user of a public water system. However, legally enforceable monitoring or regulatory requirements for PPCPs in drinking water that take into account potential human health risks have not been established in the U.S. (and most other countries). PPCPs also are not listed among the chemicals covered by the Emergency Planning and Community Right-to-Know Act of 1986 (EPCRA) (U.S. EPA 2006a).

No contaminants in drinking water are currently regulated on the basis of potential endocrine disruptive effects, and reportable quantities for release under the Federal Water Pollution Control Act (Clean Water Act), as amended, or the Comprehensive Environmental Response, Compensation, and Liability Act (CERCLA), as amended have not been published. One exception is perchlorate, a compound that at sufficient exposure levels can affect the thyroid gland (U.S. EPA, 2005c). Perchlorate occurs in drinking water in the United States, and is listed on the U.S. EPA's second Drinking Water Contaminant Candidate List (CCL 2) (U.S. EPA 2006b). The U.S. EPA has developed a reference dose for perchlorate based on its effects on the thyroid, and the resulting drinking water equivalent level (DWEL) is 24.5 μg/L. However, the DWEL is not a drinking water standard but an interim value used to develop a standard. There is no federal maximum contaminant level (MCL) for perchlorate, and it appears unlikely that a federal standard will be set within the next several years. California Department of Health Services set a public health goal (PHG) of 6 μg/L for perchlorate in drinking water and has proposed an MCL of 6 μg/L (CA DHS 2006a). Massachusetts Department of Environmental Protection promulgated a drinking water standard of 2 ppb (2 μg/L) for perchlorate; this is the first drinking water standard for perchlorate in the United States (Massachusetts DEP 2006).

To the best of the authors' knowledge, no regulatory limits have been established in the United States for EDCs in municipal WWTP effluents. For pharmaceuticals, the U.S. Food and Drug Administration (FDA) requires manufacturers to present environmental risk assessments with New Drug Applications for compounds that have expected introduction concentrations (EICs) at the point of entry into the aquatic environment greater than 1 ppb, which in practice correlates to a production rate of 44,300 kg or more per year (FDA 1998 and Velagaleti et al. 2002). If the EIC exceeds this threshold, the applicant is requested to submit an environmental assessment that includes information on mechanisms that deplete concentrations in the environment (*e.g.*, hydrolysis, aerobic biodegradation, soil biodegradation, photolysis, metabolism), microbial inhibition (to assess whether the substance has the potential to inhibit microorganisms and subsequently disrupt waste treatment processes), acute ecotoxicity, and chronic ecotoxicity (if the compound has the potential to bioaccumulate or bioconcentrate). In practice, more data on acute/ chronic toxicity in mammalian species are submitted if they are available. Because many pharmaceutical active ingredients are not produced at volumes in excess of 44,300 kg per year, they are granted categoric exclusion by the FDA (Velagaleti et al. 2002).

Some have questioned whether the FDA's EIC approach is sufficiently protective. They point out that the EIC does not accurately capture the true distribution and volume of pharmaceuticals in the environment, since the calculation does not consider inputs of the same drug from different sources (*e.g.*, other manufacturers) or the diversity of potential sources to the environment (*e.g.*, flushing of unused drugs by consumers) and does not consider local variability in volumes released resulting from variations in population density, drug usage patterns, and water treatment methods (Daughton 2003). They also note that this approach categorically assumes that exposure to any pharmaceutical is safe at a concentration of 1 ppb or less, without individual evaluation of the risks of chronic exposure to this concentration on reproductive, developmental, endocrine, immune, and other potentially sensitive endpoints. Concerns also arise about the lack of attention to risks potentially introduced by pharmaceutical metabolites and the effects of concurrent exposure to multiple compounds with different or common modes of action.

Although the U.S. EPA has not set guidance levels or limits for PPCPs in water, the Office of Research and Development (ORD) has begun gathering results from studies on the various aspects of PPCPs including sources and origins, environmental occurrence and distribution, transport and fate, exposure of biological receptors, effects, remediation (engineered treatment processes), pollution prevention, and risk communication and perception (U.S. EPA, 2007). The ORD's goals are to organize scientific information to establish whether adverse health effects from environmental exposure to PPCPs are possible, to foster research, and to educate the public on the potential impact their individual actions can have on the environment and human health (Daughton 2001). As of 2004, most research had been devoted to sources and origins, environmental occurrence and distribution, and remediation. More recently, efforts are increasingly invested toward identification of effects, pollution prevention, and risk communication and perception.

In Europe, pharmaceuticals in the environment have received regulatory attention through the European Medicines Agency (EMEA), which coordinates the evaluation and supervision of medicinal products throughout the European Union. The EMEA has released final guidelines for Environmental Risk Assessments (ERAs) to accompany Marketing Authorization Applications. Revised draft ERA guidelines were recently reviewed by various stakeholders and the final guidelines are expected to come into effect in late 2006 (EMEA 2006). These guidelines require conduct of an ERA if the predicted environmental concentration (PEC) in surface water is greater than 0.01 μg/L. In practice, environmental risk is judged to be unlikely if the maximum daily dose is less than 2 mg (EMEA 2006). Health Canada is developing Environmental Assessment Regulations (EARs) to ensure that new substances, including pharmaceuticals, meet the requirements of the Canadian Environmental Protection Act (CEPA) of 1999. Japan also is working to develop guidelines for pharmaceuticals in water systems.

## RISK ASSESSMENT OF EDCS AND PPCPS IN DRINKING WATER

As discussed in Chapters 2 and 4, EDCs and PPCPs are being increasingly detected in public water systems due to markedly enhanced analytical capabilities and increased interest in their presence. In the absence of regulatory guidelines, utilities are left with the task of establishing whether these detected concentrations of EDCs and PPCPs pose a risk to human or ecological health. In establishing guidelines for acceptable levels of contaminants in drinking and surface water, regulatory agencies use a process known as risk assessment. The following sections describe this process and how it can be used to set guideline levels for EDCs and PPCPs in drinking and surface water.

### Risk Assessment Basics

In the context of environmental contaminants, risk describes the probability of adverse effects resulting from exposure to an environmental agent or mixture of agents (U.S. EPA 2005). The process of risk assessment involves estimating the type and magnitude of adverse effects that could be caused by exposure to the substance. Due to limitations in available knowledge about how chemicals are distributed and behave and the inherently low incidence of adverse effects caused by low level exposures, thereby making direct measurement of disease difficult, the process is necessarily characterized by uncertainty. Thus, risk assessors cannot precisely characterize risks but must describe them in terms of probabilities that certain effects could occur within a large population. For example, the annual risk of death due to drowning in the U.S. is between 1 and 2 per 100,000 residents, and the annual risk of death due to long-term smoking is about 12 per 10,000 residents (Gray and Ropeik 2002).

Environmental risk assessment consists of several important steps. First, risk assessors identify substances in the environment that might cause a health risk. Second, they characterize exposure levels—the amount of a substance with which a person or an organism could come into contact and the type (dermal, oral, inhalation, etc.) and duration of contact. Usually, this question is answered by sampling environmental media and conducting analyses for chemical compounds or by using models to estimate concentrations at points of exposure. Third, risk assessors evaluate the toxicity of substances of interest—the levels of exposure required to cause an adverse effect. Finally, risk assessors characterize risks posed by the substance in its present or predicted use pattern by combining information on exposure and toxicity to predict what might happen.

The fact that something is detected in the environment does not mean that it is "bad." In general, as exposure (also known as dose) increases, the degree of response also increases. This relationship between the amount of chemical present and the effect that it causes is called "dose-response." The smaller the dose needed to cause an effect, the more potent (toxic) the substance is. For most chemicals, the rate of change in response is different at different doses. For example, the response may change slowly at low doses but rapidly at high doses. For compounds other than cancer-causing agents, there is assumed to be a threshold dose below which no effects occur, similar to a drug for which a dose that is too small has no beneficial effect. For most carcinogens, it is assumed that any exposure, no matter how small, can increase the potential for developing cancer over a lifetime, though that potential might be infinitesimal.

As discussed in Chapter 6, Human Health, sensitive individuals such as the very young (including the fetus) and the elderly are generally most susceptible to adverse effects from exposure to chemicals. The young are more susceptible because they are rapidly developing and younger cells are more easily harmed than mature cells and also because the smaller size of young animals and humans means that a given exposure could be greater on the basis of body weight. The elderly have increased sensitivity due to reduced ability to metabolize agents and to compensate for effects. Gender, genetic background, and health status also can affect the manner in which a person responds to an exposure.

In the environment, people are exposed to more than one chemical at a time. However, risks from exposure to multiple chemicals are difficult to predict, and most studies of toxicity are based on exposure to one chemical or a very small number of chemicals. When data on the toxicity of the mixture are lacking, most regulatory agencies recommend basing assessments on the toxicity of individual mixture components and assuming that the effects are additive (ACGIH 2006 and U.S. EPA 2000). This is the effect most commonly observed after simultaneous exposure to two chemicals (Eaton and Klaassen 2001). However, in any given mixture, a combination of interactive effects is likely. While assuming that the toxicity of each chemical is additive disregards possible synergistic effects, it also disregards antagonistic effects that might decrease toxicity. The appropriateness of assuming that effects are additive is supported by experimental evidence suggesting this assumption does not underestimate, and in most cases probably overestimates, risk when chemicals are administered at doses at or near their individual no observed adverse effect levels (NOAELs) (Smyth et al. 1970; Ikeda 1988 and US EPA 2000a).

## Characterization of Exposure

Chapter 4 described available data on occurrence of EDCs and PPCPs in treated wastewater effluents, drinking water sources (*i.e.*, groundwater or surface water), and finished drinking water. Direct exposure of humans to treated wastewater effluents is expected to be minimal; however, humans could be significantly exposed to EDCs and PPCPs in groundwater, surface water, and finished drinking water, depending on local concentrations. Table 7.1 lists EDCs and PPCPs that have been detected in these media.

**Table 7.1**
**EDCs and PPCPs detected in drinking water sources and raw and finished drinking water**

| Drinking Water Sources (groundwater or surface water) | | Raw Drinking Water | Finished Drinking Water |
|---|---|---|---|
| Pharmaceuticals<br>Analgesics:<br>acetaminophen<br>diclofenac<br>ibuprofen<br>ketoprofen<br>naproxen<br>Antibiotics:<br>chlorotetracycline<br>ciprofloxacin<br>clarithromycin<br>erythromycin-H2O<br>lincomycin<br>roxithromycin<br>sulfadiazine<br>sulfadimethoxine<br>sulfamethazine<br>sulfamethoxazole<br>trimethoprim<br>Antiepileptic drugs:<br>carbamazepine<br>primidone<br>Anti-hypertensive:<br>diltiazem<br>Anti-microbials:<br>triclocarban<br>triclosan<br>β-blockers:<br>atenolol<br>metoprolol<br>propranolol<br>Contrast media:<br>diatrizoate<br>iopromide<br>Lipid regulators:<br>bezafibrate<br>clofibric acid<br>gemfibrozil<br>Muscle Relaxant:<br>diazepam | Personal Care Products<br>Antioxidant:<br>butylated hydroxy-anisole (BHA)<br>Fragrances:<br>galaxolide (HHCB)<br>tonalide (AHTN)<br>Surfactants:<br>4-nonyl-phenol<br>Steroids<br>Androgen:<br>testosterone<br>Estrogens<br>17α-estradiol<br>17 α-ethynyl-estradiol (EE2)<br>17β-estradiol (E2)<br>estrone (E1)<br>Other<br>Caffeine and Metabolite:<br>Caffeine<br>1,7-dimethyl-xanthine<br>Flame Retardants:<br>tris-(2-chloro-ethyl)-phosphate (TCEP)<br>tris-(2-chloro-isopropyl)-phosphate (TCPP)<br>Insecticide:<br>n,n-Diethyl-meta-toluamide (DEET)<br>Nicotine Metabolite:<br>cotinine<br>Plasticizers:<br>bisphenol A (BPA)<br>di-n-butyl phthalate (DBP) | Pharmaceuticals<br>Analgesics:<br>diclofenac<br>ketoprofen<br>naproxen<br>Antiepileptic drugs:<br>carbamazepine<br>primidone<br>Contrast media:<br>diatrizoate<br>iopromide<br>Lipid regulators:<br>bezafibrate<br>clofibric acid<br>Muscle Relaxant:<br>diazepam<br>Personal Care Products<br>Fragrances:<br>galaxolide (HHCB)<br>tonalide (AHTN)<br>Surfactants:<br>4-nonyl-phenol<br>Steroids<br>Estrogens<br>17α-estradiol<br>17β-estradiol (E2)<br>estrone (E1)<br>Other<br>Flame Retardants:<br>tris-(2-chloro-ethyl)-phosphate (TCEP)<br>tris-(2-chloro-isopropyl)-phosphate (TCPP)<br>Plasticizers:<br>bisphenol A (BPA)<br>Stimulant:<br>caffeine | Pharmaceuticals<br>Analgesics:<br>diclofenac<br>Antiepileptic drugs:<br>carbamazepine<br>primidone<br>Contrast media:<br>diatrizoate<br>Lipid regulators:<br>bezafibrate<br>clofibric acid<br>Personal Care Products<br>Antioxidant:<br>butylated hydroxy-anisole (BHA)<br>Fragrances:<br>galaxolide (HHCB)<br>tonalide (AHTN)<br>Surfactants:<br>4-nonyl-phenol<br>Steroids<br>Estrogens<br>17α-estradiol<br>17 α-ethynyl-estradiol (EE2)<br>17β-estradiol (E2)<br>estrone (E1)<br>Other<br>Flame Retardants:<br>tris-(2-chloro-ethyl)-phosphate (TCEP)<br>tris-(2-chloro-isopropyl)-phosphate (TCPP)<br>Plasticizers:<br>bisphenol A (BPA)<br>di-n-butyl phthalate (DBP)<br>Stimulant:<br>caffeine |

While humans are expected to be exposed to waterborne EDCs and PPCPs primarily through ingesting water from the tap, they could also inhale or have skin contact with chemicals in tap water during showering and bathing, or inhale, ingest, or have skin contact with chemicals in surface water during swimming, boating, or other recreational activities. However, the contribution of these other pathways is likely to be very small. EDCs and PPCPs are generally not very volatile, limiting the potential for volatilization into household air during showering or bathing (or use of washing machines, toilet flushing, washing dishes, etc.), and the skin serves as a barrier to penetration of chemicals into the body.

As discussed in Chapter 4, although data are becoming increasingly available on pharmaceuticals in water systems, testing of water supplies for EDCs and PPCPs is not yet systematic, and available measurements are distributed widely. In general, data that are collected provide qualitative information on whether PPCPs or EDCs are present, not quantitative information on levels to which people could be exposed over time. Most sampling programs to date have not been designed to specifically assess concentrations at locations where people are known to come into contact with the water. Nonetheless, by making conservative assumptions about exposure (for example, assuming that someone could be exposed to levels measured in untreated water), possible maximum risks from exposure to these compounds can be evaluated.

**Health Effects of Concern**

(Daughton and Ternes 1999) discussed possible health effects of long-term exposure to pharmaceuticals. Based on information about the activity and potential toxicity of pharmaceuticals, they suggested that at sufficient levels of exposure, health risks could include endocrine disrupting activity, induction of antibiotic resistance, genotoxicity, carcinogenicity, allergic reactions, and effects on reproduction or fetal/ child development.

As discussed in Chapter 6, the biological activity that makes a drug beneficial for one population group can be hazardous to other population groups or at certain times—for example, during pregnancy or while taking other medications. Prior to marketing a pharmaceutical in the United States, the manufacturer is required to collect data to establish its safety. These studies, conducted to assess the drug's behavior, toxicologic effects in animals, and safety and efficacy in humans, generally use therapeutic or higher doses that are delivered for a relatively short period of time. Few data on the hazards associated with low-level exposure to PPCPs or EDCs are available, and these data largely limited to aquatic studies with fish or other aquatic species, often involving mixtures.

Historically, the susceptibility of humans to endocrine disrupting effects from exposure to EDCs became evident when it was discovered in 1971 that female children of women who took diethylstilbestrol (DES), a synthetic estrogenic medication, during pregnancy to prevent miscarriage had an increased incidence of a rare vaginal cancer (McLachlan 2006). However, DES is not an environmental contaminant, and in these incidences DES was administered at therapeutic doses, which are much greater than typical environmental exposures to trace concentrations of EDCs. According to (Ayscough et al. 2000), 17α-ethynylestradiol (EE2), a synthetic contraceptive estrogen, is the only pharmaceutical for which subtle endocrine disrupting effects have been demonstrated in laboratory studies at environmentally relevant concentrations, based on effects in fathead minnow on growth and egg production and not on potential effects on humans. Analysis of human data have so far failed to provide firm evidence of direct causal associations between low-level exposure to EDCs and adverse health outcomes (WHO 2002) (U.S. EPA 2004) (Anderson 2005).

Although no specific regulatory guidance exists prescribing how to assess human health risks of exposure to PPCPs or EDCs in drinking water, standard methods exist for determining exposure levels to environmental contaminants that are not likely to be associated with adverse health effects, as described in Chapter 6. Acceptable daily intakes (ADIs) are typically derived from animal or human toxicity data and uncertainty factors (UFs) and are commonly defined as the amount of a chemical to which a person can be exposed on a daily basis over an extended period of time (usually a lifetime) without suffering a deleterious effect (U.S. EPA 1993). ADIs are often presented in terms of dose per day (*e.g.*, mg/d) or dose per unit of body weight per day (*e.g.*, mg/kg-d). ADIs can be converted to drinking water equivalent levels (DWELs) in units of micrograms per liter (μg/L; equivalent to parts per billion, or ppb) or nanograms per liter (ng/L; equivalent to a parts per trillion, or ppt) by assuming a person (who weighs 70 kg, or 154 lbs) consumes two liters of water per day (U.S. EPA 2006c).

For drugs that show evidence of carcinogenicity in animal studies, extrapolation models can be used to predict the carcinogenic response at lower doses. Resulting "slope factors" (SFs) predict the lifetime excess risk of cancer to a person per 1 mg/kg-day of exposure to the chemical of interest. DWELs for carcinogens can be estimated assuming that a *de minimis* lifetime excess cancer risk of one additional cancer per one million lifetime exposures ($10^{-6}$) is acceptable, and that a 70-kg person consumes two liters of water daily for 30 years over a 70-year lifetime.

Some potential effects of pharmaceuticals are difficult to predict. For example, some pharmaceuticals (*e.g.*, antibiotics) are known to cause allergic reactions in humans, but predicting the likelihood of allergic effects is difficult because people sensitized to a chemical can respond to it at extremely low levels of exposure. Additionally, microbial resistance to antibiotics has been noted in surface water and sewage effluent (Ayscough, J.Fawell et al. 2000).

**Total Population and Individual Risks**

While a wide range of EDCs and PPCPs have been detected in effluent, surface, and drinking water, the number of samples overall is insufficient to adequately characterize exposure, and therefore risk, to aquatic organisms or humans. However, preliminary, screening-level risk assessments using conservative assumptions about rates and levels of exposure can be undertaken for these chemicals until more information on the duration and timing of exposure can be gathered. These assessments provide perspective on which contaminants are likely to pose the greatest risks, and what the maximal levels of these risks are likely to be.

Tables 7.2 and 7.3 lists pharmaceuticals that have been detected in source water and drinking water, respectively (see Chapter 4), their maximum detected concentrations, and their lowest therapeutic doses (*i.e.*, the lowest recommended dosage level indicated on the package labeling, assumed to be the lowest exposure level at which the chemical produces the desired pharmacologic effect). The table also shows water concentrations for each pharmaceutical that would correlate to the lowest therapeutic dose, if one assumes that a person drinks two liters of water at this concentration every day. To provide an additional margin of safety, these therapeutic dose concentrations were divided by a factor of 1,000. As shown, in each case, the highest detected concentration of a pharmaceutical in source water or drinking water is well below the concentration based on the lowest therapeutic dose divided by 1,000. Overall, the maximum detected concentrations in source water are a factor of 5 to 12,000 lower than the levels calculated as "safe" using this method and are far below unadjusted therapeutic dose levels. Assuming that people are exposed to the highest detected concentration is particularly conservative because people do not generally consume untreated surface water and all of the maximum detected concentrations presented in this table are surface water or groundwater concentrations. The maximum detected concentrations in drinking water are a factor of 53 to 91,000 lower than the levels calculated as "safe" using this method.

**Putting Risks in Perspective**

People incur a certain amount of risk every day. Whether they drive a car, cross a street, take medication, drink coffee, or breathe the air, they put themselves at some risk. "Zero risk" does not exist. When considering risks, it is important to note that everyone is at some background level of risk of developing diseases or conditions, such as cancer, during their lifetime. The causes of these conditions typically are not known but can include genetics, lifestyle factors (*e.g.*, diet, activity levels, general attention to health and well-being), voluntary choices (*e.g.*, smoking, illicit drug use, alcohol consumption), or other "environmental" exposures (*e.g.*, microbes or viruses, chemicals in the workplace or hobby-related chemical exposure, background radiation). Tobacco smoke, for example, contains about 4,000 individual compounds that have been identified as known or probable carcinogens, including benzo(a)pyrene, arsenic, cadmium, polycyclic aromatic hydrocarbons, vinyl chloride, formaldehyde, and nitrosamines (Ropeik and Gray 2002). Smoking is estimated to account for roughly 30 percent of all deaths due to cancer in the United States (Ropeik and Gray 2002). Overall, the probability that a person will develop cancer over the course of a lifetime in the U.S. is 1 in 2 for men and 1 in 3 for women (ACS 2006).

A 1 in 3 risk can also be expressed as 0.33 or $3.3 \times 10^{-1}$. For chemicals that can cause cancer, a risk of 1 in 100,000 (0.00001 or $1 \times 10^{-5}$) to 1 in 1,000,000 (0.000001 or $1 \times 10^{-6}$) of developing cancer during a lifetime is typically considered to be an acceptable risk for exposure to a chemical substance. If the average U.S. woman was exposed to an additional 1 in 100,000 risk (0.00001 or 1 × 10-5) of developing cancer in her lifetime, her total lifetime cancer risk would rise from 0.33 to 0.33001.

Risk estimates are often incorrectly assumed to define a "bright line" between harmful and safe. In general, risk estimates derived using the approach described above—which is consistent with methodologies used by U.S. EPA and other regulatory agencies—use inputs that, when combined, are intended to represent the upper bound of a possible exposure. For example, assumptions about the concentration of a contaminant to which a person could be exposed are often biased toward the high end, either by using data from samples collected at locations where higher-than-average levels are expected (*e.g.*, near an effluent outfall) or by using the maximum or near-maximum measured concentrations. Risk assessors do this because it is essentially impossible to precisely characterize actual exposure concentrations—a limitless number of samples would have to be collected. By biasing the assumed concentration toward higher estimates, risk assessors can be fairly confident that the actual average exposure concentration over time does not exceed the concentration that is assumed.

Similarly, assumptions about the amount, frequency, and duration of exposure are typically biased towards higher estimates. For example, risk assessments for drinking water assume that a person drinks two liters of tap water (about 8 ½ cups) containing the highest measured concentrations of contaminants every day, year after year. However, most people leave home for part of the day to go to work, to school, or to engage in other activities, and they typically take vacations and move about to a number of different residences during their lifetime. So, in general, people can be expected to drink water or fluids other than the drinking water available at a single residence. Also, many people drink other kinds of beverages during the day, including bottled water and prepared fruit juices and soft drinks, and do not rely on tap water for their entire daily fluid intake.

Furthermore, when human health risk assessors evaluate the toxicity of a contaminant, they use assumptions that provide a significant margin of safety between the dose that is indicated, on the basis of available data, to be toxic in animals and the dose assumed to be acceptable for a range of human populations. As discussed in Chapter 6, the lowest toxic dose (or the highest nontoxic dose) for the most sensitive endpoint in the most sensitive species is typically selected, and then uncertainty factors are applied to account for inadequacies in the available data set. The resulting "acceptable" dose is thus typically at least 30 to 1,000-fold lower than the lowest dose shown to have an adverse effect in toxicity studies, depending on the severity of the toxic effect.

Human exposures to many EDCs and PPCPs detected in water are expected to be small compared to exposures to potentially hazardous chemicals through prescription and non-prescription medications, food and beverages, occupational exposures, and residential activities (*e.g.*, cleaning products, personal care products, hobby chemicals, pesticides). When reporting the results of a risk evaluation for a chemical exposure, it can be helpful to compare the risk estimates to those for other more familiar situations to help people visualize the estimates and put the numbers in perspective. Table 7.2 presents lifetime risks associated with other activities or with exposure to other environmental agents.

As an example of chemicals to which people can be exposed "naturally" every day, plants produce natural pesticides to defend themselves against fungi, insects, and other animals that eat plants. Although about 10,000 natural pesticides and their break-down products are assumed to occur in foods that make up the human diet, only 71 have been tested adequately for carcinogenicity in rodent studies (Gold et al. 2001). Half of these have been shown to be carcinogenic in these tests: these include caffeic acid, d-limonene, and coumarin in such foods and beverages as coffee, lettuce, apples, tomatoes, orange juice, and cinnamon (Gold et al. 2001). Cooking and preparation of food also produce chemicals that are carcinogenic in rodent studies. Furfural is a chemical formed naturally when sugars are heated, and is a widespread constituent of food flavor. A variety of mutagenic and carcinogenic heterocyclic amines, such as benzo(a)pyrene, are formed when meat, chicken or fish is cooked, particularly when charred (Gold et al. 2001). Epidemiological studies indicate that all types of alcoholic beverages are associated with increased cancer risk, suggesting that ethyl alcohol produced by fermentation is the carcinogenic component rather than any particular type of beverage (Gold et al. 2001).

One way to put possible cancer risks from exposure to pharmaceuticals in drinking water in perspective is to compare the relative carcinogenic risk of daily exposure to these agents with other commonly present agents. For example, rodent testing suggests that gemfibrozil and phenytoin, two pharmaceuticals that have been detected in surface water, are carcinogenic. Risks from daily exposure to these agents in water at their maximum detected concentrations (assuming consumption of 2 liters of this water per day) can be compared to agents present in different food products or in drugs taken voluntarily for therapeutic purposes, using information on the carcinogenic potency of nearly 1,500 chemicals based on data from long-term animal testing that has been compiled in a unique database by the Carcinogenic Potency Project at the Lawrence Berkeley Laboratory of the University of California (http://potency.berkeley.edu). "TD50" values (i.e., chronic dose-rates in mg/kg body weight/day that would induce tumors in half the test animals at the end of a standard lifespan) for these compounds can be compared, incorporating assumptions about average daily human exposure rates to each chemical. Table 7.3

shows these comparisons for gemfibrozil and phenytoin in drinking water versus other chemicals to which people are typically exposed, including "naturally" present compounds in foods. These comparisons show that carcinogenic risks of exposure to gemfibrozil and phenytoin in drinking water are on the order of more than 1,000,000 times lower than carcinogenic risks from daily ingestion of alcoholic beverages and consumption of other compounds naturally present in foods.

**Table 7.2**
**Comparison of lifetime risks of different events or situations**

| Event or situation | Approximate lifetime risk |
|---|---|
| **Estimated Site-Related Excess Cancer Risks** | |
| Contact with residual contaminants in soil on Site | Less than 1 in 1,000,000* |
| Inhalation of vapor from groundwater that volatilizes into the parking structure on Site | Less than 1 in 1,000,000† |
| **Risks of Accidental Death** | |
| Dying in a motor vehicle accident | 1 in 100‡ |
| Dying in a motor vehicle accident per 100 miles traveled | 1 in 1,000,000§ |
| Dying while a pedestrian | 1 in 1,000‡ |
| Drowning | 1 in 1,000** |
| Dying from a fall at home | 3 in 1,000‡ |
| Dying while riding a bicycle | 2 in 10,000†† |
| Dying from a lightening strike | 2 in 100,000** |
| **Risks of Developing Cancer from Environmental Agents** | |
| Skin cancer from exposure to the sun | 1 in 100‡‡ |
| Cigarette smoking (a pack or more per day) | 8 in 100§§ |
| Outside radiation (radon and cosmic rays) | 1 in 1,000§§ |
| Environmental tobacco smoke (death from cancer) | 1 in 1,000*** |
| Human-made chemicals in indoor air at home | 2 in 10,000§§ |
| Outdoor air in industrialized area | 1 in 10,000§§ |
| Human-made chemicals in drinking water (chlorination) | 1 in 100,000§§ |
| Consumption of 2 oz. of peanut butter per week (naturally occurring aflatoxin) | 8 in 100,000§§ |
| Consumption of one meal per year of small Lake Michigan trout | 1 in 100,000§§ |

*Estimates assume an individual contacts contaminated soil, which is covered with at least 1 to 3 feet of clean soil or pavement, every working day for a working lifetime. Actual risks to employees on the 6th floor of the WWCC building are significantly lower, and are likely to be virtually zero.

†Estimates assume an individual breathes contaminated vapors from groundwater in an enclosed parking structure (the actual onsite parking structure is open to the outside air on two sides) for 24 hours a day, every working day for a working lifetime.

‡Actual risks to employees on the 6th floor of the WWCC building are significantly lower, and are likely to be virtually zero.

Source: National Safety Council, 2000a

§Source: U.S. DOT, 2000

**Source: National Safety Council, 2000b

††Source: IIHS, 2001

‡‡Source: AAD, 2001

§§Source: U.S. EPA, 1991

***Source: U.S. EPA, 1993b

**Table 7.3**
**Relative cancer risks from exposure to pharmaceuticals in water to other carcinogens based on the human exposure/ rodent potency index (HERP)**

| Chemical | Description | Chronic human daily exposure (mg)* | Rodent potency index ($TD_{50}$) (mg/kg-d)† | Human exposure/ rodent potency index (%)‡ | Fraction of highest food-related HERP |
|---|---|---|---|---|---|
| **Pharmaceutical in water** | | | | | |
| Clofibric acid | Lipid regulator, based on maximum-detected water concentration (7.3 ug/L) | 0.015 | 169 | 0.00013 | 0.000036 |
| Gemfibrozil | Lipid regulator, based on maximum-detected water concentration (1.55 ug/L) | 0.0031 | 125 | 0.000035 | 0.0000097 |
| **Chemicals in food** | | | | | |
| Aflatoxin (fungal toxin) | In: Peanut butter | 0.000018 | 0.0032 | 0.008 | 0.0022 |
| Caffeic acid (natural pesticide) | In: Coffee (~0.5 oz.)<br>Lettuce (14.9 g)<br>Tomato (89 g)<br>Apple (32.0 g) | 20.8<br>7.90<br>5.46<br>3.40 | 297 | 0.1<br>0.04<br>0.03<br>0.02 | 0.028<br>0.011<br>0.0083<br>0.0055 |
| Coumarin (natural pesticide) | In: Cinnamon (21.9 mg) | 0.065 | 13.9 | 0.007 | 0.0019 |
| d-Limonene (natural pesticide) | In: Orange juice (~4.5 oz.) | 4.28 | 204 | 0.03 | 0.0083 |
| Ethyl alcohol (produced from fermentation) | In: Alcoholic drinks, all types<br>Beer (~8 oz.)<br>Wine (~1 oz.) | 22,800<br>11,700<br>3,670 | 9,110 | 3.6<br>1.8<br>0.6 | 1<br>0.5<br>0.17 |
| Furfural (formed when sugars are heated) | In: Bread (~3 slices) | 0.867 | 197 | 0.004 | 0.0011 |
| PhIP (formed during cooking) | In: Hamburger, pan fried (one 3 oz. serving) | 0.000176 | 1.64 | 0.0002 | 0.000056 |
| **Chemicals in home environment** | | | | | |
| Formaldehyde | In: Conventional home air (14 hours/day) | 0.598 | 2.19 | 0.4 | 0.11 |

Source: CPDB (2005), a unique and widely used international resource of results from 6,153 chronic, long-term animal cancer tests on 1,485 chemicals.
*Average U.S. exposure or consumption data for which human exposure could be chronic for a lifetime. For pharmaceuticals the doses are recommended doses.
†A $TD_{50}$ value is the tumorigenic dose-rate for rodents, defined as the chronic dose-rate in mg/kg body weight/day that would induce tumors in half the test animals at the end of a standard lifespan for a species.
‡Human Exposure/Rodent Potency index (HERP) -- indicates the percentage of the rodent $TD_{50}$ (in mg/kg/day) a person receives from average daily exposure to a chemical over a lifetime (mg/kg/day).

## ECOLOGICAL EFFECTS: POTENTIAL EDC/PPCP EXPOSURES AND EFFECTS ON FISH

Fish and other aquatic organisms are expected to receive much greater exposure than humans to EDCs and PPCPs in the aquatic environment. Fish are constantly immersed in water and can bioconcentrate waterborne contaminants by direct uptake from water across the gill. Fish also can be exposed to lipophilic contaminants through diet or exposure to contaminated sediment or suspended particulate material. Humans are likely to receive much less exposure to wastewater-derived contaminants by these routes. In oligotrophic to mesotrophic water bodies, fish often congregate near municipal wastewater discharges where increased nutrient input can result in greater availability of food and plant cover. This behavior also results in greater exposure to wastewater-related contaminants. While people are exposed intermittently to waterborne contaminants, fish may be exposed continuously and possibly for an entire lifetime and are more likely to be exposed during critical periods of development when organisms are most sensitive to the effects of contaminants.

While there is little or no evidence that widespread endocrine disruption is occurring among humans exposed to commonly encountered concentrations of EDCs, there is substantial evidence that EDCs and PPCPs at levels found in WWTP effluents can cause endocrine disruption in fish. Studies conducted in Europe (Harries, Sheahan et al. 1997; Jobling, Nolan et al. 1998; Routledge, Sheahan et al. 1998; Rodgers-Gray, Jobling et al. 2001), Canada (Schoenfuss, Levitt et al. 2002), the United States (Hemming, Waller et al. 2001), and elsewhere have reported estrogenic effects in fish due to exposure to municipal and industrial WWTP effluents. Because WWTP effluents contain complex mixtures of EDCs and PPCPs, most of these studies have been unable to pinpoint the specific chemicals that are responsible for the observed effects. However, a few studies have identified likely causative agents including the animal hormones 17ß-estradiol (E2) and estrone, the pharmaceutical EE2, and in some cases nonylphenol and octylphenol, which are breakdown products from certain surfactants and detergents.

There is an increasing body of literature describing the effects of waterborne exposure to EDCs and PPCPs on fish and other aquatic life. However, no specific guidelines, water quality criteria, or "safe" levels of EDCs or PPCPs in wastewater have been developed in the United States for the protection of aquatic life. While surface water criteria and regulations do exist for some chemicals that might be considered to be EDCs or potential EDCs, these guidelines or criteria are not based on potential endocrine disruptive effects or on pharmaceutical mode of action, but rather on more traditional toxicity testing endpoints used for aquatic life, such as mortality, effects on reproduction, decreased growth, etc. While these test results are useful, they often do not address the potential for more subtle, long-term effects of chronic exposure to environmentally relevant concentrations of EDC and PPCPs. Basing standards or other levels of concern on acute or overt effects, which might occur at higher concentrations than effects mediated through endocrine or pharmaceutical modes of action, could lead to over-confidence in the level of protection that they afford. For most EDCs and PPCPs, more information is needed to develop concentrations in water that are protective of aquatic life.

Nonylphenol is an example of an EDC for which water quality criteria for protection of aquatic life have been developed. Nonylphenol is an organic chemical used primarily as an intermediate to produce other chemicals such as nonylphenol ethoxylates (NPEs). NPEs are nonionic surfactants produced and used in large quantities in the United States and elsewhere.

Nonylphenol is moderately soluble and persistent in water and is toxic to aquatic organisms. 4-Nonylphenol is a common contaminant of surface waters. 4-Nonylphenol in the environment occurs primarily as a result of degradation of NPEs by sewage sludge in industrial and municipal WWTP and further transformation of degradates in effluents discharged to the environment. (U.S. EPA 2005)

The U.S. EPA has set ambient water quality criteria for nonylphenol for protection of aquatic life. According to (U.S. EPA 2005), aquatic life should not be affected unacceptably if the one-hour average concentration of nonylphenol does not exceed 28 μg/L more than once every three years on the average (criteria maximum concentration (CMC), or acute criterion) and the four-day average concentration of nonylphenol does not exceed 6.6 μg/L more than once every three years on the average (criteria continuous concentration (CCC), or chronic criterion). Several review articles have described the estrogenic (endocrine disruptive) effects of nonylphenol, primarily demonstrated by effects on aquatic species using endpoints that do not meet acceptability requirements for data used to derive national ambient water quality criteria (NAWQC). Hence, U.S. EPA did not consider this information during development of the U.S. EPA criteria, except to the extent that the effects on these endpoints were integrated into whole organism endpoints such as growth and development that are deemed suitable for deriving the U.S. EPA NAWQC (U.S. EPA 2005).

According to data summarized by (U.S. EPA 2005), nonylphenol seldom has been reported to induce estrogenic effects at concentrations less than the U.S. EPA chronic aquatic criterion of 6.6 μg/L. However, one study reported effects at 1 μg/L or less. (Giesy, Pierens et al. 2000) reported that exposure to waterborne nonylphenol for 42 days at concentrations of >0.3 to 0.4 μg/L appeared to reduce fecundity in fathead minnows, while fish exposed to 0.09 and 0.1 μg/L produced more eggs than control fish, possibly indicating a hormetic response of fecundity to nonylphenol. A hormetic response is one that results in a U-shaped dose-response curve, meaning that a response occurs at a low dose, does not occur or occurs to a lesser degree at a higher dose, then reappears or increases at an even greater dose. However, sample sizes used in this study were small, and it is not clear whether the observed effects would be likely to cause population-level impacts in fathead minnows.

England and Wales are likely to be the first places in the world to target EDCs for removal from their municipal WWTP effluents. The Environment Agency of England and Wales has proposed a demonstration project to assess how estrogenic EDCs can be prevented from entering these effluents. According to the science manager for the Environment Agency's ecosystems section, the agency has set threshold exposure limits used for risk assessment for steroid estrogens at 0.1 ng/L for ethynylestradiol (EE2), 1 ng/L for 17ß-estradiol (E2), and 3 ng/L for estrone (E1). A total threshold value based on E2 also has been set because the estrogenic effects of these EDCs are expected to be additive.

## ESTABLISHING TREATMENT GOALS AND MONITORING PROGRAMS FOR EDCS AND PPCPS IN DRINKING WATER

Because "safe" levels for most PPCPs and EDCs have not been published at this time, it is challenging to determine health-based treatment goals. In the absence of regulatory guidelines and toxicity and exposure data required to set health-based goals, each community must determine whether it will wait for additional information and guidance or take proactive measures to treat and remove PPCPs and EDCs from wastewater and drinking water despite the costs and uncertain benefits.

While it might be tempting to set a goal of reducing the concentrations of PPCPs and EDCs of potential concern in wastewater or particularly in drinking water to levels below the analytical detection limits, the costs of adhering to this approach would, for many compounds, far outweigh any known benefits. Developments in analytical chemistry techniques are rapidly out-pacing improvements in treatment technologies such that it is impossible to remove all waterborne contaminants to levels below detection limits. Furthermore, analytical detection limits currently have no relationship to health-based levels of concern. It is possible for toxic effects of a chemical to occur at concentrations less than the achievable analytical detection limits for that chemical. Conversely, as analytical equipment becomes more exquisitely sensitive, it is likely that detectable concentrations for many PPCPs or EDCs will fall far below any concentrations that produce an observable toxic effect. Forcing water treatment technologies to arbitrarily reduce contaminant concentrations below analytical detection limits could result in substantial expenditures of public funds with no appreciable benefit to public or environmental health. Striving to reach an increasingly unattainable goal of non-detectable levels of contaminants should be abandoned in favor of tailoring drinking water and wastewater treatment goals to achieve concentrations of emerging contaminants that consider safety and costs.

Caution should be excercised in the the development of monitoring plans to characterize occurrence and removal of EDCs and PPCPs from source water and drinking water. Reliable laboratories that are qualified to conduct analyses and that can demonstrate appropriate analytical detection limits for trace contaminants in water and reliable quality assurance/quality control procedures should be selected. Because field and laboratory contamination is extremely difficult to avoid during sampling for trace analysis of EDCs and PPCPs, personnel who collect the samples should be thoroughly trained in proper sampling techniques for trace analyses.

Any bioassays selected for detection of EDCs and PPCPs in source water and drinking water and characterization of their potential to cause health effects should be carefully researched to identify their strengths and weaknesses prior to implementation, and the meaning and importance of bioassay results should not be extrapolated beyond the intended purposes and capabilities of the bioassays. For example, while *in vitro* bioassays are useful tools for investigating the occurrence and removal of certain EDCs from drinking water, their responses have limited revlevance (or no relevance in some cases) to the potential for contaminants in the water to cause health effects in humans who consume the water. Finally, utilities and researchers who collect occurrence and bioassay data should prepare in advance, before the first samples are collected, to disclose their monitoring or research plans and results in a responsible manner such that the media, the public and particularly drinking water consumers will understand the importance and limitations of the work. Many utilities and researchers have been unpleasantly surprised by interest and suspicion from the public and the press during sampling and monitoring before data are available for release. Having a plan in place in advance to explain ongoing activities and present finished results can prevent an air of secrecy from surrounding these activities and allow time for anticipation of concerns and preparation of carefully considered responses as well as materials that can be distributed to the public and the media. Honest, responsible, and timely disclosure of information is necessary to avoid undue stress and to maintain trust between utilities and consumers, the media, and regulators.

## RISK COMMUNICATION GUIDELINES

When individuals learn that they may face a health risk due to exposure to a chemical in the environment, their response is often emotional, and subsequent communications regarding the risk frequently are fraught with controversy. In many cases, the response to the perceived risk may be disproportionate to the magnitude of the risk itself (Sandman 1993). For those charged with addressing and, if appropriate, mitigating the chemical's presence, effectively communicating with concerned citizens about the risk can be challenging, but may be improved by focusing on respecting the audience's concerns about the perceived risks and answering questions in astraight forward manner. As described by Dr. Peter Sandman, a leading risk communication expert, "The ultimate job of risk communication is to try to produce a citizenry that has the knowledge, the power, and the will to assess its own risks rationally, decide which ones it wants to tolerate and which ones it wants to reduce or eliminate, and act accordingly" (Sandman 1983).

Typically, people tend to perceive "unnatural" and "involuntary" risks, such as those associated with chemical contaminants, as greater than "natural" or "voluntary" risks, and thus are less willing to accept them (Sandman, 1993). When faced with such risks, a common response is outrage and distrust. According to Sandman, "risk management and risk communication require differentiating between what you can and can't control, and focusing on what you can." Some recommendations for addressing concern and outrage include (Sandman 1993 and Krause 2006):

- Listen and respond accordingly, and tolerate early over-reactions
- Acknowledge uncertainty or diversity of opinion when things are not clear cut, and avoid "over-reassuring"
- Acknowledge, and apologize for, past errors, deficiencies and misbehaviors
- Avoid lies or half-truths
- Aim for total candor and transparency

With regard to responding to concerns about the occurrence of PPCPs and EDCs in water systems, utility managers will benefit from staying on top of issues related to these chemicals. Strategies that might be considered are:

- Maintaining up-to-date knowledge of "emerging" chemicals of concern and understanding how these chemicals might be relevant for a specific utility
- Initiating a monitoring program and hiring experts in toxicology or public health to characterize the potential health risks of any detected chemicals
- Understanding treatment options and the trade-offs between cost of contaminant removal and mitigated risks
- Developing a robust plan to educate management staff as well as other key internal and external parties

In the absence of regulatory guidelines, utility managers must determine whether they will wait for additional information and guidance or take proactive measures to address concerns about PPCPs and EDCs in public water supplies. Establishing a program of internal and external education and community participation will be a good step toward understanding the true and perceived magnitude of the issue and toward developing an appropriate strategy for addressing new issues as they arise.

# APPENDIX A:
# RECENT ANALYTICAL METHODS LITERATURE REVIEW

**Table A.1**
**Recent methods for the determination of potential EDCs and PPCPs**

| Compound | Matrixes | Extraction Method | Detection Method | Reporting Limit (ng/L) | Reference |
|---|---|---|---|---|---|
| **Pharmaceuticals** | | | | | |
| ***Analgesics*** | | | | | |
| Acetaminophen | WW, SW, DW | SPE | LC-MS/MS | 1.0 | (Vanderford, Pearson et al. 2003) |
| | WW, SW | SPE | LC-MS/MS | 40 – 58 | (Gros, Petrovic et al. 2006) |
| | WW | SPE | LC-MS/MS | 132 | (Gomez, Petrovic et al. 2006) |
| | WW, SW | SPE | LC-TOF-MS | 1.5 | (Benotti and Brownawell 2007) |
| Diclofenac | WW, SW, DW | SPE | LC-MS/MS | 1.0 | (Vanderford, Pearson et al. 2003) |
| | WW, SW | SPE | LC-MS/MS | 5 – 30 | (Gros, Petrovic et al. 2006) |
| | WW, SW, DW | SPE | Isotope Dilution LC-MS/MS | 0.50 | (Vanderford and Snyder 2006) |
| Ibuprofen | WW, SW, DW | SPE | LC-MS/MS | 1.0 | (Vanderford, Pearson et al. 2003) |
| | GW, SW | SPE | LC-MS | 18 | (Cahill, Furlong et al. 2004) |
| | WW, SW | SPE | LC-MS/MS | 20 – 42 | (Gros, Petrovic et al. 2006) |
| Ketoprofen | WW, SW | SPE | GC-MS/MS | 10 | (Fono, Kolodziej et al. 2006) |
| | WW, SW | SPE | LC-MS/MS | 70 – 95 | (Gros, Petrovic et al. 2006) |
| | WW, SW, DW | SPE | LC-MS/MS | 75 | (Hernando, Heath et al. 2006) |
| Naproxen | WW, SW, DW | SPE | LC-MS/MS | 1.0 | (Vanderford, Pearson et al. 2003) |
| | WW, SW | SPE | LC-MS/MS | 20 – 32 | (Gros, Petrovic et al. 2006) |
| | WW, SW, DW | SPE | Isotope Dilution LC-MS/MS | 0.50 | (Vanderford and Snyder 2006) |
| ***Antibiotics*** | | | | | |
| Chlortetracycline | WW | SPE | LC-MS/MS | 4 | (Miao, Bishay et al. 2004) |
| | WW, SW | SPE | LC-MS/MS | 40 | (Yang, Cha et al. 2004) |
| | WW, GW, SW | SPE | Ion Trap LC-MS/MS | 59 | (Batt and Aga 2005) |
| | SW | SPE | LC-MS/MS | 20 | (Hao, Lissemore et al. 2006) |
| Ciprofloxacin | WW | SPE | LC-MS/MS | 1 | (Miao, Bishay et al. 2004) |
| | WW, GW, SW | SPE | Ion Trap LC-MS/MS | 30 | (Batt and Aga 2005) |
| | DW | SPE | LC-MS/MS | 1.0 | (Ye, Weinberg et al. 2007) |
| | WW, GW, SW | SPE | LC-MS/MS | 8.4 – 163 | (Vieno, Tuhkanen et al. 2006) |
| Clarithromycin | WW | SPE | LC-MS/MS | 2 – 4 | (Gobel, McArdell et al. 2004) |
| | WW | SPE | LC-MS/MS | 1 | (Miao, Bishay et al. 2004) |
| | SW | SPE | LC-MS and LC-MS/MS | 43 – 470 | (Abuin, Codony et al. 2006) |

(continued)

**Table A.1 (Continued)**

| Compound | Matrixes | Extraction Method | Detection Method | Reporting Limit (ng/L) | Reference |
|---|---|---|---|---|---|
| Erythromycin | WW, SW, DW | SPE | LC-MS/MS | 1.0 | (Vanderford, Pearson et al. 2003) |
| | WW, GW, SW | SPE | Ion Trap LC-MS/MS | 75 | (Batt and Aga 2005) |
| | WW, SW | SPE | LC-MS/MS | 14 – 20 | (Gros, Petrovic et al. 2006) |
| | GW, SW | SPE | LC-MS/MS | 0.4 – 5 | (Xu, Zhang et al. 2007) |
| | DW | SPE | LC-MS/MS | 0.50 | (Ye, Weinberg et al. 2007) |
| Lincomycin | WW | SPE | LC-MS/MS | 0.31 | (Castiglioni, Bagnati et al. 2005) |
| | SW | SPE | LC-MS/MS | 10 | (Hao, Lissemore et al. 2006) |
| Roxithromycin | WW, GW, SW | SPE | Ion Trap LC-MS/MS | 190 | (Batt and Aga 2005) |
| | SW | SPE | LC-MS/MS | 10 | (Hao, Lissemore et al. 2006) |
| | GW, SW | SPE | LC-MS/MS | 0.4 – 5 | (Xu, Zhang et al. 2007) |
| | DW | SPE | LC-MS/MS | 0.50 | (Ye, Weinberg et al. 2007) |
| Sulfadiazine | WW | SPE | LC-MS/MS | 3 | (Miao, Bishay et al. 2004) |
| | WW | SPE/SPME | LC-MS/MS | 3.4 – 9000 | (Balakrishnan, Terry et al. 2006) |
| | WW, GW, SW | SPE | LC-DAD/FLD | 30 – 150 | (Peng, Wang et al. 2006) |
| | GW, SW | SPE | LC-MS/MS | 0.2 – 1 | (Xu, Zhang et al. 2007) |
| Sulfadimethoxine | WW, GW, SW | SPE | Ion Trap LC-MS/MS | 50 | (Batt and Aga 2005) |
| | WW | SPE/SPME | LC-MS/MS | 6.7 – 12000 | (Balakrishnan, Terry et al. 2006) |
| | DW | SPE | LC-MS/MS | 1.0 | (Ye, Weinberg et al. 2007) |
| Sulfamethazine | WW, GW, SW | SPE | Ion Trap LC-MS/MS | 53 | (Batt and Aga 2005) |
| | WW | SPE/SPME | LC-MS/MS | 3.7 – 16000 | (Balakrishnan, Terry et al. 2006) |
| | DW | SPE | LC-MS/MS | 1.0 | (Ye, Weinberg et al. 2007) |
| Sulfamethoxazole | WW, SW, DW | SPE | LC-MS/MS | 1.0 | (Vanderford, Pearson et al. 2003) |
| | WW, GW, SW | SPE | Ion Trap LC-MS/MS | 83 | (Batt and Aga 2005) |
| | WW, SW | SPE | LC-MS/MS | 16 – 120 | (Gros, Petrovic et al. 2006) |
| | WW, SW, DW | SPE | Isotope Dilution LC-MS/MS | 0.25 | (Vanderford and Snyder 2006) |
| | DW | SPE | LC-MS/MS | 2.1 | (Ye, Weinberg et al. 2007) |
| Trimethoprim | WW, SW, DW | SPE | LC-MS/MS | 1.0 | (Vanderford, Pearson et al. 2003) |
| | WW, GW, SW | SPE | Ion Trap LC-MS/MS | 91 | (Batt and Aga 2005) |
| | WW, SW | SPE | LC-MS/MS | 4 – 82 | (Gros, Petrovic et al. 2006) |
| | WW, SW, DW | SPE | Isotope Dilution LC-MS/MS | 0.25 | (Vanderford and Snyder 2006) |
| | DW | SPE | LC-MS/MS | 0.50 | (Ye, Weinberg et al. 2007) |

(continued)

**Table A.1 (Continued)**

| Compound | Matrixes | Extraction Method | Detection Method | Reporting Limit (ng/L) | Reference |
|---|---|---|---|---|---|
| ***β-Blockers*** | | | | | |
| Atenolol | WW, SW | SPE | LC-MS/MS | 30 – 141 | (Gros, Petrovic et al. 2006) |
| | WW | SPE | LC-MS/MS | 12 | (Nikolai, McClure et al. 2006) |
| | WW, SW, DW | SPE | Isotope Dilution LC-MS/MS | 0.25 | (Vanderford and Snyder 2006) |
| Metoprolol | WW, SW | SPE | LC-MS/MS | 12 – 36 | (Gros, Petrovic et al. 2006) |
| | WW, SW | SPE | GC-MS/MS | 1.0 | (Fono and Sedlak 2005) |
| | WW | SPE | LC-MS/MS | 17 | (Nikolai, McClure et al. 2006) |
| Propanolol | WW, SW | SPE | GC-MS/MS | 0.1 – 1.0 | (Fono and Sedlak 2005) |
| | WW, SW | SPE | LC-MS/MS | 7 – 40 | (Gros, Petrovic et al. 2006) |
| | WW | SPE | LC-MS/MS | 4.4 | (Nikolai, McClure et al. 2006) |
| ***Lipid regulators*** | | | | | |
| Bezafibrate | WW | SPE | LC-MS/MS | 0.1 | (Castiglioni, Bagnati et al. 2005) |
| | WW, SW | SPE | LC-MS/MS | 4 – 31 | (Gros, Petrovic et al. 2006) |
| | SW | SPE | LC-MS/MS | 2 | (Hao, Lissemore et al. 2006) |
| | SW, DW | SPE | LC-MS/MS | 0.1 | (Loos, Wollgast et al. 2007) |
| Clofibric acid | WW | SPE | LC-MS/MS | 0.36 | (Castiglioni, Bagnati et al. 2005) |
| | WW, SW | SPE | LC-MS/MS | 3 – 6 | (Gros, Petrovic et al. 2006) |
| | SW | SPE | LC-MS/MS | 3 | (Hao, Lissemore et al. 2006) |
| Gemfibrozil | WW, SW, DW | SPE | LC-MS/MS | 1.0 | (Vanderford, Pearson et al. 2003) |
| | GW, SW | SPE | LC-MS | 15 | (Cahill, Furlong et al. 2004) |
| | WW, SW | SPE | LC-MS/MS | 3 – 9 | (Gros, Petrovic et al. 2006) |
| | SW | SPE | LC-MS/MS | 5 | (Hao, Lissemore et al. 2006) |
| | WW, SW, DW | SPE | Isotope Dilution LC-MS/MS | 0.25 | (Vanderford and Snyder 2006) |
| ***Anti-hypertensive*** | | | | | |
| Diltiazem | GW, SW | SPE | LC-MS | 12 | (Cahill, Furlong et al. 2004) |
| | WW, SW | SPE | LC-TOF-MS | 0.10 | (Benotti and Brownawell 2007) |

(continued)

Table A.1 (Continued)

| Compound | Matrixes | Extraction Method | Detection Method | Reporting Limit (ng/L) | Reference |
|---|---|---|---|---|---|
| ***Anti-convulsants*** | | | | | |
| Carbamazepine | WW, SW, DW | SPE | LC-MS/MS | 1.0 | (Vanderford, Pearson et al. 2003) |
| | WW, SW | SPE | LC-MS/MS | 10 – 61 | (Gros, Petrovic et al. 2006) |
| | WW, SW, DW | SPE | Isotope Dilution LC-MS/MS | 0.50 | (Vanderford and Snyder 2006) |
| Primidone | WW, SW | SPE | LC-MS/MS | 2 – 20 | (Hummel, Loeffler et al. 2006) |
| ***X-ray contrast media*** | | | | | |
| Diatrizoate | WW, GW, SW, DW | None | LC-MS/MS | 100 | (Seitz, Schulz et al. 2006) |
| | WW, SW | SPE | LC-Ion Trap MS/MS | 40 | (Seitz, Weber et al. 2006) |
| Iopromide | WW, SW, DW | SPE | LC-MS/MS | 1.0 | (Vanderford, Pearson et al. 2003) |
| | WW, GW, SW, DW | None | LC-MS/MS | 100 | (Seitz, Schulz et al. 2006) |
| | WW, SW | SPE | LC-Ion Trap MS/MS | 40 | (Seitz, Weber et al. 2006) |
| ***Muscle relaxant*** | | | | | |
| Diazepam | WW, SW, DW | SPE | LC-MS/MS | 1.0 | (Vanderford, Pearson et al. 2003) |
| | WW, SW | SPE | LC-MS/MS | 1 – 10 | (Hummel, Loeffler et al. 2006) |
| | WW, GW, SW | SPE | GC-MS | 0.4 – 6.9 | (Rabiet, Togola et al. 2006) |
| | WW, SW, DW | SPE | Isotope Dilution LC-MS/MS | 0.25 | (Vanderford and Snyder 2006) |
| **Steroids** | | | | | |
| ***Estrogens*** | | | | | |
| 17α-Estradiol | WW, SW | SPE | LC-MS/MS | 0.2 | (Isobe, Shiraishi et al. 2003) |
| | SW | SPE | GC-MS/MS | 0.25 | (Noppe, De Wasch et al. 2005) |
| | SW | SPE | GC-MS/MS | 0.3 | (Kolodziej and Sedlak 2007) |
| 17β-Estradiol | WW, SW, DW | SPE | LC-MS/MS | 1.0 | (Vanderford, Pearson et al. 2003) |
| | WW | SPE | LC-MS/MS | 5.2 | (Castiglioni, Bagnati et al. 2005) |
| | WW, SW | SPE | LC-MS/MS | 0.7 | (Yamamoto, Kakutani et al. 2006) |
| | SW, DW | SPE | LC-MS/MS | 0.1 | (Loos, Wollgast et al. 2007) |

(continued)

**Table A.1 (Continued)**

| Compound | Matrixes | Extraction Method | Detection Method | Reporting Limit (ng/L) | Reference |
|---|---|---|---|---|---|
| Estrone | WW | SPE | LC-MS/MS | 1.5 | (Castiglioni, Bagnati et al. 2005) |
| | WW, SW | SPE | LC-MS/MS | 0.7 | (Yamamoto, Kakutani et al. 2006) |
| | SW, DW | SPE | LC-MS/MS | 0.05 | (Loos, Wollgast et al. 2007) |
| 17α-Ethynylestradiol | WW, SW, DW | SPE | LC-MS/MS | 1.0 | (Vanderford, Pearson et al. 2003) |
| | WW | SPE | LC-MS/MS | 4.6 | (Castiglioni, Bagnati et al. 2005) |
| | WW, SW | SPE | LC-MS/MS | 0.9 | (Yamamoto, Kakutani et al. 2006) |
| | SW, DW | SPE | LC-MS/MS | 0.1 | (Loos, Wollgast et al. 2007) |
| ***Androgen*** | | | | | |
| Testosterone | WW, SW, DW | SPE | LC-MS/MS | 1.0 | (Vanderford, Pearson et al. 2003) |
| | WW, SW | SPE | LC-MS/MS | 0.06 | (Yamamoto, Kakutani et al. 2006) |
| | SW | SPE | GC-MS/MS | 0.3 | (Kolodziej and Sedlak 2007) |
| **Personal Care Products** | | | | | |
| ***Antimicrobials*** | | | | | |
| Triclocarban | WW, SW, DW | SPE | LC-MS | 3 – 50 | (Halden and Paull 2004) |
| | WW, SW | SPE | Isotope Dilution LC-MS/MS | 0.9 | (Sapkota, Heidler et al. 2007) |
| Triclosan | WW, SW, DW | SPE | LC-MS/MS | 1.0 | (Vanderford, Pearson et al. 2003) |
| | WW, SW, DW | SPE | Isotope Dilution LC-MS/MS | 1.0 | (Vanderford and Snyder 2006) |
| | SW, DW | SPE | LC-MS/MS | 2.0 | (Loos, Wollgast et al. 2007) |
| ***Fragrances*** | | | | | |
| Galaxolide | WW, SW | SPE | GC-MS | 2 | (Buerge, Buser et al. 2003) |
| | WW, GW | SPE | GC-MS | 20 | (Kreuzinger, Clara et al. 2004) |
| | WW, SW | LLE | GC-MS | 20 | (Einsle, Paschke et al. 2006) |
| | WW, GW, SW, DW | SPE | GC-MS/MS | 10 | (Trenholm, Vanderford et al. 2006) |
| Tonalide | WW, SW | SPE | GC-MS | 1 | (Buerge, Buser et al. 2003) |
| | WW, GW | SPE | GC-MS | 20 | (Kreuzinger, Clara et al. 2004) |
| | WW, SW | LLE | GC-MS | 20 | (Einsle, Paschke et al. 2006) |

(continued)

Table A.1 (Continued)

| Compound | Matrixes | Extraction Method | Detection Method | Reporting Limit (ng/L) | Reference |
|---|---|---|---|---|---|
| ***Surfactant*** | | | | | |
| 4-Nonlyphenol | WW, SW | SPE | GC-MS | 0.2 | (Zhang, Hibberd et al. 2006) |
| | SW, DW | SPE | LC-MS/MS | 10 | (Loos, Wollgast et al. 2007) |
| | WW | SPE | GC-MS | 110 | (Gatidou, Thomaidis et al. 2007) |
| ***Antioxidant*** | | | | | |
| Butylated hydroxyanisole | WW, SW | LLE | GC-MS | 268 | (Pedersen, Soliman et al. 2005) |
| | WW, SW | SPE | GC-MS | 70 | (Loraine and Pettigrove 2006) |
| ***Insecticide*** | | | | | |
| DEET | WW, SW, DW | SPE | LC-MS/MS | 1.0 | (Vanderford, Pearson et al. 2003) |
| | SW | LLE | GC-MS | 80 | (Sandstrom, Kolpin et al. 2005) |
| | WW, SW, DW | SPE | GC-MS | 82 – 160 | (Loraine and Pettigrove 2006) |
| **Others** | | | | | |
| ***Plasticizers*** | | | | | |
| Bisphenol A | WW, SW, DW | SPE | Isotope Dilution LC-MS/MS | 1.0 | (Vanderford and Snyder 2006) |
| | WW, SW | SPE | GC-MS | 0.1 | (Zhang, Hibberd et al. 2006) |
| | SW, DW | SPE | LC-MS/MS | 0.2 | (Loos, Wollgast et al. 2007) |
| | WW | SPE | GC-MS | 480 | (Gatidou, Thomaidis et al. 2007) |
| Di-n-butyl phthalate | WW, SW | SPME | GC-MS | 26 | (Polo, Llompart et al. 2005) |
| | WW, SW, DW | SPE | GC-MS | 1350 – 2700 | (Loraine and Pettigrove 2006) |
| | DW | SBSE-LD | GC-MS | 600 | (Serodio and Nogueira 2006) |
| ***Flame retardants*** | | | | | |
| TCEP | WW, SW, DW | SPE | LC-MS/MS | 1.0 | (Vanderford, Pearson et al. 2003) |
| | SW | LLE/SPE | GC-MS | 12 - 20 | (Andresen, Grundmann et al. 2004) |
| | WW | SPE | GC-MS | 6.1 | (Meyer and Bester 2004) |
| | WW | LLE | GC-NPD | 0.8 | (Marklund, Andersson et al. 2005) |
| | WW | SPE | GC-MS | 760 | (Loraine and Pettigrove 2006) |

(continued)

**Table A.1 (Continued)**

| Compound | Matrixes | Extraction Method | Detection Method | Reporting Limit (ng/L) | Reference |
|---|---|---|---|---|---|
| TCPP | SW | LLE | GC-MS | 4.9 | (Andresen, Grundmann et al. 2004) |
| | WW | SPE | GC-MS | 1.0 | (Meyer and Bester 2004) |
| | WW | LLE | GC-NPD | 0.8 | (Marklund, Andersson et al. 2005) |
| TDCPP | SW | LLE | GC-MS | 14 | (Andresen, Grundmann et al. 2004) |
| | WW | SPE | GC-MS | 7.0 | (Meyer and Bester 2004) |
| | WW | LLE | GC-NPD | 0.8 | (Marklund, Andersson et al. 2005) |
| ***Stimulant*** | | | | | |
| Caffeine | WW, SW, DW | SPE | LC-MS/MS | 1.0 | (Vanderford, Pearson et al. 2003) |
| | GW, SW | SPE | LC-MS | 14 | (Cahill, Furlong et al. 2004) |
| | WW, GW, SW | SPE | Ion Trap LC-MS/MS | 71 | (Batt and Aga 2005) |
| | WW, SW | SPE | LC-TOF-MS | 12 | (Benotti and Brownawell 2007) |
| | WW, SW | SPE | LC-MS/MS | 0.96 – 5.0 | (Huerta-Fontela, Galceran et al. 2007) |

# APPENDIX B:
# TABLES OF OCCURRENCE OF EDCS AND PPCPS

## Table B.1
## Occurrence of select EDCs and PPCPs in drinking water sources

| Compounds | Water Source | Country | # of Waters | # of Sites | # of Samples* | Wastewater impact (% indicates degree) | Proximity and Distance (km) from Discharge(s) | Concentration Range† (ng/L) | Detection Frequency (%) | Median (ng/L) | References |
|---|---|---|---|---|---|---|---|---|---|---|---|
| **Pharmaceuticals** | | | | | | | | | | | |
| ***Analgesics*** | | | | | | | | | | | |
| Acetaminophen | surface | US | 9 | 9 | 9 | unknown | us, 0.006-49 | <36-1780 | 44 | <36 | (Glassmeyer et al. 2005) |
| | surface | US | 10 | 20 | 20 | yes, 10-95% | ds, 0.1-97 | <36-1720 | 40 | <36 | (Glassmeyer et al. 2005) |
| | surface | US | 10 | 23 | 76 | yes | ds, us | <9-1950 | 21 | <9 | (Kolpin et al. 2004) |
| | surface | G | 16 | 16 | 31 | yes | ds | <150 | 0 | <150 | (Ternes 1998) |
| | surface | US | 84 | 84 | 84 | yes | ds | <9-10000 | 24 | <9 | (Kolpin et al. 2002) |
| | surface | US | 2 | 3 | 12 | yes, 10-70% | ds | <9-40 | 50 | <9 | (Stackelberg et al. 2004) |
| | surface | US | 2 | 8 | 8 | yes | ds | <17 | 0 | <17 | (Barber et al. 2003) |
| | surface | US | 2 | 8 | 8 | yes | ds | <17-17 | 13 | <17 | (Barber et al. 2003) |
| Diclofenac | surface | A | 4 | 4 | 4 | yes | ds | 15.8-35.5 | 100 | 19.9 | (Ahrer, Scherwenk, and Buchberger 2001) |
| | surface | A | 2 | 5 | 5 | yes | ds | 28.3-392.1 | 100 | 161.4 | (Ahrer, Scherwenk, and Buchberger 2001) |
| | surface | Sw | 8 | 11 | 38 | yes | ds | <1-370 | ~74 | 3 | (Buser, Poiger, and Muller 1998) |
| | surface | G | 3 | 3 | 3 | unknown | - | 14-18 | 100 | 16 | (Deng et al. 2003) |
| | surface | Sp | 4 | 4 | 12 | yes | ds | <5-610 | 75 | 53 | (Farre et al. 2001) |
| | ground | G | 17 | 17 | 17 | yes | dg | n.d.-380 | - | - | (Heberer et al. 1997) |
| | surface | G | 12 | 27 | 27 | yes | ds | n.d.-960 | 93 | 60 | (Heberer, Schmidt-Baeumler, and Stan 1998) |
| | surface | G | 2 | 15 | 45 | yes | ds | n.d.-1030 | - | - | (Heberer, Reddersen, and Mechlinski 2002) |
| | surface | Sw | 2 | - | 45 | unknown | - | 4.5-150 | - | - | (Ollers et al. 2001) |
| | ground | G | - | 105 | 105 | unknown | - | <8.7-590 | 4 | <8.7 | (Sacher et al. 2001) |
| | ground | US | 1 | 1 | 1 | yes | dg | <10 | 0 | <10 | (Sedlak and Pinkston 2001) |
| | surface | Sw | 2 | 2 | 60 | yes | ds | <4.5-150 | - | - | (Tixier et al. 2003) |
| | surface | F | 1 | 1 | 3 | unknown | us | <1-2.5 | 33 | <1 | (Vieno, Tuhkanen, and Kronberg 2005) |
| | surface | F | 1 | 7 | 21 | yes | ds, 0-32 | <1-45 | 48 | <1 | (Vieno, Tuhkanen, and Kronberg 2005) |

(continued)

## Table B.1 (Continued)

| Compounds | Water Source | Country | # of Waters | # of Sites | # of Samples* | Wastewater impact (% indicates degree) | Proximity and Distance (km) from Discharge(s) | Concentration Range† (ng/L) | Detection Frequency (%) | Median (ng/L) | References |
|---|---|---|---|---|---|---|---|---|---|---|---|
| | surface | F | 3 | 3 | 3 | unknown | us | <1-6 | 33 | <1 | (Lindqvist, Tuhkanen, and Kronberg 2005) |
| | surface | F | 3 | 4 | 4 | yes | ds, 2-90 | <1-35 | 75 | 2.5 | (Lindqvist, Tuhkanen, and Kronberg 2005) |
| | surface | UK | 5 | 5 | 15 | yes | ds, 1 | <20-568 | 47 | <20 | (Ashton, Hilton, and Thomas 2004) |
| | surface | UK | 5 | 5 | 15 | unknown | us, 1 | <20 | 0 | <20 | (Ashton, Hilton, and Thomas 2004) |
| | surface | UK | 1 | 1 | 1 | yes | ds | 91 | 100 | 91 | (Hilton and Thomas 2003) |
| | surface | UK | 1 | 1 | 1 | unknown | us | <20 | 0 | <20 | (Hilton and Thomas 2003) |
| | surface | G | 22 | 22 | 43 | yes | ds | 10-1200 | 100 | 150 | (Ternes 1998) |
| | surface | G | 3 | 8 | 8 | yes | ds | 10-50 | 100 | 25 | (Wiegel et al. 2004) |
| | surface | G | 1 | 1 | 1 | yes | ds | <1 | 0 | <1 | (Wiegel et al. 2004) |
| | surface | G | 9 | 23 | 23 | yes | ds | <1-69 | 87 | 8 | (Wiegel et al. 2004) |
| Ibuprofen | ground | G | 17 | 17 | 17 | yes | dg | n.d.-200 | - | - | (Heberer et al. 1997) |
| | surface | G | 12 | 27 | 27 | yes | ds | n.d.-280 | 89 | 10 | (Heberer, Schmidt-Baeumler, and Stan 1998) |
| | surface | G | 22 | 22 | 43 | yes | ds | <10-530 | 81 | 70 | (Ternes 1998) |
| | surface | G | 2 | 15 | 45 | yes | ds | n.d.-55 | - | - | (Heberer, Reddersen, and Mechlinski 2002) |
| | surface | Sw | 6 | 14 | 32 | yes | ds | <0.2-7.8 | 94 | 4 | Buser et al., 1999 |
| | surface | Sw | 2 | - | 45 | unknown | - | 4.5-80 | - | - | (Ollers et al. 2001) |
| | surface | Sp | 4 | 4 | 12 | yes | ds | <43-2700 | 83 | 192 | (Farre et al. 2001) |
| | surface | Sw | 2 | 2 | 60 | yes | ds | <4.5-80 | - | - | (Tixier et al. 2003) |
| | surface | F | 1 | 1 | 3 | unknown | us | <1-75 | 67 | 10 | (Vieno, Tuhkanen, and Kronberg 2005) |
| | surface | F | 1 | 7 | 21 | yes | ds, 0-32 | <1-80 | 90 | 10 | (Vieno, Tuhkanen, and Kronberg 2005) |
| | surface | F | 3 | 3 | 3 | unknown | us | <1-6 | 67 | 3 | (Lindqvist, Tuhkanen, and Kronberg 2005) |
| | surface | F | 3 | 4 | 4 | yes | ds, 2-90 | <1-14 | 75 | 10 | (Lindqvist, Tuhkanen, and Kronberg 2005) |
| | surface | I | 2 | 8 | 8 | yes | ds | <4.2-78.5 | 63 | 4.5 | (Calamari et al. 2003) |
| | surface | UK | 55 | 5 | 15 | yes | ds, 1 | <20-5044 | 69 | 826 | (Ashton, Hilton, and Thomas 2004) |

(continued)

## Table B.1 (Continued)

| Compounds | Water Source | Country | # of Waters | # of Sites | # of Samples* | Wastewater impact (% indicates degree) | Proximity and Distance (km) from Discharge(s) | Concentration Range† (ng/L) | Detection Frequency (%) | Median (ng/L) | References |
|---|---|---|---|---|---|---|---|---|---|---|---|
| | surface | UK | 5 | 5 | 15 | unknown | us, 1 | <20 | 57 | <20 | (Ashton, Hilton, and Thomas 2004) |
| | surface | UK | 1 | 1 | 1 | yes | ds | <20 | 0 | <20 | (Hilton and Thomas 2003) |
| | surface | UK | 1 | 1 | 1 | unknown | us | <20 | 0 | <20 | (Hilton and Thomas 2003) |
| | surface | US | 1 | 4 | 13 | unknown | - | <280-5850 | 8 | <280 | (Loraine and Pettigrove 2006) |
| | surface | US | 10 | 23 | 76 | yes | ds, us | <18 | 0 | <18 | (Kolpin et al. 2002) |
| | surface | US | 84 | 84 | 84 | yes | ds | <18-1000 | 10 | <18 | (Kolpin et al. 2002) |
| | ground | US | 1 | 1 | 1 | yes | dg | <10 | 0 | <10 | (Sedlak and Pinkston 2001) |
| | surface | G | 3 | 8 | 8 | yes | ds | <20-70 | 50 | <20 | (Wiegel et al. 2004) |
| | surface | G | 1 | 1 | 1 | yes | ds | <2 | 0 | <2 | (Wiegel et al. 2004) |
| | surface | G | 9 | 23 | 23 | yes | ds | <2-146 | 70 | 5.5 | (Wiegel et al. 2004) |
| | surface | US | 2 | 8 | 8 | yes | ds | <36 | 0 | <36 | (Barber et al. 2003) |
| | surface | US | 2 | 8 | 8 | yes | ds | <36-108 | 13 | <36 | (Barber et al. 2003) |
| | surface | US | 2 | 3 | 12 | yes, 10-70% | ds | <18 | 0 | <18 | (Stackelberg et al. 2004) |
| Ketoprofen | surface | F | 1 | 1 | 3 | unknown | us | <25 | 67 | <25 | (Vieno, Tuhkanen, and Kronberg 2005) |
| | surface | F | 1 | 7 | 21 | yes | ds, 0-32 | <25-40 | 57 | <25 | (Vieno, Tuhkanen, and Kronberg 2005) |
| | surface | F | 3 | 3 | 3 | unknown | us | <5 | 0 | <5 | (Lindqvist, Tuhkanen, and Kronberg 2005) |
| | surface | F | 3 | 4 | 4 | yes | ds, 2-90 | <5-23 | 50 | <5 | (Lindqvist, Tuhkanen, and Kronberg 2005) |
| | surface | Sw | 2 | 2 | 60 | yes | ds | <4.5-10 | - | - | (Tixier et al. 2003) |
| | surface | Sw | 2 | - | 45 | unknown | - | 4.5.-5 | 9 | - | (Ollers et al. 2001) |
| | surface | G | 22 | 22 | 43 | yes | ds | <10-120 | 12 | <10 | (Ternes 1998) |
| | ground | US | 1 | 1 | 1 | yes | dg | <10 | 0 | <10 | (Sedlak and Pinkston 2001) |
| | surface | Sp | 4 | 4 | 12 | yes | ds | <28-300 | 25 | <28 | (Farre et al. 2001) |
| | surface | G | 2 | 15 | 45 | yes | ds | n.d.-65 | - | - | (Heberer, Reddersen, and Mechlinski 2002) |
| Naproxen | surface | F | 1 | 1 | 3 | unknown | us | 10-95 | 100 | 12 | (Vieno, Tuhkanen, and Kronberg 2005) |
| | surface | F | 1 | 7 | 21 | yes | ds, 0-32 | <5-130 | 95 | 15 | (Vieno, Tuhkanen, and Kronberg 2005) |

(continued)

## Table B.1 (Continued)

| Compounds | Water Source | Country | # of Waters | # of Sites | # of Samples* | Wastewater impact (% indicates degree) | Proximity and Distance (km) from Discharge(s) | Concentration Range† (ng/L) | Detection Frequency (%) | Median (ng/L) | References |
|---|---|---|---|---|---|---|---|---|---|---|---|
| | surface | F | 3 | 3 | 3 | unknown | us | <5 | 0 | <5 | (Lindqvist, Tuhkanen, and Kronberg 2005) |
| | surface | F | 3 | 4 | 4 | yes | ds, 2-90 | <5-45 | 75 | 5.8 | (Lindqvist, Tuhkanen, and Kronberg 2005) |
| | surface | Sw | 2 | 2 | 60 | yes | ds | <4.5-375 | - | - | (Tixier et al. 2003) |
| | surface | Sw | 2 | - | 45 | unknown | us | <4.5-400 | - | - | (Ollers et al. 2001) |
| | surface | G | 20 | 20 | 20 | yes | ds | 10-390 | 100 | 70 | (Ternes 1998) |
| | ground | US | 1 | 1 | 1 | yes | dg | <10 | 0 | <10 | (Sedlak and Pinkston 2001) |
| | surface | A | 2 | 5 | 5 | yes | ds | <13-38.2 | 20 | <13 | (Ahrer, Scherwenk, and Buchberger 2001) |
| | surface | Sp | 4 | 4 | 12 | yes | ds | <29-2000 | 58 | 67 | (Farre et al. 2001) |
| | surface | G | 1 | 1 | 1 | yes | ds | <1 | 0 | <1 | (Wiegel et al. 2004) |
| | surface | G | 9 | 23 | 23 | yes | ds | <1-32 | 61 | 1 | (Wiegel et al. 2004) |
| | surface | G | 2 | 15 | 45 | yes | ds | n.d.-95 | - | - | (Heberer, Reddersen, and Mechlinski 2002) |
| ***Antibiotics*** | | | | | | | | | | | |
| Ciprofloxacin | surface | US | 10 | 23 | 76 | yes | ds, us | <10-30 | 1 | <20 | (Kolpin et al. 2004) |
| | surface | I | 2 | 8 | 8 | yes | ds | <0.3-20.3 | 25 | <0.3 | (Calamari et al. 2003) |
| | surface | US | 115 | 115 | 115 | yes | ds | <20-30 | 3 | 20 | (Kolpin et al. 2002) |
| | surface | US | 2 | 3 | 12 | yes, 10-70% | ds | <20 | 0 | <20 | (Stackelberg et al. 2004) |
| Clarithromycin | surface | G | - | 33 | 33 | yes | ds | <20-260 | 21 | <20 | (Hirsch et al. 1999) |
| | ground | G | - | 59 | 59 | yes | dg | <20 | 0 | <20 | (Hirsch et al. 1999) |
| | surface | I | 2 | 8 | 8 | yes | ds | 0.5-20.3 | 100 | 1.6 | (Calamari et al. 2003) |
| | surface | G | 3 | 8 | 8 | yes | ds | <30-40 | 25 | <30 | (Wiegel et al. 2004) |
| Roxithromycin | surface | G | - | 52 | 52 | yes | ds | <20-560 | 44 | <20 | (Hirsch et al. 1999) |
| | ground | G | - | 59 | 59 | yes | dg | <20 | 0 | <20 | (Hirsch et al. 1999) |
| | surface | US | 10 | 23 | 76 | yes | ds, us | <10 | 0 | <10 | (Kolpin et al. 2002) |
| | surface | US | 104 | 104 | 104 | yes | ds | <30-180 | 5 | <30 | (Kolpin et al. 2002) |
| | surface | G | 3 | 8 | 8 | yes | ds | <30-40 | 50 | <30 | (Wiegel et al. 2004) |
| | surface | US | 2 | 3 | 12 | yes, 10-70% | ds | <30 | 0 | <30 | (Stackelberg et al. 2004) |

(continued)

## Table B.1 (Continued)

| Compounds | Water Source | Country | # of Waters | # of Sites | # of Samples* | Wastewater impact (% indicates degree) | Proximity and Distance (km) from Discharge(s) | Concentration Range† (ng/L) | Detection Frequency (%) | Median (ng/L) | References |
|---|---|---|---|---|---|---|---|---|---|---|---|
| Chlortetracycline | surface | US | 84 | 84 | 84 | yes | ds | < 100-690 | 2 | < 100 | (Kolpin et al. 2002) |
| | surface | G | - | 14 | 14 | yes | ds | <50 | 0 | <50 | (Hirsch et al. 1999) |
| | ground | G | - | 37 | 37 | yes | dg | <50 | 0 | <50 | (Hirsch et al. 1999) |
| | surface | US | 2 | 2 | 2 | unknown | - | <500-500 | 50 | <500 | (Meyer et al. 2000) |
| | ground | US | 1 | 1 | 1 | unknown | - | <500 | 0 | <500 | (Meyer et al. 2000) |
| | surface | US | 10 | 23 | 76 | yes | ds, us | <20-100 | 3 | <20 | (Kolpin et al. 2002) |
| | surface | US | 2 | 3 | 12 | yes, 10-70% | ds | <50 | 0 | <50 | (Stackelberg et al. 2004) |
| Sulfadiazine | surface | G | 5 | 5 | 5 | unknown | - | <0.3-7 | 20 | <0.3 | (Hartig, Storm, and Jekel 1999) |
| Sulfamethoxazole | surface | G | 5 | 5 | 5 | unknown | - | <0.9-85 | 80 | 60 | (Hartig, Storm, and Jekel 1999) |
| | surface | G | - | 52 | 52 | yes | ds | <20-480 | 50 | 30 | (Hirsch et al. 1999) |
| | ground | G | - | 59 | 59 | yes | dg | <20-470 | 3 | <20 | (Hirsch et al. 1999) |
| | surface | UK | 5 | 5 | 15 | yes | ds, 1 | <50 | 0 | <50 | (Ashton, Hilton, and Thomas 2004) |
| | surface | UK | 5 | 5 | 15 | unknown | us, 1 | <50 | 0 | <50 | (Ashton, Hilton, and Thomas 2004) |
| | surface | UK | 1 | 1 | 1 | yes | ds | <50 | 0 | <50 | (Hilton and Thomas 2003) |
| | surface | UK | 1 | 1 | 1 | unknown | us | <50 | 0 | <50 | (Hilton and Thomas 2003) |
| | surface | US | 2 | 3 | 12 | yes, 10-70% | ds | <23 | 17 | <23 | (Stackelberg et al. 2004) |
| | ground | G | - | 105 | 105 | unknown | - | <6.2-410 | 10 | <6.2 | (Sacher et al. 2001) |
| | surface | US | 104 | 104 | 104 | yes | ds | <50-1900 | 13 | <50 | (Kolpin et al. 2002) |
| | surface | US | 10 | 23 | 76 | yes | ds, us | <50-70 | 8 | <50 | (Kolpin et al. 2004) |
| | surface | US | 10 | 23 | 76 | yes | ds, us | <23-63 | 7 | <23 | (Kolpin et al. 2004) |
| | surface | US | 9 | 9 | 9 | unknown | us, 0.006-49 | <64-292 | 22 | <64 | (Glassmeyer et al. 2005) |
| | surface | US | 10 | 20 | 20 | yes, 10-95% | ds, 0.1-97 | <64-763 | 85 | 69[d] | (Glassmeyer et al. 2005) |
| | surface | G | 3 | 8 | 8 | yes | ds | <30-70 | 88 | 45 | (Wiegel et al. 2004) |
| | surface | US | 2 | 8 | 8 | yes | ds | <46-150 | 38 | <46 | (Barber et al. 2003) |
| | surface | US | 2 | 8 | 8 | yes | ds | <46-220 | 50 | <46 | (Barber et al. 2003) |

(continued)

## Table B.1 (Continued)

| Compounds | Water Source | Country | # of Waters | # of Sites | # of Samples* | Wastewater impact (% indicates degree) | Proximity and Distance (km) from Discharge(s) | Concentration Range† (ng/L) | Detection Frequency (%) | Median (ng/L) | References |
|---|---|---|---|---|---|---|---|---|---|---|---|
| Sulfamethazine | surface | G | - | 52 | 52 | yes | ds | <20 | 0 | <20 | (Hirsch et al. 1999) |
| | ground | G | - | 59 | 59 | yes | dg | <20-160 | 3 | <20 | (Hirsch et al. 1999) |
| | surface | US | 10 | 23 | 76 | yes | ds, us | <10 | 0 | <10 | (Kolpin et al. 2004) |
| | surface | US | 104 | 104 | 104 | yes | ds | <50-120 | 5 | <50 | (Kolpin et al. 2002) |
| | surface | US | 2 | 3 | 12 | yes, 10-70% | ds | <50 | 17 | <50 | (Stackelberg et al. 2004) |
| Sulfadimethoxine | surface | US | 10 | 23 | 76 | yes | ds, us | <10-10 | 1 | <10 | (Kolpin et al. 2004) |
| | surface | US | 104 | 104 | 104 | yes | ds | <50-60 | 0 | <50 | (Kolpin et al. 2002) |
| | surface | US | 2 | 3 | 12 | yes, 10-70% | ds | <50 | 17 | <50 | (Stackelberg et al. 2004) |
| Erythromycin | ground | G | - | 105 | 105 | unknown | - | <12-49 | 10 | <12 | (Sacher et al. 2001) |
| | surface | US | 10 | 23 | 76 | yes | ds, us | <20-220 | 4 | <20 | (Kolpin et al. 2004) |
| | surface | G | - | 52 | 52 | yes | ds | <20-260 | 59 | 150 | (Hirsch et al. 1999) |
| | ground | G | - | 59 | 59 | yes | dg | <20 | 0 | <20 | (Hirsch et al. 1999) |
| | surface | I | 2 | 8 | 8 | yes | ds | 1.40-15.9 | 100 | 3.2 | (Calamari et al. 2003) |
| | surface | UK | 5 | 5 | 15 | yes | ds, 1 | <10-1022 | 38 | <10 | (Ashton, Hilton, and Thomas 2004) |
| | surface | UK | 5 | 5 | 15 | unknown | us, 1 | <10-57 | 17 | <10 | (Ashton, Hilton, and Thomas 2004) |
| | surface | UK | 1 | 1 | 1 | yes | ds | 1000 | 100 | 1000 | (Hilton and Thomas 2003) |
| | surface | UK | 1 | 1 | 1 | unknown | us | 57 | 100 | 57 | (Hilton and Thomas 2003) |
| | surface | US | 2 | 3 | 12 | yes, 10-70% | ds | <30-700 | 67 | 30 | (Stackelberg et al. 2004) |
| | surface | US | 104 | 104 | 104 | yes | ds | <50-1700 | 22 | <50 | (Kolpin et al. 2002) |
| | surface | G | 3 | 8 | 8 | yes | ds | 30-70 | 100 | 40 | (Wiegel et al. 2004) |
| Trimethoprim | surface | G | - | 52 | 52 | yes | ds | <20-200 | 19 | <20 | (Hirsch et al. 1999) |
| | ground | G | - | 59 | 59 | yes | dg | <20 | 0 | <20 | (Hirsch et al. 1999) |
| | surface | UK | 5 | 5 | 15 | yes | ds, 1 | <10-42 | 36 | <10 | (Ashton, Hilton, and Thomas 2004) |
| | surface | UK | 5 | 5 | 15 | unknown | us, 1 | <10-36 | 38 | <10 | (Ashton, Hilton, and Thomas 2004) |
| | surface | UK | 1 | 1 | 1 | yes | ds | 39 | 100 | 39 | (Hilton and Thomas 2003) |
| | surface | UK | 1 | 1 | 1 | unknown | us | <10 | 0 | <10 | (Hilton and Thomas 2003) |
| | surface | US | 9 | 9 | 9 | unknown | us, 0.006-49 | <13-54 | 22 | <13 | (Glassmeyer et al. 2005) |

(continued)

## Table B.1 (Continued)

| Compounds | Water Source | Country | # of Waters | # of Sites | # of Samples* | Wastewater impact (% indicates degree) | Proximity and Distance (km) from Discharge(s) | Concentration Range† (ng/L) | Detection Frequency (%) | Median (ng/L) | References |
|---|---|---|---|---|---|---|---|---|---|---|---|
| | surface | US | 10 | 20 | 20 | yes, 10-95% | ds, 0.1-97 | <13-414 | 70 | 13[d] | (Glassmeyer et al. 2005) |
| | surface | US | 10 | 23 | 76 | yes | ds, us | <10-80 | 6 | <10 | (Kolpin et al. 2004) |
| | surface | US | 104 | 104 | 104 | yes | ds | <30-710 | 13 | <30 | (Kolpin et al. 2002) |
| | surface | G | 3 | 8 | 8 | yes | ds | <30-40 | 38 | <30 | (Wiegel et al. 2004) |
| | surface | US | 2 | 8 | 8 | yes | ds | <28-66 | 13 | <28 | (Barber et al. 2003) |
| | surface | US | 2 | 8 | 8 | yes | ds | <28-160 | 50 | <28 | (Barber et al. 2003) |
| | surface | US | 2 | 3 | 12 | yes, 10-70% | ds | <14 | 83 | <14 | (Stackelberg et al. 2004) |
| Lincomycin | surface | I | 2 | 8 | 8 | yes | ds | 3.13-249 | 100 | 24.4 | Calamari et al., 2003 |
| | surface | US | 10 | 23 | 76 | yes | ds, us | <10-10 | 1 | <10 | (Kolpin et al. 2004) |
| | surface | US | 104 | 104 | 104 | yes | ds | <50-730 | 19 | <50 | (Kolpin et al. 2002) |
| | surface | US | 2 | 3 | 12 | yes, 10-70% | ds | <50 | 0 | <50 | (Stackelberg et al. 2004) |
| ***Anti-convulsants*** | | | | | | | | | | | |
| Carbamazepine | surface | US | 9 | 9 | 9 | unknown | us, 0.006-49 | <11-158 | 33 | <11 | (Glassmeyer et al. 2005) |
| | surface | US | 10 | 20 | 20 | yes, 10-95% | ds, 0.1-97 | 11-186[c] | 100 | 76[d] | (Glassmeyer et al. 2005) |
| | surface | US | 10 | 23 | 76 | yes | ds, us | <11-263 | 30 | <11 | (Kolpin et al. 2004) |
| | surface | Sw | 2 | 2 | 60 | yes | ds | <8.7-325 | - | - | (Tixier et al. 2003) |
| | ground | G | - | 105 | 105 | unknown | - | <32-900 | 12 | <32 | (Sacher et al. 2001) |
| | surface | Sw | 2 | - | 45 | unknown | - | 30-250 | 100 | - | (Ollers et al. 2001) |
| | surface | US | 2 | 3 | 12 | yes, 10-70% | ds | 60-1500 | 100 | 200 | (Stackelberg et al. 2004) |
| | surface | G | 20 | 20 | 26 | yes | ds | <30-1100 | 92 | 250 | (Ternes 1998) |
| | surface | A | 4 | 4 | 4 | yes | ds | 23-133.1 | 100 | 47.2 | (Ahrer, Scherwenk, and Buchberger 2001) |
| | surface | G | 3 | 8 | 8 | yes | ds | <30-140 | 88 | 65 | (Wiegel et al. 2004) |
| | surface | G | 109 | 179 0 | 1790 | yes | ds | <20-7100 | n/a | 400** | (Wiegel et al. 2004) |
| | surface | G | 2 | 15 | 45 | yes | ds | 25-1075 | 100 | 550 | (Heberer, Reddersen, and Mechlinski 2002) |
| Primidone | surface | G | 2 | 15 | 45 | yes | ds | n.d.-635 | - | - | (Heberer, Reddersen, and Mechlinski 2002) |

(continued)

## Table B.1 (Continued)

| Compounds | Water Source | Country | # of Waters | # of Sites | # of Samples* | Wastewater impact (% indicates degree) | Proximity and Distance (km) from Discharge(s) | Concentration Range† (ng/L) | Detection Frequency (%) | Median (ng/L) | References |
|---|---|---|---|---|---|---|---|---|---|---|---|
| ***β-blockers*** | | | | | | | | | | | |
| Atenolol | surface | I | 2 | 8 | 8 | yes | ds | 3.44-241 | 100 | 17 | (Calamari et al. 2003) |
| Metoprolol | surface | G | 23 | 23 | 45 | yes | ds | <10-2200 | 84 | 45 | (Ternes 1998) |
| | ground | US | 1 | 1 | 1 | yes | ds | <10 | 0 | <10 | (Sedlak and Pinkston 2001) |
| | surface | G | 1 | 1 | 1 | no | us | <1 | 0 | <1 | (Wiegel et al. 2004) |
| | surface | G | 7 | 24 | 24 | yes | ds | 2-224 | 100 | 48 | (Wiegel et al. 2004) |
| Propanolol | surface | US | 1 | 7 | 16 | yes | ds | <0.1-31 | - | - | (Fono and Sedlak 2005) |
| | surface | UK | 5 | 5 | 15 | yes | ds, 1 | <10-215 | 87 | 29 | (Ashton, Hilton, and Thomas 2004) |
| | surface | UK | 5 | 5 | 15 | unknown | us, 1 | <10-115 | 14 | <10 | (Ashton, Hilton, and Thomas 2004) |
| | surface | UK | 1 | 1 | 1 | yes | ds | 37 | 100 | 37 | (Hilton and Thomas 2003) |
| | surface | UK | 1 | 1 | 1 | unknown | us | <10 | 0 | <10 | (Hilton and Thomas 2003) |
| | surface | G | 23 | 23 | 45 | yes | ds | <10-590 | 58 | 12 | (Ternes 1998) |
| ***Lipid regulators*** | | | | | | | | | | | |
| Clofibric acid | ground | G | 17 | 17 | 17 | yes | dg | 70-7300 | 100 | - | (Heberer et al. 1997) |
| | surface | G | 12 | 27 | 27 | yes | ds | n.d.-875 | 78 | 125 | (Heberer, Schmidt-Baeumler, and Stan 1998) |
| | surface | G | 2 | 15 | 45 | yes | ds | n.d.-450 | - | - | (Heberer, Reddersen, and Mechlinski 2002) |
| | surface | Sw | 2 | 2 | 60 | yes | ds | <4.5-30 | - | - | (Tixier et al. 2003) |
| | surface | I | 2 | 8 | 8 | yes | ds | <0.3-5.77 | 88 | 1.1 | Calamari et al., 2003 |
| | surface | A | 2 | 5 | 5 | yes | ds | 19.3-43.5 | 100 | 24.8 | (Ahrer, Scherwenk, and Buchberger 2001) |
| | surface | Sw | 2 | - | 45 | unknown | - | 4.5-25 | - | - | (Ollers et al. 2001) |
| | surface | UK | 1 | 1 | 1 | yes | ds | <50 | 0 | <50 | (Hilton and Thomas 2003) |
| | surface | UK | 1 | 1 | 1 | unknown | us | <50 | 0 | <50 | (Hilton and Thomas 2003) |
| | surface | US | 1 | 4 | 13 | unknown | - | <130-630 | 8 | <130 | (Loraine and Pettigrove 2006) |
| | surface | G | 22 | 22 | 43 | yes | ds | <10-550 | 81 | 66 | (Ternes 1998) |

(continued)

## Table B.1 (Continued)

| Compounds | Water Source | Country | # of Waters | # of Sites | # of Samples* | Wastewater impact (% indicates degree) | Proximity and Distance (km) from Discharge(s) | Concentration Range† (ng/L) | Detection Frequency (%) | Median (ng/L) | References |
|---|---|---|---|---|---|---|---|---|---|---|---|
| | surface | G | 3 | 8 | 8 | yes | ds | <10-40 | 88 | 15 | (Wiegel et al. 2004) |
| | surface | G | 1 | 1 | 1 | no | us | <1 | 0 | <1 | (Wiegel et al. 2004) |
| | surface | G | 9 | 23 | 23 | yes | ds | <1-22 | 74 | 4.5 | (Wiegel et al. 2004) |
| Bezafibrate | surface | F | 1 | 1 | 3 | unknown | us | <1-20 | 67 | 2 | (Vieno, Tuhkanen, and Kronberg 2005) |
| | surface | F | 1 | 7 | 21 | yes | ds, 0-32 | <1-45 | 48 | <1 | (Vieno, Tuhkanen, and Kronberg 2005) |
| | surface | F | 3 | 3 | 3 | unknown | us | <1-4.5 | 33 | <1 | (Lindqvist, Tuhkanen, and Kronberg 2005) |
| | surface | F | 3 | 4 | 4 | yes | ds, 2-90 | <1-4 | 50 | <1 | (Lindqvist, Tuhkanen, and Kronberg 2005) |
| | surface | I | 2 | 8 | 8 | yes | ds | 0.8-57 | 100 | 1.9 | (Calamari et al. 2003) |
| | surface | G | 22 | 22 | 43 | yes | ds | <25-3100 | 91 | 350 | (Ternes 1998) |
| | surface | A | 4 | 4 | 4 | yes | ds | 1.6-12.5 | 100 | 5 | (Ahrer, Scherwenk, and Buchberger 2001) |
| | surface | A | 2 | 5 | 5 | yes | ds | <4.8-20.4 | 80 | 10.5 | (Ahrer, Scherwenk, and Buchberger 2001) |
| | surface | G | 3 | 8 | 8 | yes | ds | <50-130 | 50 | <50 | (Wiegel et al. 2004) |
| | surface | G | 1 | 1 | 1 | no | us | <50 | 0 | <50 | (Wiegel et al. 2004) |
| | surface | G | 9 | 23 | 23 | yes | ds | <50-88 | 9 | <50 | (Wiegel et al. 2004) |
| Gemfibrozil | surface | US | 10 | 23 | 76 | yes | ds, us | <15 | 0 | <15 | (Kolpin et al. 2004) |
| | surface | US | 84 | 84 | 84 | yes | ds | <15-790 | 4 | <15 | (Kolpin et al. 2002) |
| | surface | G | 22 | 22 | 43 | yes | ds | <10-510 | 65 | 52 | (Ternes 1998) |
| | ground | US | 1 | 1 | 1 | yes | dg | <10 | 0 | <10 | (Sedlak and Pinkston 2001) |
| | surface | Sp | 4 | 4 | 12 | yes | ds | <56-1550 | 92 | 228 | (Farre et al. 2001) |
| | surface | G | 1 | 1 | 1 | no | us | <2 | 0 | <2 | (Wiegel et al. 2004) |
| | surface | G | 9 | 23 | 23 | yes | ds | <2-11 | 52 | <2 | (Wiegel et al. 2004) |
| | surface | US | 2 | 8 | 8 | yes | ds | <28 | 0 | <28 | (Barber et al. 2003) |
| | surface | US | 2 | 8 | 8 | yes | ds | <28 | 0 | <28 | (Barber et al. 2003) |
| | surface | US | 2 | 3 | 12 | yes, 10-70% | ds | <15 | 0 | <15 | (Stackelberg et al. 2004) |
| | surface | G | 2 | 15 | 45 | yes | ds | n.d.-35 | - | - | (Heberer, Reddersen, and Mechlinski 2002) |

(continued)

## Table B.1 (Continued)

| Compounds | Water Source | Country | # of Waters | # of Sites | # of Samples* | Wastewater impact (% indicates degree) | Proximity and Distance (km) from Discharge(s) | Concentration Range† (ng/L) | Detection Frequency (%) | Median (ng/L) | References |
|---|---|---|---|---|---|---|---|---|---|---|---|
| ***X-ray contrast media*** | | | | | | | | | | | |
| Iopromide | surface | G | 1 | 1 | 1 | yes | ds | 1600 | 100 | 1600 | (Putschew, Wischnack, and Jekel 2000) |
| | ground | G | 1 | 1 | 1 | yes | dg | <50 | 0 | <50 | (Putschew, Wischnack, and Jekel 2000) |
| | surface | G | 1 | - | - | yes | ds | <20-1200 | | 860 | (Schittko, Putschew, and Jekel 2004) |
| Diatrizoate | surface | G | 1 | 1 | 1 | yes | ds | 2000 | 100 | 2000 | (Putschew, Wischnack, and Jekel 2000) |
| | ground | G | 1 | 1 | 1 | yes | dg | 4000 | 100 | 4000 | (Putschew, Wischnack, and Jekel 2000) |
| | surface | G | 1 | - | - | yes | ds | <100-1450 | | 960 | (Schittko, Putschew, and Jekel 2004) |
| ***Muscle relaxant*** | | | | | | | | | | | |
| Diazepam | surface | G | 11 | 11 | 11 | yes | ds, 5-10 | <10-33 | 18 | <10 | (Ternes, Bonerz, and Schmidt 2001) |
| | surface | G | 20 | 20 | 30 | yes | ds | <30 | 0 | <30 | (Ternes 1998) |
| ***Anti-hypertensive*** | | | | | | | | | | | |
| Diltiazem | surface | US | 9 | 9 | 9 | unknown | us, 0.006-49 | <16-74 | 22 | <16 | (Glassmeyer et al. 2005) |
| | surface | US | 10 | 20 | 20 | yes, 10-95% | ds, 0.1-97 | <16-67 | 75 | <16 | (Glassmeyer et al. 2005) |
| | surface | US | 10 | 23 | 76 | yes | ds, us | <12-106 | 9 | <12 | (Kolpin et al. 2004) |
| | surface | US | 2 | 8 | 8 | yes | ds | <24-25 | 25 | <24 | (Barber et al. 2003) |
| | surface | US | 2 | 8 | 8 | yes | ds | <24 | 13 | <24 | (Barber et al. 2003) |
| | surface | US | 2 | 3 | 12 | yes, 10-70% | ds | <2 | 0 | <12 | (Stackelberg et al. 2004) |
| | surface | US | 84 | 84 | 84 | yes | ds | <12-49 | 13 | <12 | (Kolpin et al. 2002) |

(continued)

## Table B.1 (Continued)

| Compounds | Water Source | Country | # of Waters | # of Sites | # of Samples* | Wastewater impact (% indicates degree) | Proximity and Distance (km) from Discharge(s) | Concentration Range† (ng/L) | Detection Frequency (%) | Median (ng/L) | References |
|---|---|---|---|---|---|---|---|---|---|---|---|
| **Steroids** | | | | | | | | | | | |
| ***Estrogens*** | | | | | | | | | | | |
| 17α-Ethinylestradiol | surface | N | 11 | 11 | 16 | yes | ds | <0.3-5.5 | 25 | <0.3 | (Belfroid et al. 1999) |
| | surface | G | 15 | 15 | 15 | unknown | - | <0.5 | 0 | <0.5 | (Ternes et al. 1999) |
| | surface | G | 2 | 2 | 31 | yes | ds, 1 | <0.1-5.1 | 48 | 0.4 | (Kuch and Ballschmiter 2001) |
| | surface | G | 2 | 2 | 2 | no | us | <0.10 | 0 | <0.10 | (Kuch and Ballschmiter 2001) |
| | surface | I | 1 | 2 | 2 | yes | ds, 1-20 | <0.03-0.04 | 50 | - | (Baronti et al. 2000) |
| | surface | Fr | 2 | 7 | 7 | yes | ds | 1.1-2.9 | 100 | 1.5 | (Cargouet et al. 2004) |
| | surface | US | 3 | 3 | 9 | yes | ds | <0.05-0.07 | 33 | <0.05 | (Huang and Sedlak 2001) |
| | surface | US | 70 | 70 | 70 | yes | ds | <5-831 | 16 | <5 | (Kolpin et al. 2002) |
| | surface | N | - | - | - | unknown | - | <0.4 | - | - | ([Gwrc] 2003) |
| | surface | UK | - | - | - | unknown | - | <0.3 | - | - | ([Gwrc] 2003) |
| | surface | G | - | - | - | unknown | - | <1 | - | - | ([Gwrc] 2003) |
| | surface | US | - | - | - | unknown | - | - | - | 73 | ([Gwrc] 2003) |
| | surface | US | 2 | 13 | 14 | yes | ds | <0.05-.5 | 29 | <0.05 | (Snyder et al. 1999) |
| | surface | UK | 5 | 5 | 5 | unknown | - | <0.3 | 0 | <0.3 | (Fawell et al. 2001) |
| | surface | US | 2 | 8 | 8 | yes | ds | <0.5 | 0 | <0.5 | (Barber et al. 2003) |
| | surface | US | 2 | 8 | 8 | yes | ds | <0.5 | 0 | <0.5 | (Barber et al. 2003) |
| 17β-Estradiol | surface | N | 11 | 11 | 16 | yes | ds | <0.6-5.5 | 50 | <0.6 | (Belfroid et al. 1999) |
| | surface | US | 3 | 3 | 9 | yes | ds | 0.05-0.8 | 100 | 0.08 | (Huang and Sedlak 2001) |
| | surface | G | 15 | 15 | 15 | unknown | - | <0.5 | 0 | <0.5 | (Ternes et al. 1999) |
| | surface | G | 2 | 2 | 31 | yes | ds, 1 | <0.15-3.6 | 45 | <0.15 | (Kuch and Ballschmiter 2001) |
| | surface | G | 2 | 2 | 2 | no | us | - | - | 0.2 | (Kuch and Ballschmiter 2001) |
| | Surface | US | 2 | 13 | 14 | yes | ds | <0.1-2.7 | 71 | 0.7 | (Snyder et al. 1999) |
| | surface | I | 1 | 2 | 2 | yes | ds, 1-20 | <0.02-0.11 | 50 | - | (Baronti et al. 2000) |
| | surface | Fr | 2 | 7 | 7 | yes | ds | 1.4-3.2 | 100 | 2.1 | (Cargouet et al. 2004) |
| | surface | US | 70 | 70 | 70 | yes | ds | <5-93 | 10 | <5 | (Kolpin et al. 2002) |

(continued)

## Table B.1 (Continued)

| Compounds | Water Source | Country | # of Waters | # of Sites | # of Samples* | Wastewater impact (% indicates degree) | Proximity and Distance (km) from Discharge(s) | Concentration Range† (ng/L) | Detection Frequency (%) | Median (ng/L) | References |
|---|---|---|---|---|---|---|---|---|---|---|---|
| | surface | N | - | - | - | unknown | - | <0.4 | - | - | ([Gwrc] 2003) |
| | surface | UK | - | - | - | unknown | - | <0.2 | - | - | ([Gwrc] 2003) |
| | surface | G | - | - | - | unknown | - | <1 | - | - | ([Gwrc] 2003) |
| | surface | US | - | - | - | unknown | - | - | - | 9 | ([Gwrc] 2003) |
| | surface | UK | 5 | 5 | 5 | unknown | - | <0.3-25 | 40 | <0.3 | (Fawell et al. 2001) |
| | surface | US | 2 | 8 | 8 | yes | ds | <0.5 | 0 | <0.5 | (Barber et al. 2003) |
| | surface | US | 2 | 8 | 8 | yes | ds | <0.5 | 0 | <0.5 | (Barber et al. 2003) |
| Estrone | surface | G | 2 | 2 | 31 | yes | ds, 1 | <0.1-4.1 | 94 | 0.4 | (Kuch and Ballschmiter 2001) |
| | surface | G | 2 | 2 | 2 | no | us | - | - | 0.1 | (Kuch and Ballschmiter 2001) |
| | surface | G | 15 | 15 | 15 | unknown | - | <0.5-1.6 | 0 | <0.5 | (Ternes et al. 1999) |
| | surface | N | 11 | 11 | 16 | yes | ds | <0.3-3.4 | 69 | 0.3 | (Belfroid et al. 1999) |
| | surface | US | 70 | 70 | 70 | yes | ds | <5-112 | 7 | <5 | (Kolpin et al. 2002) |
| | surface | I | 1 | 2 | 2 | yes | ds, 1-20 | <0.008-1.5 | 50 | - | (Baronti et al. 2000) |
| | surface | Fr | 2 | 7 | 7 | yes | ds | 1.1-3 | 100 | 1.8 | (Cargouet et al. 2004) |
| | surface | N | - | - | - | unknown | - | <0.4-3.4 | - | - | ([Gwrc] 2003) |
| | surface | UK | - | - | - | unknown | - | <0.8-7.1 | - | - | ([Gwrc] 2003) |
| | surface | G | - | - | - | unknown | - | <1 | - | - | ([Gwrc] 2003) |
| | surface | US | - | - | - | unknown | - | - | - | 27 | ([Gwrc] 2003) |
| | Surfac e | UK | 5 | 5 | 5 | unknown | - | 0.8-7.1 | 100 | 2.5 | (Fawell et al. 2001) |
| | surface | US | 2 | 8 | 8 | yes | ds | <0.5 | 0 | <0.5 | (Barber et al. 2003) |
| | surface | US | 2 | 8 | 8 | yes | ds | <0.5 | 0 | <0.5 | (Barber et al. 2003) |
| 17α-Estradiol | surface | G | 2 | 2 | 31 | yes | ds, 1 | <0.15-2 | 26 | 0.4 | (Kuch and Ballschmiter 2001) |
| | surface | G | 2 | 2 | 2 | no | us | <0.15 | 0 | <0.15 | (Kuch and Ballschmiter 2001) |
| | surface | N | 11 | 11 | 16 | yes | ds | <0.1-3 | 44 | <0.1 | (Belfroid et al. 1999) |
| | surface | US | 70 | 70 | 70 | yes | ds | <5-74 | 6 | <5 | (Kolpin et al. 2002) |
| | surface | N | - | - | - | unknown | - | <0.4 | - | - | ([Gwrc] 2003) |
| | surface | US | 2 | 8 | 8 | yes | ds | <0.5 | 0 | <0.5 | (Barber et al. 2003) |
| | surface | US | 2 | 8 | 8 | yes | ds | <0.5 | 0 | <0.5 | (Barber et al. 2003) |

(continued)

## Table B.1 (Continued)

| Compounds | Water Source | Country | # of Waters | # of Sites | # of Samples* | Wastewater impact (% indicates degree) | Proximity and Distance (km) from Discharge(s) | Concentration Range† (ng/L) | Detection Frequency (%) | Median (ng/L) | References |
|---|---|---|---|---|---|---|---|---|---|---|---|
| ***Androgen*** | | | | | | | | | | | |
| Testosterone | surface | US | 70 | 70 | 70 | yes | ds | <5-214 | 3 | <5 | (Kolpin et al. 2002) |
| **Personal Care Products** | | | | | | | | | | | |
| ***Surfactant*** | | | | | | | | | | | |
| 4-Nonylphenol | surface | G | 2 | 2 | 31 | yes | ds, 1 | 6.7-134 | 100 | 23 | (Kuch and Ballschmiter 2001) |
| | surface | G | 2 | 2 | 2 | no | us | - | - | 15 | (Kuch and Ballschmiter 2001) |
| | surface | US | 10 | 23 | 76 | yes | ds, us | <5000 | 4 | <5000 | (Kolpin et al. 2004) |
| | surface | US | 85 | 85 | 85 | yes | ds | <500-40000 | 51 | <500 | (Kolpin et al. 2002) |
| | surface | N | - | - | - | unknown | - | <100-710 | - | - | ([Gwrc] 2003) |
| | surface | UK | - | - | - | unknown | - | <200 | - | - | ([Gwrc] 2003) |
| | surface | G | - | - | - | unknown | - | <5-3300 | - | - | ([Gwrc] 2003) |
| | surface | US | - | - | - | unknown | - | - | - | 800 | ([Gwrc] 2003) |
| | surface | US | 2 | 13 | 14 | yes | ds | <11-1190 | 79 | 693 | (Snyder et al. 1999) |
| | surface | US | | | 30 | unknown | - | <0.11-0.64 | | - | (Naylor et al. 1992) |
| | surface | G | 5 | 8 | 23 | yes | ds | <50-485 | 70 | - | (Bolz, Hagenmaier, and Körner 2001) |
| | surface | J | 2 | 2 | 20 | yes | ds | <20-300 | 98 | - | (Tsuda et al. 2002) |
| | surface | C | 6 | 35 | 35 | unknown | - | <10-920 | 24 | <10 | (Bennie et al. 1997) |
| | surface | Sw | 1 | 1 | 4 | yes | ds | 700-9000 | 100 | 3775 | (Ahel, Schaffner, and Giger 1996) |
| | ground | Sw | 1 | 4 | 15 | yes | dg, 0-0.013 | <100-3500 | 80 | 500 | (Ahel, Schaffner, and Giger 1996) |
| | surface | Sw | 1 | 1 | 12 | yes | ds | 700-2700 | 100 | 1800 | (Ahel, Schaffner, and Giger 1996) |
| | ground | Sw | 1 | 1 | 12 | yes | dg, 0.05 | <100-200 | 100 | <100 | (Ahel, Schaffner, and Giger 1996) |
| | surface | UK | 5 | 29 | 29 | yes | ds | <1900-53000 | 59 | <1900 | (Blackburn and Waldock 1995) |
| | surface | UK | 5 | 5 | 5 | unknown | - | <200 | 0 | <200 | (Fawell et al. 2001) |
| | surface | Sp | 2 | 2 | 2 | unknown | us, 1.5-5 | 18000-51000 | 100 | 35000 | (Sole et al. 2000) |
| | surface | Sp | 2 | 4 | 4 | yes | ds, 4-27 | <150-644000 | 75 | 220000 | (Sole et al. 2000) |

(continued)

## Table B.1 (Continued)

| Compounds | Water Source | Country | # of Waters | # of Sites | # of Samples* | Wastewater impact (% indicates degree) | Proximity and Distance (km) from Discharge(s) | Concentration Range† (ng/L) | Detection Frequency (%) | Median (ng/L) | References |
|---|---|---|---|---|---|---|---|---|---|---|---|
| | surface | US | 2 | 8 | 8 | yes | ds | 12-340 | 100 | 43 | (Barber et al. 2003) |
| | surface | US | 2 | 8 | 8 | yes | ds | 11-180 | 100 | 53 | (Barber et al. 2003) |
| ***Antioxidant*** | | | | | | | | | | | |
| BHA | surface | US | 1 | 4 | 13 | unknown | - | <70-3520 | 15 | <70 | (Loraine and Pettigrove 2006) |
| | surface | US | 2 | 8 | 8 | yes | ds | <0.5 | 0 | <0.5 | (Barber et al. 2003) |
| | surface | US | 2 | 8 | 8 | yes | ds | <0.5 | 0 | <0.5 | (Barber et al. 2003) |
| | surface | US | 2 | 3 | 12 | yes, 10-70% | ds | <5000 | 0 | <5000 | (Stackelberg et al. 2004) |
| ***Antimicrobials*** | | | | | | | | | | | |
| Triclosan | surface | US | 9 | 9 | 9 | unknown | us, 0.006-49 | <1000 | 11 | <1000 | (Glassmeyer et al. 2005) |
| | surface | US | 10 | 20 | 20 | yes, 10-95% | ds, 0.1-97 | <1000 | 65 | <1000 | (Glassmeyer et al. 2005) |
| | surface | UK | 1 | 1 | 3 | unknown | us, 0.05 | - | 100 | 19 | (Sabaliunas et al. 2003) |
| | surface | UK | 1 | 4 | 12 | yes | ds, 0.02-3.5 | - | 100 | 55 | (Sabaliunas et al. 2003) |
| | surface | US | 10 | 23 | 76 | yes | ds, us | <1000 | 4 | <1000 | (Kolpin et al. 2004) |
| | surface | US | 85 | 85 | 85 | yes | ds | <50-2300 | 58 | <50 | (Kolpin et al. 2002) |
| | surface | Sw | 2 | 2 | 24 | yes | ds | 11-98 | 100 | 20 | (Singer et al. 2002) |
| | surface | Sw | 1 | 7 | 26 | yes | ds | 4-12 | 100 | 7 | (Tixier et al. 2002) |
| | surface | Sw | 3 | 3 | 12 | yes | ds | 2.3-74 | 100 | 10 | (Lindstrom et al. 2002) |
| | surface | US | 2 | 3 | 12 | yes, 10-70% | ds | <1000 | 67 | <1000 | (Stackelberg et al. 2004) |
| | surface | US | 1 | 4 | 13 | unknown | - | <125-734 | 8 | <125 | (Loraine and Pettigrove 2006) |
| | surface | US | - | - | - | unknown | - | - | - | 140 | ([Gwrc] 2003) |
| | surface | US | 2 | 8 | 8 | yes | ds | <0.5-170 | 63 | 10 | (Barber et al. 2003) |
| | surface | US | 2 | 8 | 8 | yes | ds | <0.5-45 | 38 | <0.5 | (Barber et al. 2003) |
| Triclocarban | surface | US | 6 | - | 28 | unknown | us | <30-9000 | 68 | 100 | (Halden and Paull 2004) |

(continued)

## Table B.1 (Continued)

| Compounds | Water Source | Country | # of Waters | # of Sites | # of Samples* | Wastewater impact (% indicates degree) | Proximity and Distance (km) from Discharge(s) | Concentration Range† (ng/L) | Detection Frequency (%) | Median (ng/L) | References |
|---|---|---|---|---|---|---|---|---|---|---|---|
| ***Insecticide*** | | | | | | | | | | | |
| DEET | surface | US | 2 | 3 | 12 | yes, 10-70% | ds | <500 | 25 | <500 | (Stackelberg et al. 2004) |
| | surface | US | 1 | 4 | 13 | unknown | - | <82-131 | 8 | <82 | (Loraine and Pettigrove 2006) |
| | surface | US | 9 | 9 | 9 | unknown | us, 0.006-49 | <500 | 33 | <500 | (Glassmeyer et al. 2005) |
| | surface | US | 10 | 20 | 20 | yes, 10-95% | ds, 0.1-97 | <500-640 | 81 | 500 | (Glassmeyer et al. 2005) |
| | surface | US | 10 | 23 | 76 | yes | ds, us | nd-130 | 4 | - | (Kolpin et al. 2004) |
| | surface | G | 1 | 1 | 1 | no | us | <20 | 0 | <20 | (Wiegel et al. 2004) |
| | surface | G | 9 | 15 | 15 | yes | ds | <20-506 | 13 | <20 | (Wiegel et al. 2004) |
| ***Fragrances*** | | | | | | | | | | | |
| Galaxolide | surface | US | 2 | 3 | 12 | yes, 10-70% | ds | <500 | 92 | <500 | (Stackelberg et al. 2004) |
| | surface | US | 9 | 9 | 9 | unknown | us, 0.006-49 | <500 | 11 | <500 | (Glassmeyer et al. 2005) |
| | surface | US | 10 | 20 | 20 | yes, 10-95% | ds, 0.1-97 | <500 | 55 | <500 | (Glassmeyer et al. 2005) |
| | surface | G/Fr/Sp | 4 | 4 | 4 | unknown | - | 36-329 | 100 | 55 | (Mitjans and Ventura 2004) |
| | surface | US | 10 | 23 | 76 | yes | ds, us | <500 | 9 | <500 | (Kolpin et al. 2004) |
| | surface | G | 12 | 27 | 27 | yes | ds | 20-12500 | 100 | 825 | (Heberer, Gramer, and Stan 1999) |
| Tonalide | surface | US | 9 | 9 | 9 | unknown | us, 0.006-49 | <500 | 22 | <500 | (Glassmeyer et al. 2005) |
| | surface | US | 10 | 20 | 20 | yes, 10-95% | ds, 0.1-97 | <500-2100 | 95 | <500 | (Glassmeyer et al. 2005) |
| | surface | US | 2 | 3 | 12 | yes, 10-70% | ds | <500 | 100 | <500 | (Stackelberg et al. 2004) |
| | surface | US | 10 | 23 | 76 | yes | ds, us | <500-1200 | 20 | <500 | (Kolpin et al. 2004) |
| | surface | G/Fr/Sp | 4 | 4 | 4 | unknown | - | <7-62 | 25 | <7 | (Mitjans and Ventura 2004) |
| | surface | G | 12 | 27 | 27 | yes | ds | 30-6800 | 100 | 500 | (Heberer, Gramer, and Stan 1999) |

(continued)

## Table B.1 (Continued)

| Compounds | Water Source | Country | # of Waters | # of Sites | # of Samples* | Wastewater impact (% indicates degree) | Proximity and Distance (km) from Discharge(s) | Concentration Range† (ng/L) | Detection Frequency (%) | Median (ng/L) | References |
|---|---|---|---|---|---|---|---|---|---|---|---|
| **Other** | | | | | | | | | | | |
| ***Flame retardants*** | | | | | | | | | | | |
| TCEP | surface | US | 9 | 9 | 9 | unknown | us, 0.006-49 | <500 | 33 | <500 | (Glassmeyer et al. 2005) |
| | surface | US | 10 | 20 | 20 | yes, 10-95% | ds, 0.1-97 | <500 | 80 | <500 | (Glassmeyer et al. 2005) |
| | surface | US | 2 | 3 | 12 | yes, 10-70% | ds | <500 | 100 | <500 | (Stackelberg et al. 2004) |
| | surface | US | 10 | 23 | 76 | yes | ds, us | <500 | 15 | <500 | (Kolpin et al. 2004) |
| | surface | US | 85 | 85 | 85 | yes | ds | <40-540 | 58 | <40 | (Kolpin et al. 2002) |
| | surface | G | 6 | 9 | 11 | yes | ds | <1-220 | 82 | 24 | (Fries and Puettmann 2001) |
| | ground | G | 1 | 1 | 45 | yes | dg | <1-195 | n/a | 9 | (Fries and Puettmann 2001) |
| | ground | G | 1 | 1 | 45 | yes | dg | <1-754 | n/a | 50 | (Fries and Puettmann 2001) |
| | surface | G | 4 | 4 | 9 | yes | ds | <1-1236 | 89 | 220 | (Fries and Puettmann 2003) |
| | ground | G | 2 | 5 | 6 | yes | dg | <1-312 | 67 | 84 | (Fries and Puettmann 2003) |
| TCPP | surface | US | 9 | 9 | 9 | unknown | us, 0.006-49 | <500 | 22 | <500 | (Glassmeyer et al. 2005) |
| | surface | US | 10 | 20 | 20 | yes, 10-95% | ds, 0.1-97 | <500 | 90 | <500 | (Glassmeyer et al. 2005) |
| | surface | G | 11 | 11 | 11 | yes | ds, 5-10 | 150-880 | 100 | 530 | (Ternes, Bonerz, and Schmidt 2001) |
| | surface | US | 10 | 23 | 76 | yes | ds, us | <500 | 7 | <500 | (Kolpin et al. 2004) |
| | surface | US | 85 | 85 | 85 | yes | ds | <100-160 | 13 | <100 | (Kolpin et al. 2002) |
| | Surfac e | US | 2 | 3 | 12 | yes, 10-70% | ds | <500 | 100 | <500 | (Stackelberg et al. 2004) |
| ***Stimulant*** | | | | | | | | | | | |
| Caffeine | surface | US | 9 | 9 | 9 | unknown | us, 0.006-49 | <16-807 | 67 | 40 | (Glassmeyer et al. 2005) |
| | surface | US | 10 | 20 | 20 | yes, 10-95% | ds, 0.1-97 | <16-2600 | 77 | 46[d] | (Glassmeyer et al. 2005) |
| | surface | Sw | 2 | 2 | 2 | no | us | <2 | 0 | <2 | (Buerge et al. 2003) |
| | surface | Sw | 14 | 19 | 25 | yes | ds | 6-250 | 100 | 47 | (Buerge et al. 2003) |
| | surface | US | 2 | 3 | 12 | yes, 10-70% | ds | 40-250 | 100 | 100 | (Stackelberg et al. 2004) |
| | surface | US | 10 | 23 | 76 | yes | ds, us | <14-1390 | 72 | - | (Kolpin et al. 2004) |
| | surface | US | 84 | 84 | 84 | yes | ds | <23-6000 | 62 | 81 | (Kolpin et al. 2002) |

(continued)

## Table B.1 (Continued)

| Compounds | Water Source | Country | # of Waters | # of Sites | # of Samples* | Wastewater impact (% indicates degree) | Proximity and Distance (km) from Discharge(s) | Concentration Range† (ng/L) | Detection Frequency (%) | Median (ng/L) | References |
|---|---|---|---|---|---|---|---|---|---|---|---|
| | ground | US | 2 | 12 | 12 | unknown | - | <5000 | 0 | <5000 | (Seiler et al. 1999) |
| | ground | US | 3 | 14 | 14 | unknown | - | <40 | 0 | <40 | (Seiler et al. 1999) |
| | surface | US | 2 | 8 | 8 | yes | ds | 11-170 | 100 | 32 | (Barber et al. 2003) |
| | surface | US | 2 | 8 | 8 | yes | ds | <0.5-350 | 100 | 28 | (Barber et al. 2003) |
| | surface | G | 2 | 15 | 45 | yes | ds | 80-265 | 100 | - | (Heberer, Reddersen, and Mechlinski 2002) |
| ***Plasticizers*** | | | | | | | | | | | |
| Bisphenol A | surface | US | 85 | 85 | 85 | yes | ds | <90-12000 | 41 | <90 | (Kolpin et al. 2002) |
| | surface | US | 10 | 23 | 76 | yes | ds, us | <1000 | 8 | <1000 | (Kolpin et al. 2004) |
| | surface | US | 9 | 9 | 9 | unknown | us, 0.006-49 | <1000 | 11 | <1000 | (Glassmeyer et al. 2005) |
| | surface | US | 10 | 20 | 20 | yes, 10-95% | ds, 0.1-97 | <1000 | 25 | <1000 | (Glassmeyer et al. 2005) |
| | surface | J | 2 | 2 | 14 | no | us | nd-3 | 14 | - | (Suzuki et al. 2004) |
| | surface | J | 2 | 4 | 35 | yes | ds | nd-230 | 97 | - | (Suzuki et al. 2004) |
| | surface | G | 2 | 2 | 31 | yes | ds, 1 | 0.5-14 | 100 | 3.8 | (Kuch and Ballschmiter 2001) |
| | surface | G | 2 | 2 | 2 | no | us | <0.04 | 0 | <0.04 | (Kuch and Ballschmiter 2001) |
| | surface | US | 2 | 3 | 12 | yes, 10-70% | ds | <1000 | 100 | <1000 | (Stackelberg et al. 2004) |
| | surface | N | - | - | - | unknown | - | <10-2200 | - | - | ([Gwrc] 2003) |
| | surface | UK | - | - | - | unknown | - | <5100 | - | - | ([Gwrc] 2003) |
| | surface | G | - | - | - | unknown | - | 5-200 | - | - | ([Gwrc] 2003) |
| | surface | US | - | - | - | unknown | - | - | - | 140 | ([Gwrc] 2003) |
| | surface | G | 5 | 8 | 23 | yes | ds | <50-272 | 100 | <50 | (Bolz, Hagenmaier, and Körner 2001) |
| | surface | J | 2 | 2 | 20 | yes | ds | <10-100 | 27 | - | (Tsuda et al. 2002) |
| | surface | G | - | 116 | 116 | yes | ds | 5-229 | 100 | 10 | (Fromme et al. 2002) |
| | surface | UK | 5 | 5 | 5 | unknown | - | <5100 | 0 | <5100 | (Fawell et al. 2001) |
| | surface | G | 1 | 1 | 1 | no | us | 16 | 100 | 16 | (Wiegel et al. 2004) |
| | surface | G | 7 | 13 | 13 | yes | ds | 16-100 | 100 | 26 | (Wiegel et al. 2004) |
| | surface | US | 2 | 8 | 8 | yes | ds | <0.5-150 | 85 | 11 | (Barber et al. 2003) |

(continued)

## Table B.1 (Continued)

| Compounds | Water Source | Country | # of Waters | # of Sites | # of Samples* | Wastewater impact (% indicates degree) | Proximity and Distance (km) from Discharge(s) | Concentration Range† (ng/L) | Detection Frequency (%) | Median (ng/L) | References |
|---|---|---|---|---|---|---|---|---|---|---|---|
| Di-n-butyl phthalate | surface | US | 2 | 8 | 8 | yes | ds | <0.5-72 | 70 | 5.6 | (Barber et al. 2003) |
| | surface | N | - | - | - | unknown | - | 10-1900 | - | - | ([Gwrc] 2003) |
| | surface | UK | - | - | - | unknown | - | <200 | - | - | ([Gwrc] 2003) |
| | surface | G | - | - | - | unknown | - | 30-700 | - | - | ([Gwrc] 2003) |
| | surface | SA | - | - | - | unknown | - | 40-76000 | - | - | ([Gwrc] 2003) |
| | surface | G | - | 116 | 116 | yes | ds | <20-8800 | 100 | 500 | (Fromme et al. 2002) |
| | surface | US | 1 | 4 | 13 | unknown | - | <1350-8340 | 31 | <1350 | (Loraine and Pettigrove 2006) |

*Total number of samples, including temporal samples

†Lower limit is based on the minimum concentration or reported reporting limit, minimum detection limit, limit of detection or limit of quantification

‡100% detection; minimum not given

§Average of two medians

**Average median

us – upstream; ds- downstream; dg – downgradient; US – United States; G – Germany; A- Austria; Sp – Spain; Sw – Switzerland; F – Finland; UK – United Kingdom; I – Italy; SA – South Africa; Fr – France; N – Netherlands; J – Japan; n.d. – not detected

**Table B.2**
**Occurrence of select EDCs and PPCPs in raw and finished drinking water**

| Compounds | Treatment Train | Raw Water (ng/L) | # of Finished Water Samples | Finished Water (ng/L) | Median (ng/L) | Method | References |
|---|---|---|---|---|---|---|---|
| **Pharmaceuticals** | | | | | | | |
| ***Analgesics*** | | | | | | | |
| Diclofenac | Bank filtration | n.d.-500 | 25 | n.d.-25 | n.d. | - | (Kuehn and Mueller 2000) |
| | Bank filtration | 50 | 1 | n.d. | n.d. | GC-MS w/ der. | (Heberer et al. 2001) |
| | Pre$O_3$/Coag./$O_3$/GAC | 35 | 1 | - | <2 | GC-MS | (Ternes et al. 2002) |
| | Coag./GAC | 65 | 1 | - | <2 | GC-MS | (Ternes et al. 2002) |
| | Bank filtration | 65 | 1 | - | <2 | GC-MS | (Ternes et al. 2002) |
| | Coag./GAC/$Cl_2$ | 8.5 | 2 | | <1 | LC-MS-MS | (Vieno, Tuhkanen, and Kronberg 2005) |
| | 3 Coag. & 1 Coag./$O_3$/Filt. | - | 15 | <280-930 | <280 | GC-MS | (Loraine and Pettigrove 2006) |
| Naproxen | Coag./GAC/$Cl_2$ | 7.25 | 2 | | <5 | LC-MS-MS | (Vieno, Tuhkanen, and Kronberg 2005) |
| Ketoprofen | Coag./GAC/$Cl_2$ | 9.5 | 2 | | <5 | LC-MS-MS | (Vieno, Tuhkanen, and Kronberg 2005) |
| ***Anti-convulsants*** | | | | | | | |
| Carbamazepine | Bank filtration | n.d.-550 | 25 | n.d.-200 | 50 | - | (Kuehn and Mueller 2000) |
| | Bank filtration | 235 | 1 | 20 | 20 | GC-MS w/ der. | (Heberer et al. 2001) |
| | PAC/Coag./$Cl_2$/Filt. | 60-400 | 4 | 60-250 | 120 | GC-MS | (Stackelberg et al. 2004) |

(continued)

**Table B.2 (Continued)**

| Compounds | Treatment Train | Raw Water (ng/L) | # of Finished Water Samples | Finished Water (ng/L) | Median (ng/L) | Method | References |
|---|---|---|---|---|---|---|---|
| | $PreO_3$/Coag./$O_3$/GAC | 80 | 1 | - | <2 | GC-MS | (Ternes et al. 2002) |
| | Coag./GAC | 140-180 | 2 | - | <2 | GC-MS | (Ternes et al. 2002) |
| | Bank filtration | 140-180 | 1 | - | 20-50 | GC-MS | (Ternes et al. 2002) |
| Primidone | Bank filtration | 105 | 1 | 15 | 15 | GC-MS w/ der. | (Heberer et al. 2001) |
| | Coag./GAC | 15 | 1 | - | 5 | GC-MS | (Ternes et al. 2002) |
| | Bank filtration | 15 | 1 | - | 10 | GC-MS | (Ternes et al. 2002) |
| ***Lipid regulators*** | | | | | | | |
| Clofibric acid | $PreO_3$/Coag./$O_3$/GAC | 10 | 1 | - | <2 | GC-MS | (Ternes et al. 2002) |
| | Coag./GAC | 10 | 1 | - | 5 | GC-MS | (Ternes et al. 2002) |
| | Bank filtration | 10 | 1 | - | <2 | GC-MS | (Ternes et al. 2002) |
| | 3 Coag. & 1 Coag./$O_3$/Filt. | - | 15 | <130 | <130 | GC-MS | (Loraine and Pettigrove 2006) |
| | Bank filtration | 50 | 1 | 70 | 70 | GC-MS w/ der. | (Heberer et al. 2001) |
| Bezafibrate | Coag./GAC | 80 | 1 | - | <10 | GC-MS | (Ternes et al. 2002) |
| | Bank filtration | 80 | 1 | - | <10 | GC-MS | (Ternes et al. 2002) |
| | Bank filtration | 170 | 1 | 100 | 100 | GC-MS w/ der. | (Heberer et al. 2001) |

(continued)

**Table B.2 (Continued)**

| Compounds | Treatment Train | Raw Water (ng/L) | # of Finished Water Samples | Finished Water (ng/L) | Median (ng/L) | Method | References |
|---|---|---|---|---|---|---|---|
| ***X-ray contrast media*** | | | | | | | |
| Iopromide | Bank filtration | 1600 | 1 | - | <50 | LC-MS | (Putschew, Wischnack, and Jekel 2000) |
| | Bank filtration | 860 | 1 | - | <20 | LC-MS | (Schittko, Putschew, and Jekel 2004) |
| Diatrizoate | Bank filtration | 2000 | 1 | - | 4000 | LC-MS | (Putschew, Wischnack, and Jekel 2000) |
| | Bank filtration | 960 | 1 | - | 166 | LC-MS | (Schittko, Putschew, and Jekel 2004) |
| **Steroids** | | | | | | | |
| ***Estrogens*** | | | | | | | |
| 17α-Ethinylestradiol | unknown | - | 10 | 0.15-0.5 | 0.35 | GC-MS w/ der. | (Kuch and Ballschmiter 2001) |
| | unknown | - | - | <1 | - | - | ([Gwrc] 2003) |
| | unknown | <0.2 | 5 | <.2 | <.2 | LC-MS-MS | (Verstraeten et al. 2003) |
| 17β-Estradiol | unknown | - | 10 | 0.2-2.1 | 0.3 | GC-MS w/ der. | (Kuch and Ballschmiter 2001) |
| | unknown | - | - | <1 | - | - | ([Gwrc] 2003) |
| | unknown | 25 | 1 | - | <0.3 | GC-MS-MS | (Fawell et al. 2001) |
| | unknown | <.2 | 5 | <.2 | <.2 | LC-MS-MS | (Verstraeten et al. 2003) |

(continued)

**Table B.2 (Continued)**

| Compounds | Treatment Train | Raw Water (ng/L) | # of Finished Water Samples | Finished Water (ng/L) | Median (ng/L) | Method | References |
|---|---|---|---|---|---|---|---|
| Estrone | unknown | - | 10 | 0.2-0.6 | 0.4 | GC-MS w/ der. | (Kuch and Ballschmiter 2001) |
| | unknown | - | - | <1 | - | - | ([Gwrc] 2003) |
| | unknown | 0.86 | 5 | - | 0.16 | LC-MS-MS | (Verstraeten et al. 2003) |
| 17α-Estradiol | unknown | - | 10 | 0.3-0.3 | 0.3 | GC-MS w/ der. | (Kuch and Ballschmiter 2001) |
| | unknown | - | - | <0.4 | - | - | ([Gwrc] 2003) |
| | unknown | 0.8 | 1 | - | <0.3 | GC-MS-MS | (Fawell et al. 2001) |
| | unknown | 1.8 | 1 | - | <0.3 | GC-MS-MS | (Fawell et al. 2001) |
| **Personal Care Products** | | | | | | | |
| ***Surfactant*** | | | | | | | |
| 4-Nonylphenol | unknown | - | 10 | 2.5-16 | 6.6 | GC-MS w/ der. | (Kuch and Ballschmiter 2001) |
| | unknown | - | - | <200 | - | - | ([Gwrc] 2003) |
| | $PreCl_2$/Coag./Filt./$O_3$/GAC/$Cl_2$ | 1100-2200 | 4 | 25-90 | 55 | LC-MS-MS | (Petrovic et al. 2003) |
| ***Fragrances*** | | | | | | | |
| HHCB | PAC/Coag./$Cl_2$/Filt. | 62-105 | 4 | 60-80 | 75 | GC-MS | (Stackelberg et al. 2004) |
| | unknown | - | 4 | <6 | <6 | CLSA-GC-MS | (Mitjans and Ventura 2004) |

(continued)

**Table B.2 (Continued)**

| Compounds | Treatment Train | Raw Water (ng/L) | # of Finished Water Samples | Finished Water (ng/L) | Median (ng/L) | Method | References |
|---|---|---|---|---|---|---|---|
| AHTN | PAC/Coag./$Cl_2$/Filt. | 320-620 | 4 | 105-320 | 110 | GC-MS | (Stackelberg et al. 2004) |
| | unknown | - | 4 | <7 | <7 | CLSA-GC-MS | (Mitjans and Ventura 2004) |
| ***Antioxidant*** | | | | | | | |
| BHA | 3 Coag. & 1 Coag/$O_3$/Filt. | - | 15 | <70-3450 | <70 | GC-MS | (Loraine and Pettigrove 2006) |
| ***Insecticide*** | | | | | | | |
| DEET | 3 Coag. & 1 Coag/$O_3$/Filt. | - | 15 | <82 | <82 | GC-MS | (Loraine and Pettigrove 2006) |
| **Other** | | | | | | | |
| ***Flame retardants*** | | | | | | | |
| TCEP | PAC/Coag./$Cl_2$/Filt. | 82-100 | 4 | 70-100 | 90 | GC-MS | (Stackelberg et al. 2004) |
| | Filtration/Bank Filtration | 12-130 | 5 | 0.74-1.7 | 1.2 | GC-MS | (Andersen and Bester 2006) |
| | $O_3$/Filt./$O_3$/GAC/UV | 35 | 1 | <0.3 | <0.3 | GC-MS | (Andersen and Bester 2006) |
| | Coag./$O_3$/Filt./GAC | 72 | 1 | <0.3 | <0.3 | GC-MS | (Andersen and Bester 2006) |
| TCPP | PAC/Coag./$Cl_2$/Filt. | 180-210 | 4 | 110-230 | 150 | GC-MS | (Stackelberg et al. 2004) |
| | Filtration/Bank Filtration | 47-65 | 5 | <1-4.1 | 2.3 | GC-MS | (Andersen and Bester 2006) |

(continued)

Table B.2 (Continued)

| Compounds | Treatment Train | Raw Water (ng/L) | # of Finished Water Samples | Finished Water (ng/L) | Median (ng/L) | Method | References |
|---|---|---|---|---|---|---|---|
| | $O_3$/Filt./$O_3$/GAC/UV | 95 | 1 | <1 | <1 | GC-MS | (Andersen and Bester 2006) |
| | Coag./$O_3$/Filt./GAC | 72 | 1 | <1 | <1 | GC-MS | (Andersen and Bester 2006) |
| ***Stimulant*** | | | | | | | |
| Caffeine | PAC/Coag./$Cl_2$/Filt. | 98-220 | 4 | 48-105 | 60 | LC-MS | (Stackelberg et al. 2004) |
| ***Plasticizers*** | | | | | | | |
| Bisphenol A | PAC/Coag./$Cl_2$/Filt. | 130-320 | 4 | 300-500 | 400 | GC-MS | (Stackelberg et al. 2004) |
| | unknown | - | 10 | 0.5-2 | 1.1 | GC-MS w/ der. | (Kuch and Ballschmiter 2001) |
| | unknown | - | - | <5.1 | - | - | ([Gwrc] 2003) |
| Di-n-butyl phthalate | unknown | - | 1 | - | 1040 | GC-MS | (Psillakis and Kalogerakis 2003) |
| | 3 Coag. & 1 Coag/$O_3$/Filt. | - | 15 | <1350-2730 | <1350 | GC-MS | (Loraine and Pettigrove 2006) |
| | unknown | - | 2 | 380-640 | 510 | GC-MS | (Luks-Betlej et al. 2001) |
| | unknown | - | - | <10-1100 | - | - | ([Gwrc] 2003) |

## REFERENCES


(Gwrc), G.W.R.C. 2003. Endocrine disrupting compounds: Occurrence of EDC in Water Systems. London, United Kingdom: Alliance House.

Adams, C., Y. Wang, K. Loftin, and M. Meyer. 2002. Removal of antibiotics from surface and distilled water in conventional water treatment processes. Journal of Environmental Engineering, 128(3):253-260.

Agenson, K.O., J.-I. Oh, and T. Urase. 2003. Retention of a wide variety of organic pollutants by different nanofiltration/reverse osmosis membranes: controlling parameters of process. Journal of Membrane Science, 225(1-2):91-103.

Ahel, M., C. Schaffner, and W. Giger. 1996. Behavior of alkylphenol polyethoxylate surfactants in the aquatic environment - III. Occurrence and elimination of their persistent metabolites during infiltration of river water to groundwater. Water Research, 30(1): 37-46.

Ahrer, W., E. Scherwenk, and W. Buchberger. 2001. Determination of drug residues in water by the combination of liquid chromatography or capillary electrophoresis with electrospray mass spectrometry. Journal of Chromatography A, 910:69-78.

Andersen, J., and K. Bester. 2006. Elimination of organophosphate ester flame retardants and plasticizers in drinking water purification. Water Research, 40:621-629.

Ashton, D., M. Hilton, and K. Thomas. 2004. Investigating the environmental transport of human pharmaceuticals to streams in the United Kingdom. The Science of the Total Environment, 333:167-184.

Barber, L.B., E.T. Furlong, S.H. Keefe, G.M. Brown, and J.D. Cahill. 2003. Natural and contaminant organic compounds in the Boulder Creek watershed, Colorado during high-flow and low-flow conditions, 2000. In Comprehensive water quality of the Boulder Creek watershed, Colorado during high-flow and low-flow conditions, 2000. Edited by Denver, Colorado: USGS.

Baronti, C., R. Curini, G. D'ascenzo, A. Di Corcia, A. Gentili, and R. Samperi. 2000. Monitoring natural and synthetic estrogens at activated sludge sewage treatment plants and in a receiving river water. Environmental Science & Technology, 34(24):5059-5066.

Belfroid, A.C., A. Van Der Horst, A.D. Vethaak, A.J. Schafer, G.B.J. Rijs, J. Wegener, and W.P. Cofino. 1999. Analysis and occurrence of estrogenic hormones and their glucuronides in surface water and waste water in The Netherlands. The Science of the Total Environment, 225:101-108.

Bennie, D.T., C.A. Sullivan, H.B. Lee, T.E. Peart, and R.J. Maguire. 1997. Occurence of alkylphenols and alkylphenol mono- and diethoxylates in natural waters of the Laurentian Great Lakes basin and the upper St. Lawrence River. The Science of the Total Environment, 193:263-275.

Blackburn, M.A., and M.J. Waldock. 1995. Concentrations of alkylphenols in rivers and estuaries in England and Wales. Water Research, 29(7):1623-1629.

Bolz, U., H. Hagenmaier, and W. Körner. 2001. Phenolic xenoestrogens in surface water, sediments, and sewage sludge from Baden-Württemberg, south-west Germany. Environmental Pollution, 115(2):291-301.

Buerge, I.J., T. Poiger, M.D. Mueller, and H.-R. Buser. 2003. Caffeine, an Anthropogenic Marker for Wastewater Contamination of Surface Waters. Environmental Science and Technology, 37(4):691-700.

Buser, H.R., T. Poiger, and M.D. Muller. 1998. Occurrence and fate of the pharmaceutical drug diclofenac in surface waters: Rapid photodegradation in a lake. Environmental Science & Technology, 32(22):3449-3456.

Calamari, D., E. Zuccato, S. Castiglioni, R. Bagnati, and R. Fanelli. 2003. Strategic Survey of Therapeutic Drugs in the Rivers Po and Lambro in Northern Italy. Environmental Science and Technology, 37(7):1241-1248.

Cargouet, M., D. Perdiz, A. Mouatassim-Souali, S. Tamisier-Karolak, and Y. Levi. 2004. Assessment of river contamination by estrogenic compounds in Paris area (France). Science Of The Total Environment, 324(1-3):55-66.

Deng, A., M. Himmelsbach, Q.-Z. Zhu, S. Frey, M. Sengl, W. Buchberger, R. Niessner, and D. Knopp. 2003. Residue Analysis of the Pharmaceutical Diclofenac in Different Water Types Using ELISA and GC-MS. Environmental Science and Technology, 37(15):3422-3429.

Farre, M., I. Ferrer, A. Ginebreda, M. Figueras, L. Olivella, L. Tirapu, M. Vilanova, and D. Barcelo. 2001. Determination of drugs in surface water and wastewater samples by liquid chromatography-mass spectrometry: methods and preliminary results including toxicity studies with Vibrio fischeri. Journal of Chromatography, A, 938(1-2):187-197.

Fawell, J.K., D. Sheahan, H.A. James, M. Hurst, and S. Scott. 2001. Oestrogens and oestrogenic activity in raw and treated Severn Trent water. Water Research, 35(5):1240-1244.

Fono, L.J., and D.L. Sedlak. 2005. Use of the Chiral Pharmaceutical Propranolol to Identify Sewage Discharges into Surface Waters. Environmental Science & Technology, 39:9244-9252.

Fries, E., and W. Puettmann. 2001. Occurrence of organophosphate esters in surface water and ground water in Germany. Journal of Environmental Monitoring, 3(6):621-626.

Fries, E., and W. Puettmann. 2003. Monitoring of the three organophosphate esters TBP, TCEP and TBEP in river water and ground water (Oder, Germany). Journal of Environmental Monitoring, 5(2):346-352.

Fromme, H., T. Kuchler, T. Otto, K. Pilz, J. Muller, and A. Wenzel. 2002. Occurrence of phthalates and bisphenol A and F in the environment. Water Research, 36(6):1429-1438.

Glassmeyer, S.T., E.T. Furlong, D.W. Kolpin, J.D. Cahill, S.D. Zaugg, S.L. Werner, M.T. Meyer, and D.D. Kryak. 2005. Transport of chemical and microbial compounds from known wastewater discharges: potential for use as indicators of human fecal contamination. Environmental Science & Technology, 39:5157-5169.

Halden, R., and D. Paull. 2004. Analysis of triclocarban in aquatic samples by liquid chromatography electrospray ionization mass spectrometry. Environmental Science & Technology, 38(18):4849-4855.

Hartig, C., T. Storm, and M. Jekel. 1999. Detection and identification of sulphonamide drugs in municipal waste water by liquid chromatography coupled with electrospray ionisation tandem mass spectrometry. Journal of Chromatography A, 854(1-2):163-173.

Heberer, T., U. Dunnbier, C. Reilich, and H.J. Stan. 1997. Detection of drugs and drug metabolites in groundwater samples of a drinking water treatment plant. Fresenius Environmental Bulletin, 6(7-8):438-443.

Heberer, T., K. Schmidt-Baeumler, and H.-J. Stan. 1998. Occurrence and distribution of organic contaminants in the aquatic system in Berlin, Germany. Part 1. Drug residues and other polar contaminants in Berlin surface and groundwater. Acta Hydrochimica Et Hydrobiologica, 26(5):272-278.

Heberer, T., S. Gramer, and H.-J. Stan. 1999. Occurrence and distribution of organic contaminants in the aquatic system in Berlin, Germany. Part 3. Determination of synthetic musks in Berlin surface water applying solid-phase microextraction (SPME) and gas chromatography-mass spectrometry (GC-MS). Acta Hydrochimica Et Hydrobiologica, 27(3):150-156.

Heberer, T., I.M. Verstraeten, M.T. Meyer, A. Mechlinski, and K. Reddersen. 2001. Occurrence and fate of pharmaceuticals during bank filtration-preliminary results from investigations in Germany and the United States. Water Resources Update, 120:4-17.

Heberer, T., K. Reddersen, and A. Mechlinski. 2002. From municipal sewage to drinking water: fate and removal of pharmaceutical residues in the aquatic environment in urban areas. Water Science and Technology, 46(3):81-88.

Heberer, T. 2002. Occurrence, fate, and removal of pharmaceutical residues in the aquatic environment: a review of recent research data. Toxicology Letters, 131(1-2):5-17.

Hilton, M.J., and K.V. Thomas. 2003. Determination of selected human pharmaceutical compounds in effluent and surface water samples by high-performance liquid chromatography-electrospray tandem mass spectrometry. Journal of Chromatography A, 1015(1-2):129-141.

Hirsch, R., T. Ternes, K. Haberer, and K.-L. Kratz. 1999. Occurrence of antibiotics in the aquatic environment. The Science of the Total Environment, 225:109-118.

Huang, C.-H., and D.L. Sedlak. 2001. Analysis of estrogenic hormones in municipal wastewater effluent and surface water using enzyme-linked immunosorbent assay and gas chromatography/tandem mass spectrometry. Environmental Toxicology and Chemistry, 20(1):133-139.

Huber, M.M., S. Canonica, G.-Y. Park, and U. Von Gunten. 2003. Oxidation of pharmaceuticals during ozonation and advanced oxidation processes. Environmental Science & Technology, 37(5):1016-1024.

Kimura, K., G. Amy, J.E. Drewes, T. Heberer, T.-U. Kim, and Y. Watanabe. 2003. Rejection of organic micropollutants (disinfection by-products, endocrine disrupting compounds, and pharmaceutically active compounds) by NF/RO membranes. Journal of Membrane Science, 227(1-2):113-121.

Kimura, K., S. Toshima, G. Amy, and Y. Watanabe. 2004. Rejection of neutral endocrine disprupting compounds (EDCs) and pharmaceutical active compounds (PhACs) by RO membranes. Journal of Membrane Science, 245:71-78.

Kolpin, D.W., E.T. Furlong, M.T. Meyer, E.M. Thurman, S.D. Zaugg, L.B. Barber, and H.T. Buxton. 2002. Pharmaceuticals, hormones, and other organic waste contaminants in U.S. streams, 1999-2000: a national reconnaissance. Environmental Science & Technology, 36(6):1202-1211.

Kolpin, D.W., M. Skopec, M.T. Meyer, E.T. Furlong, and S.D. Zaugg. 2004. Urban contribution of pharmaceuticals and other organic wastewater contaminants to streams during differing flow conditions. Science Of The Total Environment, 328(1-3):119-130.

Kuch, H.M., and K. Ballschmiter. 2001. Determination of endocrine-disrupting phenolic compounds and estrogens in surface and drinking water by HRGC-(NCI)-MS in the picogram per liter range. Environmental Science & Technology, 35(15):3201-3206.

Kuehn, W., and U. Mueller. 2000. Riverbank filtration. An overview. Journal - American Water Works Association, 92(12):60-69.

Lindqvist, N., T. Tuhkanen, and L. Kronberg. 2005. Occurrence of acidic pharmaceuticals in raw and treated sewages and in receiving waters. Water Research, 39:2219-2228.

Lindstrom, A., I.J. Buerge, T. Poiger, P.A. Bergqvist, M.D. Muller, and H.R. Buser. 2002. Occurrence and environmental behavior of the bactericide triclosan and its methyl derivative in surface waters and in wastewater. Environmental Science & Technology, 36(11):2322-2329.

Loraine, G.A., and M.E. Pettigrove. 2006. Seasonal variations in concentrations of pharmaceuticals and personal care products in drinking water and reclaimed wastewater in southern California. Environmental Science & Technology, 40:687-695.

Luks-Betlej, K., P. Popp, B. Janoszka, and H. Paschke. 2001. Solid-phase microextraction of phthalates from water. Journal of Chromatography A, 938:93-101.

Meyer, M., B. J.E., V. J.L., D. J.V., T. E.M., and H. K.A. 2000. Use of radioimmunoassay as a screen for antibiotics in confined animal feeding operations and confirmation by liquid chromatography/mass spectrometry. The Science of the Total Environment, 248:181-187.

Mitjans, D., and F. Ventura. 2004. Determination of musks and other fragrance compounds at ng/L levels using CLSA (closed loop stripping analysis) and GC/MS detection. Water Science and Technology(50):119-123.

Naylor, C.G., J.P. Mieure, W.J. Adams, J.A. Weeks, F.J. Castaldi, L.D. Ogle, and R.R. Romano. 1992. Alkylphenol ethoxylates in the environment. JAOCS, 69(7):695-703.

Ng, H.Y., and M. Elimelech. 2004. Influence of colloidal fouling on rejection of trace organic contaminants by reverse osmosis. Journal of Membrane Science, 244:215-226.

Ngheim, L.D., A.I. Schaefer, and M. Elimelech. 2004. Removal of natural hormones by nanofiltration membranes: measurement, modeling, and mechanisms Environmental Science & Technology, 38:1888-1896.

Nghiem, L., A. Schaefer, and M. Elimelech. 2005a. Nanofiltration of hormone mimicking trace organic contaminants. Separation Science and Technology, 40(13):2633-2649.

Nghiem, L.D., A.I. Schaefer, and M. Elimelech. 2005b. Pharmaceutical retention mechanisms by nanofiltration membranes. Environmental Science & Technology, 39:7698-7705.

Ollers, S., H. Singer, P. Fässler, and S. Müller. 2001. Simultaneous quantification of neutral and acidic pharmaceuticals and pesticides at the low-ng/l level in surface and waste water. Journal of Chromatography A, 911(2):225-234.

Petrovic, M., A. Diaz, F. Ventura, and D. Barceló. 2003. Occurrence and removal of estrogenic short-chain ethoxy nonylphenolic compounds and their halogenated derivatives during drinking water production. Environmental Science & Technology, 37(19):4442-4448.

Psillakis, E., and N. Kalogerakis. 2003. Hollow-fibre liquid-phase microextraction of phthalate esters from water. Journal of Chromatography A, 999:145-153.

Putschew, A., S. Wischnack, and M. Jekel. 2000. Occurrence of triiodinated X-ray contrast agents in the aquatic environment. Science Of The Total Environment, 255(1-3):129-134.

Rosenfeldt, E., and K.G. Linden. 2004. Degradation of endocrine disrupting chemicals bisphenol A, ethinyl estradiol, and estradiol during UV photolysis and advanced oxidation processes. Environmental Science & Technology, 38(20):5476-5483.

Sabaliunas, D., S. Webb, A. Hauk, M. Jacob, and W. Eckhoff. 2003. Environmental fate of Triclosan in the River Aire Basin, UK. Water Resources, 37(13):3145-3154.

Sacher, F., F. Lange, H.-J. Brauch, and I. Blankenhorn. 2001. Pharmaceuticals in groundwaters. Analytical methods and results of a monitoring program in Baden-Wurtemberg. Journal of Chromatography A, 938:199-210.

Schaefer, A.I., L.D. Nghiem, and T.D. Waite. 2003. Removal of the natural hormone estrone from aqueous solutions using nanofiltration and reverse osmosis. Environmental Science & Technology, 37(1):182-188.

Schittko, S., A. Putschew, and M. Jekel. 2004. Bank filtration: a suitable process for the removal of iodinated X-ray contrast media? Water Science and Technology, 50(5):261-268.

Sedlak, D., and K.E. Pinkston. 2001. Factors affecting the concentrations of pharmaceuticals released to the aquatic environment. Water Resources Update, 120:56-64.

Seiler, R.L., S.D. Zaugg, J.M. Thomas, and D.L. Howcroft. 1999. Caffeine and pharmaceuticals as indicators of waste water contamination in wells. Ground Water, 37(3):405-410.

Singer, H., S. Mueller, C. Tixier, and L. Pillonel. 2002. Triclosan: occurrence and fate of a widely used biocide in the aquatic environment: field measurements in wastewater treatment plants, surface waters, and lake sediments. Environmental Science & Technology, 36(23):4998-5004.

Snyder, S.A., T.L. Keith, D.A. Verbrugge, E.M. Snyder, T.S. Gross, K. Kannan, and J.P. Giesy. 1999. Analytical methods for detection of selected estrogenic compounds in aqueous mixtures. Environmental Science & Technology, 33(16):2814-2820.

Sole, M., M. Castillo, M.J. Lopez De Alda, C. Porte, K. Ladegaard-Pedersen, and D. Barcelo. 2000. Estrogenicity determination in sewage treatment plants and surface waters from the Catalonian area (NE Spain). Environmental Science & Technology, 34(24):5076-5083.

Stackelberg, P.E., E.T. Furlong, M.T. Meyer, S.D. Zaugg, A.K. Henderson, and D.B. Reissman. 2004. Persistence of pharmaceutical compounds and other organic wastewater contaminants in a conventional drinking-water-treatment plant. Science Of The Total Environment, 329(1-3):99-113.

Suzuki, T., Y. Nakagawa, I. Takano, K. Yaguchi, and K. Yasuda. 2004. Environmental fate of bisphenol A and its biological metabolites in river water and their xeno-estrogenic activity. Environmental Science & Technology, 38(8):2389-2396.

Ternes, T., M. Bonerz, and T. Schmidt. 2001. Determination of neutral pharmaceuticals in wastewater and rivers by liquid chromatography-electrospray tandem mass spectrometry. Journal of Chromatography A, 938(1-2):175-185.

Ternes, T.A. 1998. Occurrence of drugs in German sewage treatment plants and rivers. Water Research, 32(11):3245-3260.

Ternes, T.A., M. Stumpf, J. Mueller, K. Haberer, R.D. Wilken, and M. Servos. 1999. Behavior and occurence of estrogens in municipal sewage treatment plants - I. Investigations in Germany, Canada and Brazil. The Science of the Total Environment, 225:81-90.

Ternes, T.A., M. Meisenheimer, D. Mcdowell, F. Sacher, H.-J. Brauch, B. Haist-Gulde, G. Preuss, U. Wilme, and N. Zulei-Seibert. 2002. Removal of pharmaceuticals during drinking water treatment. Environmental Science & Technology, 36(17):3855-3863.

Tixier, C., H.P. Singer, S. Canonica, and S.R. Mueller. 2002. Phototransformation of triclosan in surface waters: a relevant elimination process for this widely used biocide-laboratory studies, field measurements, and modeling. Environmental Science & Technology, 36(16):3482-3489.

Tixier, C., H.P. Singer, S. Oellers, and S.R. Mueller. 2003. Occurrence and fate of carbamazepine, clofibric acid, diclofenac, ibuprofen, ketoprofen, and naproxen in surface waters. Environmental Science & Technology, 37(6):1061-1068.

Tsuda, T., K. Suga, E. Kaneda, and M. Ohsuga. 2002. 4-Nonylphenol, 4-nonylphenol mono- and diethoxylates, and other 4-alkylphenols in water and shellfish from rivers flowing into Lake Biwa. Bulletin of Environmental Contamination and Toxicology, 68(1):126-131.

Verstraeten, I.I., T. Heberer, J. Vogel, T.F. Speth, S. Zuehlke, and U. Duennbier. 2003. Occurrence of endocrine-disrupting and other wastewater compounds during water treatment with case studies from Lincoln, Nebraska and Berlin, Germany. Practice Periodical of Hazardous,Toxic, and Radioactive Waste Management, 7(4):253-263.

Vieno, N.M., T. Tuhkanen, and L. Kronberg. 2005. Seasonal variation in the occurrence of pharmaceuticals in effluents from a sewage treatment plant and in the recipient water. Environmental Science & Technology, 39(21):8220-8226.

Wiegel, S., A. Aulinger, R. Brockmeyer, H. Harms, J. Löffler, H. Reincke, R. Schmidt, B. Stachel, W. Von Tümpling, and A. Wanke. 2004. Pharmaceuticals in the river Elbe and its tributaries. Chemosphere, 57(2):107-126.

Xu, P., J.E. Drewes, C. Bellona, G. Amy, T.-U. Kim, M. Adam, and T. Heberer. 2005. Rejection of emerging organic micropollutants in nanofiltration-reverse osmosis membrane applications. Water Environment Research, 77(1):40-48.

Yoon, Y., P. Westerhoff, S.A. Snyder, and E.C. Wert. 2006. Nanofiltration and ultrafiltration of endocrine disrupting compounds,pharmaceuticals and personal care products. Journal of Membrane Science, 270:88-100.

Zwiener, C., and F.H. Frimmel. 2000. Oxidative treatment of pharmaceuticals in water. Water Research, 34(6):1881-1885.

# APPENDIX C
# TABLES OF WATER TREATMENT

**Table C.1**
**Removal of select EDCs and PPCPs during coagulation/flocculation**

| Compounds | Water Source Type | Primary Coagulant/pH/Dose | Influent (ng/L) | Percent Removal | Reference |
|---|---|---|---|---|---|
| **Pharmaceuticals** | | | | | |
| ***Analgesics*** | | | | | |
| Acetaminophen | surface | Alum, pH 6.8, 78 mg /L | - | <20 | (Westerhoff et al. 2005) |
| Diclofenac | surface | Ferric, pH 8 | 1000 | 6 | (Ternes et al. 2002) |
| | surface | Ferric, pH 8, 10.4-12.6 mg $Fe^{3+}$/L | 65-140 | 8-25 | (Ternes et al. 2002) |
| | surface | Alum, pH 6.8, 78 mg /L | - | <20 | (Westerhoff et al. 2005) |
| Ibuprofen | surface | Ferric, acidic pH | 8.5-17.5 | 4 | (Vieno, Tuhkanen, and Kronberg 2005) |
| | surface | Alum, pH 6.8, 78 mg /L | - | <20 | (Westerhoff et al. 2005) |
| Naproxen | surface | Ferric, acidic pH | 7-7.5 | 20 | (Vieno, Tuhkanen, and Kronberg 2005) |
| | surface | Alum, pH 6.8, 78 mg /L | - | <20 | (Westerhoff et al. 2005) |
| Ketoprofen | surface | Ferric, acidic pH | 7-11.5 | 13 | (Vieno, Tuhkanen, and Kronberg 2005) |
| Erythromycin | surface | Alum, pH 6.8, 78 mg /L | 30 | 33 | (Westerhoff et al. 2005) |
| | surface | Ferric, pH 6.8, 13.1 mg $Fe^{3+}$/L | 30 | 35 | (Westerhoff et al. 2005) |
| ***Antibiotics*** | | | | | |
| Sulfamethazine | surface | Alum, 20-107 mg/L | 50000 | 0 | (Adams et al. 2002) |
| | surface | Ferric, 25-169 mg/L | 50000 | 0 | (Adams et al. 2002) |

(continued)

**Table C.1 (Continued)**

| Compounds | Water Source Type | Primary Coagulant/pH/Dose | Influent (ng/L) | Percent Removal | Reference |
|---|---|---|---|---|---|
| Trimethoprim | surface | Alum, 20-107 mg/L | 50000 | 0 | (Adams et al. 2002) |
| | surface | Ferric, 25-169 mg/L | 50000 | 0 | (Adams et al. 2002) |
| | surface | Alum, pH 6.8, 78 mg /L | - | <20 | (Westerhoff et al. 2005) |
| ***Anticonvulsants*** | | | | | |
| Carbamazepine | surface | Ferric, pH 8 | 1000 | 13 | (Ternes et al. 2002) |
| | surface | Ferric, pH 8, 10.4-12.6 mg $Fe^{3+}$/L | 80-180 | 0-6 | (Ternes et al. 2002) |
| | surface | Alum, pH 6.8, 78 mg /L | - | <20 | (Westerhoff et al. 2005) |
| | surface | Ferric, pH 6.8, 13.1 mg $Fe^{3+}$/L | 30 | 35 | (Westerhoff et al. 2005) |
| Primidone | surface | Ferric, pH 8 | 1000 | 0 | (Ternes et al. 2002) |
| ***Lipid regulators*** | | | | | |
| Clofibric acid | surface | Ferric, pH 8 | 1000 | 13 | (Ternes et al. 2002) |
| Bezafibrate | surface | Ferric, pH 8 | 1000 | 0 | (Ternes et al. 2002) |
| | surface | Ferric, pH 8, 10.4-12.6 mg $Fe^{3+}$/L | 10-15 | 0 | (Ternes et al. 2002) |
| Gemfibrozil | surface | Alum, pH 6.8, 78 mg /L | - | <20 | (Westerhoff et al. 2005) |
| ***X-ray contrast media*** | | | | | |
| Iopromide | surface | Alum, pH 6.8, 78 mg /L | - | <20 | (Westerhoff et al. 2005) |

(continued)

**Table C.1 (Continued)**

| Compounds | Water Source Type | Primary Coagulant/pH/Dose | Influent (ng/L) | Percent Removal | Reference |
|---|---|---|---|---|---|
| **Steroids** | | | | | |
| ***Estrogens*** | | | | | |
| 17α-Ethinylestradiol | surface | Alum, pH 6.8, 78 mg /L | 95 | 0 | (Westerhoff et al. 2005) |
| | surface | Ferric, pH 6.8, 13.1 mg $Fe^{3+}$/L | 95 | 0 | (Westerhoff et al. 2005) |
| | surface | Alum, pH 5.4, 57-126 mg/L | 5000-50000 | 0 | (Ballard and Mackay 2005) |
| | surface | Alum, pH 6.5, 42.5 mg/L | 5000-50000 | - | (Ballard and Mackay 2005) |
| 17β-Estradiol | surface | Alum, pH 6.8, 78 mg /L | 95 | 2 | (Westerhoff et al. 2005) |
| | surface | Ferric, pH 6.8, 13.1 mg $Fe^{3+}$/L | 95 | 0 | (Westerhoff et al. 2005) |
| Estrone | surface | Alum, pH 6.8, 78 mg /L | 95 | 5 | (Westerhoff et al. 2005) |
| | surface | Ferric, pH 6.8, 13.1 mg $Fe^{3+}$/L | 95 | 0 | (Westerhoff et al. 2005) |
| **Personal Care Products** | | | | | |
| ***Antimicrobial*** | | | | | |
| Triclosan | surface | Alum, pH 6.8, 78 mg /L | - | <20 | (Westerhoff et al. 2005) |
| ***Insecticide*** | | | | | |
| DEET | surface | Alum, pH 6.8, 78 mg /L | 78 | 0 | (Westerhoff et al. 2005) |
| | surface | Ferric, pH 6.8, 13.1 mg $Fe^{3+}$/L | 78 | 0 | (Westerhoff et al. 2005) |

(continued)

**Table C.1 (Continued)**

| Compounds | Water Source Type | Primary Coagulant/pH/Dose | Influent (ng/L) | Percent Removal | Reference |
|---|---|---|---|---|---|
| ***Fragrance*** | | | | | |
| Galaxolide | surface | Alum, pH 6.8, 78 mg /L | - | <20 | (Westerhoff et al. 2005) |
| **Other** | | | | | |
| ***Flame retardants*** | | | | | |
| TCEP | surface | Alum, pH 6.8, 78 mg /L | 30 | 0 | (Westerhoff et al. 2005) |
| | surface | Ferric, pH 6.8, 13.1 mg $Fe^{3+}$/L | 30 | 0 | (Westerhoff et al. 2005) |
| ***Plasticizer*** | | | | | |
| Bisphenol A | surface | Alum, pH 5.4, 57-126 mg/L | 5000-50000 | 0 | (Ballard and Mackay 2005) |
| | surface | Alum, pH 6.5, 42.5 mg/L | 5000-50000 | - | (Ballard and Mackay 2005) |
| ***Stimulant*** | | | | | |
| Caffeine | surface | Alum, pH 6.8, 78 mg /L | 120 | 0 | (Westerhoff et al. 2005) |
| | surface | Ferric, pH 6.8, 13.1 mg $Fe^{3+}$/L | 120 | 0 | (Westerhoff et al. 2005) |

**Table C.2**
**Removal of select EDCs and PPCPs during disinfection processes**

| Compounds | Water Source Type | Primary Disinfectant/Dose/ Contact Time | Influent (ng/L) | Percent Removal | Reference |
|---|---|---|---|---|---|
| **Pharmaceuticals** | | | | | |
| ***Analgesics*** | | | | | |
| Acetaminophen | wastewater | $Cl_2$, 4 mg/L, 1 h | 150000 0 | 88 | (Bedner and Maccrehan 2006) |
| | surface | $Cl_2$, 3.5-3.8 mg/L, 24 h | - | 85 | (Westerhoff et al. 2005) |
| | surface | $O_3$, 2.5-3 mg/L, 8.4 min | - | 94 | (Westerhoff et al. 2005) |
| Diclofenac | wastewater | $O_3$, 0.5-5 mg/L, 8.4 min | 2000 | 55-100 | (Huber et al. 2005a) |
| | flocculated | $O_3$, 0.5-3.0 mg/L, 20 min | 1000 | 100 | (Ternes et al. 2002) |
| | surface | $Cl_2O$, 0.1-1 mg/L, 10 min | 88860 | 50-100 | (Huber et al. 2005b) |
| | bank filtrate | $O_3$, 0.1-2 mg/L | 148100 | 75-99 | (Huber et al. 2003) |
| | flocculated | $O_3$, 0.1-2 mg/L | 148100 | 25-99 | (Huber et al. 2003) |
| | surface | $Cl_2$, 3.5-3.8 mg/L, 24 h | - | 95 | (Westerhoff et al. 2005) |
| | surface | $O_3$, 2.5-3 mg/L, 8.4 min | - | 95 | (Westerhoff et al. 2005) |
| Ibuprofen | surface | $Cl_2$ | 9 | 5 | (Vieno, Tuhkanen, and Kronberg 2005) |
| | wastewater | $O_3$, 5 mg/L, 8.4 min | 2000 | 80 | (Huber et al. 2005a) |
| | surface | $O_3$, 2 mg/L, 10 min | 103150 | 41 | (Huber et al. 2003) |
| | bank filtrate | $O_3$, 2 mg/L, 10 min | 103150 | 40 | (Huber et al. 2003) |
| | ground | $O_3$, 2 mg/L, 10 min | 103150 | 62 | (Huber et al. 2003) |
| | flocculated | $O_3$, 2 mg/L, 10 min | 103150 | 77 | (Huber et al. 2003) |
| | surface | $Cl_2$, 3.5-3.8 mg/L, 24 h | - | 30-80 | (Westerhoff et al. 2005) |
| | surface | $O_3$, 2.5-3 mg/L, 8.4 min | - | 60-100 | (Westerhoff et al. 2005) |
| Ketoprofen | surface | $Cl_2$ | 5 | 0 | (Vieno, Tuhkanen, and Kronberg 2005) |
| | bank filtrate | $Cl_2O$, 0.95-11.5 mg/L, 30 min | - | 0 | (Huber et al. 2005b) |

(continued)

**Table C.2 (Continued)**

| Compounds | Water Source Type | Primary Disinfectant/Dose/ Contact Time | Influent (ng/L) | Percent Removal | Reference |
|---|---|---|---|---|---|
| Naproxen | wastewater | $O_3$, 2-5 mg/L, 8.4 min | - | >90 | (Huber et al. 2005a) |
| | bank filtrate | $Cl_2O$, 0.95-11.5 mg/L, 30 min | - | 0 | (Huber et al. 2005b) |
| | surface | $Cl_2$, 3.5-3.8 mg/L, 24 h | - | 95 | (Westerhoff et al. 2005) |
| | surface | $O_3$, 2.5-3 mg/L, 8.4 min | - | 91 | (Westerhoff et al. 2005) |
| ***Antibiotics*** | | | | | |
| Sulfamethoxazole | wastewater | $O_3$, 0.5-5 mg/L, 8.4 min | 2000 | 15-100 | (Huber et al. 2005a) |
| | surface | $Cl_2O$, 0.1-1 mg/L, 10 min | 76020 | 30-100 | (Huber et al. 2005b) |
| | ground | $Cl_2O$, 0.1 mg/L, 5-180 min | 1000 | 90-100 | (Huber et al. 2005b) |
| | bank filtrate | $O_3$, 0.1-2 mg/L | 126700 | 65-99 | (Huber et al. 2003) |
| | flocculated | $O_3$, 0.1-2 mg/L | 126700 | 25-99 | (Huber et al. 2003) |
| | surface | $Cl_2$, 3.5-3.8 mg/L, 24 h | - | 92 | (Westerhoff et al. 2005) |
| | surface | $O_3$, 2.5-3 mg/L, 8.4 min | - | 88 | (Westerhoff et al. 2005) |
| Erythromycin | wastewater | $O_3$, 0.5-5 mg/L, 8.4 min | 2000 | 30-100 | (Huber et al. 2005a) |
| | ground | $Cl_2O$, 0.1 mg/L, 5-180 min | 100 | 40-95 | (Huber et al. 2005b) |
| | surface | $Cl_2$, 3.5-3.8 mg/L, 24 h | - | 100 | (Westerhoff et al. 2005) |
| | surface | $O_3$, 2.5-3 mg/L, 8.4 min | - | 97 | (Westerhoff et al. 2005) |
| Sulfamethazine | surface | $Cl_2$, 1 mg/L, 3-10 min | 50000 | 50-90 | (Adams et al. 2002) |
| | surface | UV, LP, 1000-3000 $mJ/cm^2$ | 50000 | 25-60 | (Adams et al. 2002) |
| | surface | $O_3$, 0.25-1.3 min contact | 50000 | 50-100 | (Adams et al. 2002) |
| | surface | $Cl_2O$, 0.1 mg/L, 5-180 min | 1000 | 95-100 | (Huber et al. 2005b) |
| Sulfadiazine | wastewater | $O_3$, 0.5-5 mg/L, 8.4 min | 2000 | 45-100 | (Huber et al. 2005a) |
| Trimethoprim | surface | $Cl_2$, 1 mg/L, 4-10 min | 50000 | 50-90 | (Adams et al. 2002) |
| | surface | UV, LP, 1000-3000 $mJ/cm^2$ | 50000 | 15-50 | (Adams et al. 2002) |
| | surface | $O_3$, 0.25-1.3 min contact | 50000 | 50-100 | (Adams et al. 2002) |
| | surface | $Cl_2$, 3.5-3.8 mg/L, 24 h | - | 100 | (Westerhoff et al. 2005) |
| | surface | $O_3$, 2.5-3 mg/L, 8.4 min | - | 98 | (Westerhoff et al. 2005) |

(continued)

**Table C.2 (Continued)**

| Compounds | Water Source Type | Primary Disinfectant/Dose/ Contact Time | Influent (ng/L) | Percent Removal | Reference |
|---|---|---|---|---|---|
| Clarithromycin | wastewater | $O_3$, 0.5-5 mg/L, 8.4 min | 2000 | 20-98 | (Huber et al. 2005a) |
| | ground | $Cl_2O$, 0.1 mg/L, 5-180 min | 100 | 35-85 | (Huber et al. 2005b) |
| Roxithromycin | wastewater | $O_3$, 0.5-5 mg/L, 8.4 min | 2000 | 30-100 | (Huber et al. 2005a) |
| | ground | $Cl_2O$, 0.1 mg/L, 5-180 min | 100 | 35-85 | (Huber et al. 2005b) |
| Ciprofloxacin | wastewater | $Cl_2$, 11.6 mg/L, 5-60 min | 496950 | 30-90 | (Dodd et al. 2005) |
| | drinking water | $Cl_2$, 2 mg/L, 5-60 min | 496950 | 20-90 | (Dodd et al. 2005) |
| ***Anti-convulsants*** | | | | | |
| Carbamazepine | bi-distilled | $O_3$, 1 mg/L, 10 min | 779790 | 100 | (Andreozzi et al. 2002) |
| | flocculated | $O_3$, 0.2-20 mg/L, 20 min | 500 | 100 | (Mcdowell et al. 2005) |
| | flocculated | $O_3$, 0.5-3.0 mg/L, 20 min | 1000 | 100 | (Ternes et al. 2002) |
| | bank filtrate | $O_3$, 0.1-2 mg/L | 118150 | 75-99 | (Huber et al. 2003) |
| | flocculated | $O_3$, 0.1-2 mg/L | 118150 | 15-99 | (Huber et al. 2003) |
| | surface | $Cl_2$, 3.5-3.8 mg/L, 24 h | - | 95 | (Westerhoff et al. 2005) |
| | surface | $O_3$, 2.5-3 mg/L, 8.4 min | - | 99 | (Westerhoff et al. 2005) |
| Primidone | flocculated | $O_3$, 0.5-3.0 mg/L, 20 min | 1000 | 15-90 | (Ternes et al. 2002) |
| ***Lipid regulators*** | | | | | |
| Bezafibrate | wastewater | $O_3$, 0.5-5 mg/L, 8.4 min | 2000 | 40-100 | (Huber et al. 2005a) |
| | flocculated | $O_3$, 0.5-3.0 mg/L, 20 min | 1000 | 10-80 | (Ternes et al. 2002) |
| | surface | $O_3$, 2 mg/L, 10 min | 180912 | >99 | (Huber et al. 2003) |
| | bank filtrate | $O_3$, 2 mg/L, 10 min | 180912 | >99 | (Huber et al. 2003) |
| | ground | $O_3$, 2 mg/L, 10 min | 180912 | >99 | (Huber et al. 2003) |
| | flocculated | $O_3$, 2 mg/L, 10 min | 180912 | 98 | (Huber et al. 2003) |
| | bank filtrate | $O_3$, 0.1-2 mg/L | 180912 | 8-99 | (Huber et al. 2003) |
| | flocculated | $O_3$, 0.1-2 mg/L | 180912 | 98 | (Huber et al. 2003) |

(continued)

**Table C.2 (Continued)**

| Compounds | Water Source Type | Primary Disinfectant/Dose/ Contact Time | Influent (ng/L) | Percent Removal | Reference |
|---|---|---|---|---|---|
| Clofibric acid | flocculated | $O_3$, 0.5-3.0 mg/L, 20 min | 1000 | 15-40 | (Ternes et al. 2002) |
| | bank filtrate | $Cl_2O$, 0.95-11.5 mg/L, 30 min | - | 7-41 | (Huber et al. 2005b) |
| Gemfibrozil | bank filtrate | $Cl_2O$, 0.95-11.5 mg/L, 30 min | - | 7-41 | (Huber et al. 2005b) |
| | surface | $Cl_2$, 3.5-3.8 mg/L, 24 h | - | 99 | (Westerhoff et al. 2005) |
| | surface | $O_3$, 2.5-3 mg/L, 8.4 min | - | 98 | (Westerhoff et al. 2005) |
| ***X-ray contrast media*** | | | | | |
| Iopromide | wastewater | $O_3$, 0.5-5 mg/L, 8.4 min | 5000 | 0-60 | (Huber et al. 2005a) |
| | surface | $Cl_2O$, 0.1-1 mg/L, 10 min | 237330 | 0 | (Huber et al. 2005b) |
| | surface | $O_3$, 2 mg/L, 10 min | 395550 | 24 | (Huber et al. 2003) |
| | bank filtrate | $O_3$, 2 mg/L, 10 min | 395550 | 27 | (Huber et al. 2003) |
| | ground | $O_3$, 2 mg/L, 10 min | 395550 | 58 | (Huber et al. 2003) |
| | flocculated | $O_3$, 2 mg/L, 10 min | 395550 | 68 | (Huber et al. 2003) |
| | surface | $Cl_2$, 3.5-3.8 mg/L, 24 h | - | 1-40 | (Westerhoff et al. 2005) |
| | surface | $O_3$, 2.5-3 mg/L, 8.4 min | - | 50-75 | (Westerhoff et al. 2005) |
| Diatrizoate | wastewater | $O_3$, 0.5-5 mg/L, 8.4 min | 5000 | 0 | (Huber et al. 2005a) |
| ***Muscle relaxant*** | | | | | |
| Diazepam | flocculated | $O_3$, 0.2-20 mg/L, 20 min | 500 | 30-70 | (Mcdowell et al. 2005) |
| | surface | $O_3$, 2 mg/L, 10 min | 142350 | 24 | (Huber et al. 2003) |
| | bank filtrate | $O_3$, 2 mg/L, 10 min | 142350 | >99 | (Huber et al. 2003) |
| | ground | $O_3$, 2 mg/L, 10 min | 142350 | 65 | (Huber et al. 2003) |
| | flocculated | $O_3$, 2 mg/L, 10 min | 142350 | >99 | (Huber et al. 2003) |
| | surface | $Cl_2$, 3.5-3.8 mg/L, 24 h | - | 77 | (Westerhoff et al. 2005) |
| | surface | $O_3$, 2.5-3 mg/L, 8.4 min | - | 65-95 | (Westerhoff et al. 2005) |

(continued)

### Table C.2 (Continued)

| Compounds | Water Source Type | Primary Disinfectant/Dose/ Contact Time | Influent (ng/L) | Percent Removal | Reference |
|---|---|---|---|---|---|
| **Steroids** | | | | | |
| ***Estrogens*** | | | | | |
| 17α-Ethinylestradiol | wastewater | $O_3$, 0.5-5 mg/L, 8.4 min | 1000 | 30-100 | (Huber et al. 2005a) |
| | surface | $Cl_2O$, 0.1-1 mg/L, 10 min | 1000 | 50-100 | (Huber et al. 2005b) |
| | ground | $Cl_2O$, 0.1 mg/L, 5-180 min | 1000 | 100 | (Huber et al. 2005b) |
| | ultra-pure | $O_3$, 0.2-0.9 mg/L | 2964000 | 30-100 | (Huber, Ternes, and Von Gunten 2004) |
| | bank filtrate | $O_3$, 0.1-2 mg/L | 148200 | 99 | (Huber et al. 2003) |
| | flocculated | $O_3$, 0.1-2 mg/L | 148200 | 85-99 | (Huber et al. 2003) |
| | ultra-pure | UV, LP & MP, 1000 $mJ/cm^2$ | - | 2-22 | (Rosenfeldt and Linden 2004) |
| | surface | $Cl_2$, 3.5-3.8 mg/L, 24 h | - | 99 | (Westerhoff et al. 2005) |
| | surface | $O_3$, 2.5-3 mg/L, 8.4 min | - | 99 | (Westerhoff et al. 2005) |
| | ultra-pure | $Cl_2$, 1 mg/L, 1-24 h | 29640 | 92-100 | (Alum et al. 2004) |
| | ultra-pure | $O_3$, 1.5 mg/L, 1-120 min | 29640 | 100 | (Alum et al. 2004) |
| 17β-Estradiol | ground | $Cl_2O$, 0.1 mg/L, 5-180 min | 1000 | 100 | (Huber et al. 2005b) |
| | wastewater | $O_3$, 0.5-5 mg/L, 8.4 min | 500 | 60-100 | (Huber et al. 2005a) |
| | ultra-pure | UV, LP & MP, 1000 $mJ/cm^2$ | - | 5-18 | (Rosenfeldt and Linden 2004) |
| | surface | $Cl_2$, 3.5-3.8 mg/L, 24 h | - | 98 | (Westerhoff et al. 2005) |
| | surface | $O_3$, 2.5-3 mg/L, 8.4 min | - | 98 | (Westerhoff et al. 2005) |
| | ultra-pure | $Cl_2$, 1 mg/L, 1-24 h | 27240 | 92-100 | (Alum et al. 2004) |
| | ultra-pure | $O_3$, 1.5 mg/L, 1-120 min | 27240 | 100 | (Alum et al. 2004) |
| Estrone | wastewater | $O_3$, 0.5-5 mg/L, 8.4 min | 500 | 55-95 | (Huber et al. 2005a) |
| | ground | $Cl_2O$, 0.1 mg/L, 5-180 min | 1000 | 100 | (Huber et al. 2005b) |
| | surface | $Cl_2$, 3.5-3.8 mg/L, 24 h | - | 99 | (Westerhoff et al. 2005) |
| | surface | $O_3$, 2.5-3 mg/L, 8.4 min | - | 99 | (Westerhoff et al. 2005) |

(continued)

**Table C.2 (Continued)**

| Compounds | Water Source Type | Primary Disinfectant/Dose/ Contact Time | Influent (ng/L) | Percent Removal | Reference |
|---|---|---|---|---|---|
| ***Androgen*** | | | | | |
| Testosterone | surface | $Cl_2$, 3.5-3.8 mg/L, 24 h | - | 55 | (Westerhoff et al. 2005) |
| | surface | $O_3$, 2.5-3 mg/L, 8.4 min | - | 94 | (Westerhoff et al. 2005) |
| **Personal Care Products** | | | | | |
| ***Insecticide*** | | | | | |
| DEET | surface | $Cl_2$, 3.5-3.8 mg/L, 24 h | - | 5-35 | (Westerhoff et al. 2005) |
| | surface | $O_3$, 2.5-3 mg/L, 8.4 min | - | 65-95 | (Westerhoff et al. 2005) |
| ***Fragrance*** | | | | | |
| Galaxolide | surface | $Cl_2$, 3.5-3.8 mg/L, 24 h | - | 50 | (Westerhoff et al. 2005) |
| | surface | $O_3$, 2.5-3 mg/L, 8.4 min | - | 90 | (Westerhoff et al. 2005) |
| ***Antimicrobial*** | | | | | |
| Triclosan | surface | $Cl_2$, 3.5-3.8 mg/L, 24 h | - | 98 | (Westerhoff et al. 2005) |
| | surface | $O_3$, 2.5-3 mg/L, 8.4 min | - | 92 | (Westerhoff et al. 2005) |
| **Other** | | | | | |
| ***Plasticizer*** | | | | | |
| Bisphenol A | synthetic | $Cl_2$, 0.54-1.05 mg/L, 1-5 h | 114150 | 65-98 | (Gallard, Leclercq, and Croué 2004) |
| | ultra-pure | $Cl_2$, 2.7 mg/L, 5-60 min | 1141500 | 80-100 | (Gallard, Leclercq, and Croué 2004) |
| | ultra-pure | UV, LP & MP, 1000 $mJ/cm^2$ | - | 2-15 | (Rosenfeldt and Linden 2004) |
| | ultra-pure | $Cl_2$, 1 mg/L, 1-24 h | 22830 | 92-100 | (Alum et al. 2004) |
| | ultra-pure | $O_3$, 1.5 mg/L, 1-120 min | 22830 | 100 | (Alum et al. 2004) |

(continued)

**Table C.2 (Continued)**

| Compounds | Water Source Type | Primary Disinfectant/Dose/ Contact Time | Influent (ng/L) | Percent Removal | Reference |
|---|---|---|---|---|---|
| ***Stimulant*** | | | | | |
| Caffeine | flocculated | $O_3$, 0.2-20 mg/L, 20 min | 500 | 100 | (Mcdowell et al. 2005) |
| | bank filtrate | $Cl_2O$, 0.95-11.5 mg/L, 30 min | - | 0-1 | (Huber et al. 2005b) |
| | surface | $Cl_2$, 3.5-3.8 mg/L, 24 h | - | 57 | (Westerhoff et al. 2005) |
| | surface | $O_3$, 2.5-3 mg/L, 8.4 min | - | 90 | (Westerhoff et al. 2005) |
| ***Flame retardants*** | | | | | |
| TCEP | surface | $Cl_2$, 3.5-3.8 mg/L, 24 h | - | 0-10 | (Westerhoff et al. 2005) |
| | surface | $O_3$, 2.5-3 mg/L, 8.4 min | - | 0 | (Westerhoff et al. 2005) |

**Table C.3**
**Removal of select EDCs and PPCPs through activated carbon adsorption**

| Compounds | Water Source Type | Adsorbent/Dose/Contact time | Influent (ng/L) | Percent Removal | Reference |
|---|---|---|---|---|---|
| **Pharmaceuticals** | | | | | |
| ***Analgesics*** | | | | | |
| Acetaminophen | surface | PAC, 5 mg/L, 4 h contact | 25 | 69 | (Westerhoff et al. 2005) |
| Diclofenac | ground | GAC, EBCT = 10 min, h = 160 cm, ST =50 $m^3$/kg | 40 | 100 | (Ternes et al. 2002) |
| | surface | GAC, Load 27.8-39.9 $m^3$/kg | 60 | 100 | (Ternes et al. 2002) |
| | surface | PAC, 5 mg/L, 4 h contact | 35 | 48 | (Westerhoff et al. 2005) |
| Ibuprofen | surface | GAC | 8-13.25 | 33-100 | (Vieno, Tuhkanen, and Kronberg 2005) |
| | surface | PAC, 5 mg/L, 4 h contact | 20 | 21 | (Westerhoff et al. 2005) |
| Ketoprofen | surface | GAC | 6.5-10.5 | 23-100 | (Vieno, Tuhkanen, and Kronberg 2005) |
| Naproxen | surface | PAC, 5 mg/L, 4 h contact | 20 | 47 | (Westerhoff et al. 2005) |
| Erythromycin | surface | PAC, 5 mg/L, 4 h contact | 35 | 65 | (Westerhoff et al. 2005) |
| ***Antibiotics*** | | | | | |
| Sulfamethazine | surface | PAC, 5-50 mg/L, 4 h contact | 50000 | 35-90 | (Adams et al. 2002) |
| Trimethoprim | surface | PAC, 5-50 mg/L, 4 h contact | 50000 | 35-90 | (Adams et al. 2002) |
| | surface | PAC, 5 mg/L, 4 h contact | 65 | 93 | (Westerhoff et al. 2005) |
| Sulfamethoxazole | surface | PAC, 5 mg/L, 4 h contact | 20 | 20 | (Westerhoff et al. 2005) |

(continued)

Table C.3 (Continued)

| Compounds | Water Source Type | Adsorbent/Dose/Contact time | Influent (ng/L) | Percent Removal | Reference |
|---|---|---|---|---|---|
| ***Antiepileptics*** | | | | | |
| Carbamazepine | ground | GAC, EBCT = 10 min, h = 160 cm, ST =50 $m^3$/kg | 1000 | 100 | (Ternes et al. 2002) |
| | surface | GAC, Load 27.8-39.9 $m^3$/kg | 105-180 | 100 | (Ternes et al. 2002) |
| | surface | PAC, 5 mg/L, 4 h contact | 50 | 81 | (Westerhoff et al. 2005) |
| Primidone | surface | GAC, Load 27.8-39.9 $m^3$/kg | 15 | 67 | (Ternes et al. 2002) |
| ***Lipid regulators*** | | | | | |
| Clofibric acid | ground | GAC, EBCT = 10 min, h = 160 cm, ST =50 $m^3$/kg | 1800 | 75 | (Ternes et al. 2002) |
| | surface | GAC, Load 27.8-39.9 $m^3$/kg | 10 | 50 | (Ternes et al. 2002) |
| Bezafibrate | ground | GAC, EBCT = 10 min, h = 160 cm, ST =50 $m^3$/kg | 260 | 100 | (Ternes et al. 2002) |
| | surface | GAC, Load 27.8-39.9 $m^3$/kg | 70 | 100 | (Ternes et al. 2002) |
| Gemfibrozil | surface | PAC, 5 mg/L, 4 h contact | 70 | 49 | (Westerhoff et al. 2005) |
| ***Contrast Agent*** | | | | | |
| Iopromide | surface | PAC, 5 mg/L, 4 h contact | 50 | 14 | (Westerhoff et al. 2005) |
| ***Muscle Relaxant*** | | | | | |
| Diazepam | surface | PAC, 5 mg/L, 4 h contact | 40 | 73 | (Westerhoff et al. 2005) |

(continued)

Table C.3 (Continued)

| Compounds | Water Source Type | Adsorbent/Dose/Contact time | Influent (ng/L) | Percent Removal | Reference |
|---|---|---|---|---|---|
| **Steroids** | | | | | |
| ***Estrogens*** | | | | | |
| 17α-Ethinylestradiol | surface | PAC, 5-15 mg/L, 4 h contact | 29620 | 50-100 | (Yoon et al. 2003) |
| | surface | PAC, 5 mg/L, 4 h contact | 160 | 88 | (Westerhoff et al. 2005) |
| 17β-Estradiol | surface | PAC, 5-15 mg/L, 4 h contact | 27230 | 87-100 | (Yoon et al. 2003) |
| | deionized | GAC, 2000 mg/L, 50-180 min contact | 1-100 | 49-81 | (Fuerhacker, Dürauer, and Jungbauer 2001) |
| | surface | PAC, 1 mg/L, 1 h contact | 6.8-1360 | 39-51 | (Westerhoff et al. 2005) |
| ***Androgen*** | | | | | |
| Testosterone | surface | PAC, 5 mg/L, 4 h contact | 50 | 88 | (Westerhoff et al. 2005) |
| **Personal Care Products** | | | | | |
| ***Surfactant*** | | | | | |
| 4-Nonylphenol | deionized | GAC, 100-1000 mg/L, 1 and 24 h contact | 5000 | 70-100 | (Tanghe and Verstraete 2001) |
| ***Insecticide*** | | | | | |
| DEET | surface | PAC, 5 mg/L, 4 h contact | 50 | 52 | (Westerhoff et al. 2005) |
| ***Fragrance*** | | | | | |
| Galaxolide | surface | PAC, 5 mg/L, 4 h contact | 140 | 65 | (Westerhoff et al. 2005) |

(continued)

**Table C.3 (Continued)**

| Compounds | Water Source Type | Adsorbent/Dose/Contact time | Influent (ng/L) | Percent Removal | Reference |
|---|---|---|---|---|---|
| ***Antimicrobial*** | | | | | |
| Triclosan | surface | PAC, 5 mg/L, 4 h contact | 40 | 98 | (Westerhoff et al. 2005) |
| **Other** | | | | | |
| ***Flame Retardant*** | | | | | |
| TCEP | surface | PAC, 5 mg/L, 4 h contact | 30 | 50 | (Westerhoff et al. 2005) |
| ***Plasticizer*** | | | | | |
| Bisphenol A | surface | PAC, 5-15 mg/L, 4 h contact | 22810 | 33-99 | (Yoon et al. 2003) |
| ***Stimulant*** | | | | | |
| Caffeine | surface | PAC, 5 mg/L, 4 h contact | 80 | 78 | (Westerhoff et al. 2005) |

## Table C.4
## Removal of select EDCs and PPCPs during advanced oxidation processes

| Compounds | Water Source Type | AOP/Dose | Influent (ng/L) | Percent Removal | Reference |
|---|---|---|---|---|---|
| **Pharmaceuticals** | | | | | |
| ***Analgesics*** | | | | | |
| Diclofenac | ultra-pure | $O_3/H_2O_2$, 1 mg/L $O_3$, 0.4 mg/L $H_2O_2$ | 2000 | 99 | (Zwiener and Frimmel 2000) |
| | surface | $O_3/H_2O_2$, 1 mg/L $O_3$, 0.4 mg/L $H_2O_2$ | 2000 | 99 | (Zwiener and Frimmel 2000) |
| | surface | $O_3/H_2O_2$, 3.7mg/L $O_3$, 1.4 mg/L $H_2O_2$ | 2000 | 99 | (Zwiener and Frimmel 2000) |
| | surface | $O_3/H_2O_2$, 5.0 mg/L $O_3$, 1.8 mg/L $H_2O_2$ | 2000 | 99 | (Zwiener and Frimmel 2000) |
| Ibuprofen | | | | | |
| | surface | $O_3/H_2O_2$, 2 mg/L $O_3$, 0.7 mg/L $H_2O_2$ | 103150 | 84 | (Huber et al. 2003) |
| | bank filtrate | $O_3/H_2O_2$, 2 mg/L $O_3$, 0.7 mg/L $H_2O_2$ | 103150 | 78 | (Huber et al. 2003) |
| | ground | $O_3/H_2O_2$, 2 mg/L $O_3$, 0.7 mg/L $H_2O_2$ | 103150 | 90 | (Huber et al. 2003) |
| | flocculated | $O_3/H_2O_2$, 2 mg/L $O_3$, 0.7 mg/L $H_2O_2$ | 103150 | >99 | (Huber et al. 2003) |
| | ultra-pure | $O_3/H_2O_2$, 1 mg/L $O_3$, 0.4 mg/L $H_2O_2$ | 2000 | 50 | (Zwiener and Frimmel 2000) |
| | surface | $O_3/H_2O_2$, 1 mg/L $O_3$, 0.4 mg/L $H_2O_2$ | 2000 | 30 | (Zwiener and Frimmel 2000) |
| | surface | $O_3/H_2O_2$, 3.7mg/L $O_3$, 1.4 mg/L $H_2O_2$ | 2000 | 93 | (Zwiener and Frimmel 2000) |
| | surface | $O_3/H_2O_2$, 5.0 mg/L $O_3$, 1.8 mg/L $H_2O_2$ | 2000 | 98 | (Zwiener and Frimmel 2000) |
| ***Lipid regulators*** | | | | | |
| Bezafibrate | surface | $O_3/H_2O_2$, 2 mg/L $O_3$, 0.7 mg/L $H_2O_2$ | 103150 | >99 | (Huber et al. 2003) |
| | bank filtrate | $O_3/H_2O_2$, 2 mg/L $O_3$, 0.7 mg/L $H_2O_2$ | 103150 | >99 | (Huber et al. 2003) |
| | ground | $O_3/H_2O_2$, 2 mg/L $O_3$, 0.7 mg/L $H_2O_2$ | 103150 | 97 | (Huber et al. 2003) |
| | flocculated | $O_3/H_2O_2$, 2 mg/L $O_3$, 0.7 mg/L $H_2O_2$ | 103150 | >99 | (Huber et al. 2003) |
| Clofibric acid | ultra-pure | $O_3/H_2O_2$, 1 mg/L $O_3$, 0.4 mg/L $H_2O_2$ | 2000 | 50 | (Zwiener and Frimmel 2000) |
| | surface | $O_3/H_2O_2$, 1 mg/L $O_3$, 0.4 mg/L $H_2O_2$ | 2000 | 10 | (Zwiener and Frimmel 2000) |
| | surface | $O_3/H_2O_2$, 3.7mg/L $O_3$, 1.4 mg/L $H_2O_2$ | 2000 | 92 | (Zwiener and Frimmel 2000) |
| | surface | $O_3/H_2O_2$, 5.0 mg/L $O_3$, 1.8 mg/L $H_2O_2$ | 2000 | 97 | (Zwiener and Frimmel 2000) |

(continued)

**Table C.4 (Continued)**

| Compounds | Water Source Type | AOP/Dose | Influent (ng/L) | Percent Removal | Reference |
|---|---|---|---|---|---|
| **Steroids** | | | | | |
| ***Estrogens*** | | | | | |
| 17α-Ethinylestradiol | ultra-pure | $UV/H_2O_2$, 1000 mJ/$cm^2$, 15 mg/L H2O2 | – | 98 | (Rosenfeldt and Linden 2004) |
| 17β-Estradiol | ultra-pure | $UV/H_2O_2$, 1000 mJ/$cm^2$, 15 mg/L H2O2 | – | 95 | (Rosenfeldt and Linden 2004) |
| **Other** | | | | | |
| ***Plasticizer*** | | | | | |
| Bisphenol A | ultra-pure | $UV/H_2O_2$, 1000 mJ/$cm^2$, 15 mg/L H2O2 | – | 90 | (Rosenfeldt and Linden 2004) |

– = Not reported

**Table C.5**
**Removal of select EDCs and PPCPs during nanofiltration/reverse osmosis**

| Compounds | Water Source Type | Membrane Type | Influent (ng/L) | Percent Removal | Reference |
|---|---|---|---|---|---|
| **Pharmaceuticals** | | | | | |
| ***Analgesics*** | | | | | |
| Acetaminophen | Surface | NF,ESNA, 6.8 | 50 | 30 | (Yoon et al. 2006) |
| Diclofenac | DI/EfOM | NF, TFC-SR2, pH 6 | 300 | 60 | (Xu et al. 2005) |
| | DI/EfOM | RO, TFC-HR, pH 6 | 300 | 97 | (Xu et al. 2005) |
| | Surface | NF,ESNA, 6.8 | 15 | 0 | (Yoon et al. 2006) |
| | ultra-pure | NF-ESNA, pH 7 | 100000 | 93 | (Kimura et al. 2003) |
| | ultra-pure | RO-XLE, pH 7 | 100000 | 95 | (Kimura et al. 2003) |
| | ultra-pure | RO-XLE, pH 7 | 100 | 65 | (Kimura et al. 2003) |
| | surface | RO | 329 | >99.7 | (Heberer 2002) |
| Ibuprofen | DI/EfOM | NF, TFC-SR2, pH 6 | 300 | 26 | (Xu et al. 2005) |
| | DI/EfOM | RO, TFC-HR, pH 6 | 300 | 97 | (Xu et al. 2005) |
| | DI | NF-90, pH 7 | 500 | >99 | (Nghiem, Schaefer, and Elimelech 2005b) |
| | DI | NF-90, pH 5 | 500 | >99 | (Nghiem, Schaefer, and Elimelech 2005b) |
| | DI | NF-270, pH 7 | 500 | 98 | (Nghiem, Schaefer, and Elimelech 2005b) |
| | DI | NF-270, pH 5 | 500 | 90 | (Nghiem, Schaefer, and Elimelech 2005b) |
| | Surface | NF-270, pH 5 | 35 | 0 | (Yoon et al. 2006) |
| Naproxen | DI/EfOM | NF, TFC-SR2, pH 6 | 300 | 23 | (Xu et al. 2005) |
| | DI/EfOM | RO, TFC-HR, pH 6 | 300 | 97 | (Xu et al. 2005) |
| | Surface | NF,ESNA, pH 6.8 | 20 | 0 | (Yoon et al. 2006) |
| | surface | RO | 38 | >95.0 | (Heberer 2002) |
| Ketoprofen | DI/EfOM | NF, TFC-SR2, pH 6 | 300 | 25 | (Xu et al. 2005) |
| | DI/EfOM | RO, TFC-HR, pH 6 | 300 | 98 | (Xu et al. 2005) |

(continued)

**Table C.5 (Continued)**

| Compounds | Water Source Type | Membrane Type | Influent (ng/L) | Percent Removal | Reference |
|---|---|---|---|---|---|
| Gemfibrozil | DI/EfOM | NF, TFC-SR2, pH 6 | 300 | 65 | (Xu et al. 2005) |
| | DI/EfOM | RO, TFC-HR, pH 6 | 300 | 95 | (Xu et al. 2005) |
| | Surface | NF,ESNA, 6.8 | 60 | 0 | (Yoon et al. 2006) |
| ***Antibiotics*** | | | | | |
| Sulfamethazine | DI/Surface water | RO, D2731 | 50000 | 90 | (Adams et al. 2002) |
| Trimethoprim | DI/Surface water | RO, D2731 | 50000 | 90 | (Adams et al. 2002) |
| | Surface | NF,ESNA, 6.8 | 80 | 22 | (Yoon et al. 2006) |
| Sulfamethoxazole | DI | NF-90, pH 7 | 500 | >99 | (Nghiem, Schaefer, and Elimelech 2005b) |
| | DI | NF-90, pH 5 | 500 | 98 | (Nghiem, Schaefer, and Elimelech 2005b) |
| | DI | NF-270, pH 7 | 500 | 80 | (Nghiem, Schaefer, and Elimelech 2005b) |
| | DI | NF-270, pH 5 | 500 | 30 | (Nghiem, Schaefer, and Elimelech 2005b) |
| | Surface | NF,ESNA, pH 6.8 | 10 | 18 | (Yoon et al. 2006) |
| | ultra-pure | RO-XLE, pH 7 | 100000 | 70 | (Kimura et al. 2004) |
| | ultra-pure | RO-SC-3100, pH 7 | 100000 | 82 | (Kimura et al. 2004) |
| Erythromycin | Surface | NF,ESNA, pH 6.8 | 45 | 0 | (Yoon et al. 2006) |
| ***Antiepileptics*** | | | | | |
| Primidone | DI/EfOM | NF, TFC-SR2, pH 6 | 300 | 5 | (Xu et al. 2005) |
| | DI/EfOM | RO, TFC-HR, pH 6 | 300 | 91 | (Xu et al. 2005) |
| | ultra-pure | RO-XLE, pH 7 | 100000 | 87 | (Kimura et al. 2004) |
| | ultra-pure | RO-SC-3100, pH 7 | 100000 | 85 | (Kimura et al. 2004) |
| | ultra-pure | NF-ESNA, pH 7 | 100000 | 87 | (Kimura et al. 2003) |
| | ultra-pure | RO-XLE, pH 7 | 100000 | 84 | (Kimura et al. 2003) |

(continued)

Table C.5 (Continued)

| Compounds | Water Source Type | Membrane Type | Influent (ng/L) | Percent Removal | Reference |
|---|---|---|---|---|---|
| | ultra-pure | NF-ESNA, pH 7 | 100 | 72 | (Kimura et al. 2003) |
| | ultra-pure | RO-XLE, pH 7 | 100 | 78 | (Kimura et al. 2003) |
| Carbamazepine | DI | NF-90, pH 7 | 500 | 98 | (Nghiem, Schaefer, and Elimelech 2005b) |
| | DI | NF-90, pH 5 | 500 | 98 | (Nghiem, Schaefer, and Elimelech 2005b) |
| | DI | NF-270, pH 7 | 500 | 80 | (Nghiem, Schaefer, and Elimelech 2005b) |
| | DI | NF-270, pH 5 | 500 | 80 | (Nghiem, Schaefer, and Elimelech 2005b) |
| | Surface | NF,ESNA, pH 6.8 | 125 | 20 | (Yoon et al. 2006) |
| | ultra-pure | RO-XLE, pH 7 | 100000 | 91 | (Kimura et al. 2004) |
| | ultra-pure | RO-SC-3100, pH 7 | 100000 | 85 | (Kimura et al. 2004) |
| | surface | RO | 330 | >99.7 | (Heberer 2002) |
| ***Lipid Regulators*** | | | | | |
| Clofibric acid | surface | RO | 155 | >99.4 | (Heberer 2002) |
| ***Contrast media*** | | | | | |
| Iopromide | Surface | NF,ESNA, pH 6.8 | 105 | 40 | (Yoon et al. 2006) |
| ***Muscle relaxant*** | | | | | |
| Diazepam | Surface | NF,ESNA, pH 6.8 | 55 | 15 | (Yoon et al. 2006) |
| **Steroids** | | | | | |
| ***Estrogens*** | | | | | |
| 17β-Estradiol | DI | NF-90, pH 6 | 100 | 88 | (Ngheim, Schaefer, and Elimelech 2004) |
| | DI | NF-270, pH 6 | 100 | 88 | (Ngheim, Schaefer, and Elimelech 2004) |

(continued)

**Table C.5 (Continued)**

| Compounds | Water Source Type | Membrane Type | Influent (ng/L) | Percent Removal | Reference |
|---|---|---|---|---|---|
| | DI | RO, LFC-1, pH 6.8 | 100 | 90 | (Ng and Elimelech 2004) |
| | DI w/ colloidal fouling | RO, LFC-1, pH 6.8 | 100 | 79 | (Ng and Elimelech 2004) |
| | Surface | NF,ESNA, pH 6.8 | 105 | 20 | (Yoon et al. 2006) |
| | ultra-pure | RO-XLE, pH 7 | 100000 | 83 | (Kimura et al. 2004) |
| | ultra-pure | RO-SC-3100, pH 7 | 100000 | 29 | (Kimura et al. 2004) |
| 17α-Ethinylestradiol | Surface | NF,ESNA, pH 6.8 | 105 | 22 | (Yoon et al. 2006) |
| Estrone | DI | NF-90, pH 6 | 100 | 92 | (Ngheim, Schaefer, and Elimelech 2004) |
| | DI | NF-270, pH 6 | 100 | 88 | (Ngheim, Schaefer, and Elimelech 2004) |
| | Surface | NF,ESNA, 6.8 | 105 | 20 | (Yoon et al. 2006) |
| | ultra-pure | NF,ESNA, pH 6.8 | 100 | 99 | (Schaefer, Nghiem, and Waite 2003) |
| | ultra-pure | RO-TFC-S | 100 | 99 | (Schaefer, Nghiem, and Waite 2003) |
| | ultra-pure | NF-SR1 | 100 | 98 | (Schaefer, Nghiem, and Waite 2003) |
| | ultra-pure | NF-SR2 | 100 | 98 | (Schaefer, Nghiem, and Waite 2003) |
| | ultra-pure | RO-X-20 | 100 | 98 | (Schaefer, Nghiem, and Waite 2003) |
| | ultra-pure | NF-ACM-4 | 100 | 98 | (Schaefer, Nghiem, and Waite 2003) |
| | ultra-pure | NF-XN-40 | 100 | 80 | (Schaefer, Nghiem, and Waite 2003) |
| | ultra-pure | NF-TS-80 | 100 | 98 | (Schaefer, Nghiem, and Waite 2003) |
| ***Androgen*** | | | | | |
| Testosterone | DI | NF-90, pH 6 | 100 | 95 | (Ngheim, Schaefer, and Elimelech 2004) |
| | DI | NF-270, pH 6 | 100 | 85 | (Ngheim, Schaefer, and Elimelech 2004) |
| | Surface | NF,ESNA, pH 6.8 | 100 | 60 | (Yoon et al. 2006) |

(continued)

**Table C.5 (Continued)**

| Compounds | Water Source Type | Membrane Type | Influent (ng/L) | Percent Removal | Reference |
|---|---|---|---|---|---|
| **Personal Care Products** | | | | | |
| ***Surfactants*** | | | | | |
| Nonylphenol | DI | NF-270, pH 6 | 500 | >99 | (Nghiem, Schaefer, and Elimelech 2005a) |
| | DI | NF-90, pH 6 | 500 | >99 | (Nghiem, Schaefer, and Elimelech 2005a) |
| ***Antimicrobial*** | | | | | |
| Triclosan | Surface | NF,ESNA, pH 6.8 | 45 | 96 | (Yoon et al. 2006) |
| ***Insecticide*** | | | | | |
| DEET | Surface | NF,ESNA, pH 6.8 | 75 | 20 | (Yoon et al. 2006) |
| ***Fragrance*** | | | | | |
| Galaxolide | Surface | NF,ESNA, pH 6.8 | 145 | 70 | (Yoon et al. 2006) |
| **Other** | | | | | |
| ***Plasticizer*** | | | | | |
| Bisphenol A | DI/NOM | NF-270, pH 6 | 500 | 50 | (Nghiem, Schaefer, and Elimelech 2005a) |
| | DI | NF-90, pH 6 | 500 | 92 | (Nghiem, Schaefer, and Elimelech 2005a) |
| | ultra-pure/NOM | NF-UTC60 | 50000 | 47 | (Agenson, Oh, and Urase 2003) |
| | ultra-pure/NOM | RO-NTR 729HF | 50000 | 99.8 | (Agenson, Oh, and Urase 2003) |
| | ultra-pure/NOM | RO-UTC70 | 50000 | 99.8 | (Agenson, Oh, and Urase 2003) |
| | ultra-pure/NOM | RO-ES10C | 50000 | >99.9 | (Agenson, Oh, and Urase 2003) |
| | ultra-pure/NOM | RO-LF10 | 50000 | >99.9 | (Agenson, Oh, and Urase 2003) |
| | ultra-pure | RO-XLE, pH 7 | 100000 | 83 | (Kimura et al. 2004) |
| | ultra-pure | RO-SC-3100, pH 7 | 100000 | 18 | (Kimura et al. 2004) |

(continued)

**Table C.5 (Continued)**

| Compounds | Water Source Type | Membrane Type | Influent (ng/L) | Percent Removal | Reference |
|---|---|---|---|---|---|
| | ultra-pure | NF-ESNA, pH 7 | 100000 | 45 | (Kimura et al. 2003) |
| | ultra-pure | RO-XLE, pH 7 | 100000 | 99 | (Kimura et al. 2003) |
| ***Flame Retardants*** | | | | | |
| TCEP | Surface | NF,ESNA, pH 6.8 | 25 | 10 | (Yoon et al. 2006) |
| | ultra-pure/NOM | NF-UTC60 | 50000 | 81 | (Agenson, Oh, and Urase 2003) |
| | ultra-pure/NOM | RO-NTR 729HF | 50000 | >99.9 | (Agenson, Oh, and Urase 2003) |
| | ultra-pure/NOM | RO-UTC70 | 50000 | >99.9 | (Agenson, Oh, and Urase 2003) |
| | ultra-pure/NOM | RO-ES10C | 50000 | >99.9 | (Agenson, Oh, and Urase 2003) |
| | ultra-pure/NOM | RO-LF10 | 50000 | >99.9 | (Agenson, Oh, and Urase 2003) |
| | surface | RO | 360 | >97.2 | (Heberer 2002) |
| TCPP | surface | RO | 945 | >98.9 | (Heberer 2002) |
| ***Stimulant*** | | | | | |
| Caffeine | Surface | NF,ESNA, pH 6.8 | 55 | 10 | (Yoon et al. 2006) |
| | ultra-pure | RO-XLE, pH 7 | 100000 | 70 | (Kimura et al. 2004) |
| | ultra-pure | RO-SC-3100, pH 7 | 100000 | 44 | (Kimura et al. 2004) |
| | surface | RO | 430 | >99.8 | (Heberer 2002) |

## Table C.6
## Removal of select EDCs and PPCPs during riverbank filtration

| Compounds | Water Source Type | Retention time/redox conditions | Influent (ng/L) | Filtrate (ng/L) | Reference |
|---|---|---|---|---|---|
| **Pharmaceuticals** | | | | | |
| ***Analgesics*** | | | | | |
| Diclofenac | surface | 3 weeks | ND – 500 | ND | (Kuehn and Mueller 2000) |
| | surface | – | 50 | ND | (Heberer et al. 2001) |
| | surface | anaerobic | 65 | <LOQ | (Ternes et al. 2002) |
| | wastewater | 18 d | 221 | 177 | (Kreuzinger et al. 2004) |
| | wastewater | 140 d | 221 | 21 | (Kreuzinger et al. 2004) |
| Ibuprofen | wastewater | 18 d | 2 | ND | (Kreuzinger et al. 2004) |
| | wastewater | 140 d | 2 | 2 | (Kreuzinger et al. 2004) |
| | 2° effluent | 2.5 d | 175 | 200 | (Sedlak and Pinkston 2005) |
| | 2° effluent | 15 d | 175 | 10 | (Sedlak and Pinkston 2005) |
| Naproxen | 2° effluent | 2.5 d | 200 | 490 | (Sedlak and Pinkston 2005) |
| | 2° effluent | 15 d | 200 | 10 | (Sedlak and Pinkston 2005) |
| ***Antibiotics*** | | | | | |
| Sulfamethoxazole | surface | – | 7 | ND | (Heberer et al. 2001) |
| Roxithromycin | wastewater | 18 d | 12 | 50 | (Kreuzinger et al. 2004) |
| | wastewater | 140 d | 12 | 2 | (Kreuzinger et al. 2004) |
| ***Antiepileptics*** | | | | | |
| Carbamazepine | surface | 3 weeks | ND – 550 | ND – 200 | (Kuehn and Mueller 2000) |
| | surface | – | 235 | 20 | (Heberer et al. 2001) |
| | surface | anaerobic | 140 – 180 | 25 | (Ternes et al. 2002) |
| | wastewater | 18 d | 1070 | 990 | (Kreuzinger et al. 2004) |
| | wastewater | 140 d | 1070 | 782 | (Kreuzinger et al. 2004) |
| Primidone | surface | – | 105 | 15 | (Heberer et al. 2001) |
| | surface | anaerobic | 15 | 8 | (Ternes et al. 2002) |

(continued)

Table C.6 (Continued)

| Compounds | Water Source Type | Retention time/redox conditions | Influent (ng/L) | Filtrate (ng/L) | Reference |
|---|---|---|---|---|---|
| ***β-blockers*** | | | | | |
| Metoprolol | 2° effluent | 2.5 d | 90 | 25 | (Sedlak and Pinkston 2005) |
| | 2° effluent | 15 d | 90 | 15 | (Sedlak and Pinkston 2005) |
| Propranolol | 2° effluent | 2.5 d | 18 | 15 | (Sedlak and Pinkston 2005) |
| | 2° effluent | 15 d | 18 | 15 | (Sedlak and Pinkston 2005) |
| ***Contrast media*** | | | | | |
| Iopromide | wastewater | 18 d | 3 | 5 | (Kreuzinger et al. 2004) |
| | wastewater | 140 d | 3 | 5 | (Kreuzinger et al. 2004) |
| | surface | anoxic | 860 | <20 | (Schittko, Putschew, and Jekel 2004) |
| Diatrizoate | | | | | (Putschew, Wischnack, and Jekel 2000) |
| | surface | anoxic | 960 | 166 | (Schittko, Putschew, and Jekel 2004) |
| ***Lipid regulators*** | | | | | |
| Clofibric acid | surface | anaerobic | 80 | <LOQ | (Ternes et al. 2002) |
| | surface | – | 50 | 70 | (Heberer et al. 2001) |
| Bezafibrate | surface | anaerobic | 10 | <LOQ | (Ternes et al. 2002) |
| | surface | 3 weeks | - | ND | (Kuehn and Mueller 2000) |
| | surface | – | 170 | ND | (Heberer et al. 2001) |
| | wastewater | 18 d | 14 | 38 | (Kreuzinger et al. 2004) |
| | wastewater | 140 d | 14 | ND | (Kreuzinger et al. 2004) |
| Gemfibrozil | 2° effluent | 2.5 d | 1500 | 600 | (Sedlak and Pinkston 2005) |
| | 2° effluent | 15 d | 1500 | 20 | (Sedlak and Pinkston 2005) |
| **Steroids** | | | | | |
| ***Estrogens*** | | | | | |
| Estrone | surface | – | 0.9 – 1.1 | <0.1 | (Verstraeten et al. 2003) |
| 17β-Estradiol | 3° effluent | 5 d | 4.2 | 0.5 | (Mansell and Drewes 2004) |
| | 3° effluent | 12 mo. | 4.2 | <0.4 | (Mansell and Drewes 2004) |

(continued)

**Table C.6 (Continued)**

| Compounds | Water Source Type | Retention time/redox conditions | Influent (ng/L) | Filtrate (ng/L) | Reference |
|---|---|---|---|---|---|
| | 2° effluent | 1 d | 7.2 | 1.8 | (Mansell and Drewes 2004) |
| | 2° effluent | 2 wk | 7.2 | <0.4 | (Mansell and Drewes 2004) |
| | 2° effluent | 6 d anoxic | 285 | 1.1 | (Mansell and Drewes 2004) |
| | 2° effluent | 23 d anoxic | 285 | <0.4 | (Mansell and Drewes 2004) |
| ***Androgen*** | | | | | |
| Testosterone | 3° effluent | 5 d | 3 | <0.5 | (Mansell and Drewes 2004) |
| | 3° effluent | 12 mo. | 3 | <0.5 | (Mansell and Drewes 2004) |
| | 2° effluent | 1 d | 11.5 | <0.5 | (Mansell and Drewes 2004) |
| | 2° effluent | 2 wk | 11.5 | <0.5 | (Mansell and Drewes 2004) |
| | 2° effluent | 6 d anoxic | 218 | <0.5 | (Mansell and Drewes 2004) |
| | 2° effluent | 23 d anoxic | 218 | <0.5 | (Mansell and Drewes 2004) |
| **Personal Care Products** | | | | | |
| ***Fragrances*** | | | | | |
| Galaxolide | surface | 18 d | 60 | 48 | (Kreuzinger et al. 2004) |
| | wastewater | 140 d | 60 | 56 | (Kreuzinger et al. 2004) |
| Tonalide | wastewater | 18 d | 23 | 39 | (Kreuzinger et al. 2004) |
| | wastewater | 140 d | 23 | 36 | (Kreuzinger et al. 2004) |
| ***Surfactant*** | | | | | |
| Nonylphenol | 3° effluent | 8-14 d | 20 | 8.9 | (Montgomery-Brown et al. 2003) |
| | 3° effluent | 12-18 mo | 20 | ND | (Montgomery-Brown et al. 2003) |
| **Other** | | | | | |
| ***Flame Retardants*** | | | | | |
| TCEP | surface | – | 250 | 105 | (Heberer et al. 2001) |
| TCPP | surface | – | 510 | 180 | (Heberer et al. 2001) |

ND = Not detected; <LOQ = <Limit of quantification; – = Not reported

## REFERENCES


Adams, C., Y. Wang, K. Loftin, and M. Meyer. 2002. Removal of antibiotics from surface and distilled water in conventional water treatment processes. *Journal of Environmental Engineering*, 128(3):253-260.

Agenson, K.O., J.-I. Oh, and T. Urase. 2003. Retention of a wide variety of organic pollutants by different nanofiltration/reverse osmosis membranes: controlling parameters of process. *Journal of Membrane Science*, 225(1-2):91-103.

Alum, A., Y. Yoon, P. Westerhoff, and M. Abbaszadegan. 2004. Oxidation of bisphenol A, 17.beta.-estradiol, and 17.alpha.-ethynyl estradiol and byproduct estrogenicity. *Environmental Toxicology*, 19(3):257-264.

Andreozzi, R., R. Marotta, G. Pinto, and A. Pollio. 2002. Carbamazepine in water: persistence in the environment, ozonation treatment and preliminary assessment on algal toxicity. *Water Research*, 36(11):2869-2877.

Ballard, B.D., and A.A. Mackay. 2005. Estimating the Removal of Anthropogenic Organic Chemicals from Raw Drinking Water by Coagulation Flocculation. *Journal of Environmental Engineering (Reston, VA, United States)*, 131(1):108-118.

Bedner, M., and W.A. Maccrehan. 2006. Reactions of the amine-containing drugs fluoxetine and metoprolol during chlorination and dechlorination processes used in wastewater treatment. *Chemosphere*, 65:2130-2137.

Dodd, M.C., A.D. Shah, U. Von Gunten, and C.-H. Huang. 2005. Interactions of fluoroquinolone antibacterial agents with aqueous chlorine: reaction kinetics, mechanisms, and transformation pathways. *Environmental Science & Technology*, 39(18):7065-7076.

Fuerhacker, M., A. Dürauer, and A. Jungbauer. 2001. Adsorption isotherms on 17ß-estradiol on granular activated carbon (GAC). *Chemosphere*, 44:1573-1579.

Gallard, H., A. Leclercq, and J. Croué. 2004. Chlorination of bisphenol A: kinetics and by-products formation. *Chemosphere.*, 56(5):465-473.

Heberer, T., I.M. Verstraeten, M.T. Meyer, A. Mechlinski, and K. Reddersen. 2001. Occurrence and fate of pharmaceuticals during bank filtration-preliminary results from investigations in Germany and the United States. *Water Resources Update*, 120:4-17.

Heberer, T. 2002. Occurrence, fate, and removal of pharmaceutical residues in the aquatic environment: a review of recent research data. *Toxicology Letters*, 131(1-2):5-17.

Huber, M., T. Ternes, and U. Von Gunten. 2004. Removal of estrogenic activity and formation of oxidation products during ozonation of 17alpha-ethinylestradiol. *Envrironmental Science and Technology*(38):5177-5186.

Huber, M.M., S. Canonica, G.-Y. Park, and U. Von Gunten. 2003. Oxidation of pharmaceuticals during ozonation and advanced oxidation processes. *Environmental Science & Technology*, 37(5):1016-1024.

Huber, M.M., A. Göbel, A. Joss, N. Hermann, D. Löffler, C.S. Mcardell, A. Reid, H. Siegrist, T.A. Ternes, and U. Von Gunten. 2005a. Oxidation of pharmaceuticals during ozonation of municipal wastewater effluents: a pilot study. *Environmental Science & Technology*, 39(11):4290-4299.

Huber, M.M., S. Korhonen, T.A. Ternes, and U. Von Gunten. 2005b. Oxidation of pharmaceuticals during water treatment with chlorine dioxide. *Water Research Water Research*, 39:3607-3617.

Kimura, K., G. Amy, J.E. Drewes, T. Heberer, T.-U. Kim, and Y. Watanabe. 2003. Rejection of organic micropollutants (disinfection by-products, endocrine disrupting compounds, and pharmaceutically active compounds) by NF/RO membranes. *Journal of Membrane Science*, 227(1-2):113-121.

Kimura, K., S. Toshima, G. Amy, and Y. Watanabe. 2004. Rejection of neutral endocrine disprupting compounds (EDCs) and pharmaceutical active compounds (PhACs) by RO membranes. *Journal of Membrane Science*, 245:71-78.

Kreuzinger, N., M. Clara, B. Strenn, and B. Vogel. 2004. Investigation on the behaviour of selected pharmaceuticals in the groundwater after infiltration of treated wastewater. *Water Science and Technology*, 50(2):221-228.

Kuehn, W., and U. Mueller. 2000. Riverbank filtration. An overview. *Journal - American Water Works Association*, 92(12):60-69.

Mansell, J., and J. Drewes. 2004. Fate of steroidal hormones during soil-aquifer treatment. *Ground Water Monitoring & Remediation*, 24(2):94-101.

Mcdowell, D.C., M.M. Huber, M. Wagner, U. Von Gunten, and T.A. Ternes. 2005. Ozonation of carbamazepine in drinking water: identification and kinetic study of major oxidation products. *Environmental Science & Technology*, 39(20):8014-8022.

Montgomery-Brown, J., J. Drewes, P. Fox, and M. Reinhard. 2003. Behavior of alkylphenol polyethoxylate metabolites during soil aquifer treatment. *Water Research*, 37:3672-3681.

Ng, H.Y., and M. Elimelech. 2004. Influence of colloidal fouling on rejection of trace organic contaminants by reverse osmosis. *Journal of Membrane Science*, 244:215-226.

Ngheim, L.D., A.I. Schaefer, and M. Elimelech. 2004. Removal of natural hormones by nanofiltration membranes: measurement, modeling, and mechanisms *Environmental Science & Technology*, 38:1888-1896.

Nghiem, L., A. Schaefer, and M. Elimelech. 2005a. Nanofiltration of hormone mimicking trace organic contaminants. *Separation Science and Technology*, 40(13):2633-2649.

Nghiem, L.D., A.I. Schaefer, and M. Elimelech. 2005b. Pharmaceutical retention mechanisms by nanofiltration membranes. *Environmental Science & Technology*, 39:7698-7705.

Putschew, A., S. Wischnack, and M. Jekel. 2000. Occurrence of triiodinated X-ray contrast agents in the aquatic environment. *Science Of The Total Environment*, 255(1-3):129-134.

Rosenfeldt, E., and K.G. Linden. 2004. Degradation of endocrine disrupting chemicals bisphenol A, ethinyl estradiol, and estradiol during UV photolysis and advanced oxidation processes. *Environmental Science & Technology*, 38(20):5476-5483.

Schaefer, A.I., L.D. Nghiem, and T.D. Waite. 2003. Removal of the natural hormone estrone from aqueous solutions using nanofiltration and reverse osmosis. *Environmental Science & Technology*, 37(1):182-188.

Schittko, S., A. Putschew, and M. Jekel. 2004. Bank filtration: a suitable process for the removal of iodinated X-ray contrast media? *Water Science and Technology*, 50(5):261-268.

Sedlak, D., and K.E. Pinkston. 2005. *Occurrence survey of pharmaceutically active compounds*. Denver, Colorado: AwwaRF and WERF.

Tanghe, T., and W. Verstraete. 2001. Adsorption of nonylphenol onto granular activated carbon. *Water Air and Soil Pollution*, 131:61-72.

Ternes, T.A., M. Meisenheimer, D. Mcdowell, F. Sacher, H.-J. Brauch, B. Haist-Gulde, G. Preuss, U. Wilme, and N. Zulei-Seibert. 2002. Removal of pharmaceuticals during drinking water treatment. *Environmental Science & Technology*, 36(17):3855-3863.

Verstraeten, I.I., T. Heberer, J. Vogel, T.F. Speth, S. Zuehlke, and U. Duennbier. 2003. Occurrence of endocrine-disrupting and other wastewater compounds during water treatment with case studies from Lincoln, Nebraska and Berlin, Germany. *Practice Periodical of Hazardous,Toxic, and Radioactive Waste Management*, 7(4):253-263.

Vieno, N.M., T. Tuhkanen, and L. Kronberg. 2005. Seasonal variation in the occurrence of pharmaceuticals in effluents from a sewage treatment plant and in the recipient water. *Environmental Science & Technology*, 39(21):8220-8226.

Westerhoff, P., Y. Yoon, S. Snyder, and E. Wert. 2005. Fate of endocrine-disruptor, pharmaceutical, and personal care product chemicals during simulated drinking water treatment processes *Environmental Science & Technology*, 39(17):6649-6663.

Xu, P., J.E. Drewes, C. Bellona, G. Amy, T.-U. Kim, M. Adam, and T. Heberer. 2005. Rejection of emerging organic micropollutants in nanofiltration-reverse osmosis membrane applications. *Water Environment Research*, 77(1):40-48.

Yoon, Y., P. Westerhoff, S.A. Snyder, and M. Esparza. 2003. HPLC-fluorescence detection and adsorption of bisphenol A, 17β-estradiol, and 17α-ethynyl estradiol on powdered activated carbon. *Water Research*, 37(14):3530-3537.

Yoon, Y., P. Westerhoff, S.A. Snyder, and E.C. Wert. 2006. Nanofiltration and ultrafiltration of endocrine disrupting compounds,pharmaceuticals and personal care products. *Journal of Membrane Science*, 270:88-100.

Zwiener, C., and F.H. Frimmel. 2000. Oxidative treatment of pharmaceuticals in water. *Water Research*, 34(6):1881-1885.

# APPENDIX D
# SELECTED STUDIES EMPLOYING BIOLOGICALLY-BASED ASSAYS DIRECTED AT EDCS AND PPCPS

## Table D.1
## Selected studies employing biologically-based assays directed at EDCs and PPCPs

| Chemical Category | Chemical or Mode of Action | Assay Class | Assay | Detection Limit | Matrix | Reference |
|---|---|---|---|---|---|---|
| EDCs | Thyroid | In vivo | T4 immunoreactivity assay | 0.003% PTurea | Water | Elsalini and Rohr 2002 |
| EDCs | Androgenic | In vitro | Rainbow trout androgen receptor binding assay | <6.5 ng/L testosterone EQ | Raw and treated sewage | Leusch et al. 2005 |
| EDCs | Androgenicity | In vitro | A-screen | <0.5 pM AEq | Surface water | Soto et al. 2004 |
| EDCs | Carcinogenicity/ reproductive toxicity | In vivo | OECD TG 407 assay | Tox endpoints | Diet | Fukushima and Freyberger 2003 |
| EDCs | dieldrin, pyriproxyfen, fenoxycarb, methoprene; chemicals: trans-retinoic acid, phytol, phytanic acid, 20-hydroxyecdysone, fenarimol, azadirachtin,amyl phenol, chlordane, piperonyl butoxide, 4-nonylphenol, methoprene, kinoprene, bisphenol A | In vitro | Production of male offspring | 1 µg/L fenoxycarb, 0.1 µg/L pyriproxyfen, 310 µg/L methoprene, 100 µg/L dieldrin (all others did not show significant results experimentally) | Water | Wang et al. 2005 |
| EDCs | Estrogenic | In vitro | E-screen | 0.05 pm EEQ l (0.014 ng EEQ l) | Sewage treatment plant effluent/ influent | Korner et al. 1999 |
| EDCs | Estrogenic | In vitro | E-screen, YES, Ishikawas cell ALP | Cells can respond to 10 pM E2 and 1 pM E1 | River water | Matusuoka et al. 2005 |

(continued)

**Table D.1 (Continued)**

| Chemical Category | Chemical or Mode of Action | Assay Class | Assay | Detection Limit | Matrix | Reference |
|---|---|---|---|---|---|---|
| EDCs | Estrogenic | In vitro | Estrogen receptor binding assay, YES, and ER-CALUX | | WWTP influent and effluent, river water, suspended matter, sludge | Murk et al. 2002 |
| EDCs | Estrogenic | In vitro | E-screen, ER binding assay, aromatase assay, EROD assay | | River water | Oh et al. 2006 |
| EDCs | Estrogenic | In vitro | Estrogen receptor competitive binding assay | ~1 nM | River water, groundwater | Quanrud et al. 2004 |
| EDCs | Estrogenic | In vitro | YES | 2 ng 17ß-estradiol/L | Sewage treatment plant effluent | Routledge 2003 |
| EDCs | Estrogenic | In vitro | ER and DR CALUX | Followed by GC-MS/MS: 68 pg α-E2/g dw, 37 pg E1/g dw, 47 pg E2/g dw, 157 pg E3/g dw and 55 pg EE2 pg/g dw | Harbor surface sediments | Houtman et al. 2006 |
| EDCs | Estrogenic | In vitro | E-screen | HPLC ~1 nM | Simulated drinking water | Alum et al. 2004 |
| EDCs | Estrogenic | In vitro | YES | | Simulated drinking water | Hu et al. 2003 |
| EDCs | Estrogenic | In vitro | YES | 15 ng/L E2 | Raw and drinking water | Fawall et al. 2000 |
| EDCs | Estrogenic | In vitro | Sheep estrogen receptor binding assay | <1 ng/L EEQ | Raw and treated sewage | Leusch et al. 2005 |
| EDCs | Estrogenic | In vivo | ELISA for vitellogenin | 0.8 ng EE2/L | diet | Liao et al. 2006 |

(continued)

**Table D.1 (Continued)**

| Chemical Category | Chemical or Mode of Action | Assay Class | Assay | Detection Limit | Matrix | Reference |
|---|---|---|---|---|---|---|
| EDCs | Estrogenic | In vivo | Trout vitellogenin | 0.2 ng/µL for xeno estrogens and 0.05 ng/mL for steroids | Sewage treatment plant effluent | Petterson et al. 2006. |
| EDCs | Estrogenic | In vitro | Human ER a test in yeast | 0.1 ng EEQ/L | Raw and treated sewage | Svenson et al. 2003 |
| EDCs | Estrogenic | In vitro | E-screen | 1-3 pg/L | Surface water | Soto et al. 2004 |
| EDCs | Thyroid | In vivo | Xenopus tail resorption | Lowest doses tested (mg/L): Methimazole 6.25, 6-propylthiouracil 1.25, Thyroxine 0.25 | Water | Degitz et al. 2005 |
| EDCs | Thyroid ($KClO_4$) | In vitro | uptake and organification by the larval lamprey endostyle | 0.72 mM | Water | Manzon and Youson 2002 |
| EDCs (nonylphenol) | Estrogenic | In vitro | ELISA | 2.3 ± 0.9 µg/L | Water | Estevez et al 2006 |
| EDCs, PCBs, PCDD/Fs | Estrogenic | In vitro | E-screen assay (MCF-7 proliferation) | | Leachate | Behnisch et al 200 |
| EDCs Pesticides, phytoestrogens | Estrogenic | In vitro | *β*-gal reporter assay in yeast | | Culture media | Graumann et al. 199 |
| Pharmaceutical | Carbamazepine | In vivo | Multiple endpoints (EC50, growth, mitotic index, micronuclei, bioluminescence, immobilization, protein content, LDH leakage, MTT, neutral red, LDH activitiy, G6PDH activity | 19 µM in Vero cells was the most sensitive for EC50 | Carbamazepine added to water/media | Jos et al. 2003 |

(continued)

**Table D.1 (Continued)**

| Chemical Category | Chemical or Mode of Action | Assay Class | Assay | Detection Limit | Matrix | Reference |
|---|---|---|---|---|---|---|
| Pharmaceuticals | Diclofenac | In vitro | ELISA | 0.020 μg/L | Water | Richardson and Ternes, 2005, citing Deng et al. 2003 |
| Pharmaceuticals | Tetracyclines, sulfonamides | In vitro | Radioimmunoassay | 0.05 μg/L | Water | Richardson and Ternes, 2005, citing Yang and Carlson, 2004 |

Aeq- androgen equivalents; dw – dry weight; E1 – estrone; E2 – 17ß-estradiol; E3 – estriol; EC50 – effective concentration 50%; EE2 – 17α-ethynyl estradiol; EEQ – estradiol equivalents; ER – estrogen receport; ER-CALUX assay – estrogen receptor-mediated chemically activated luciferase gene expression assay; WWTP - wastewater treatment plant; YES – yeast estrogen screen

**Alum et al. 2004.** Used E-screen to determine estrogenicity of BPA, E2 and EE2 oxidation products.

**Barnes et al. 2004.** Ground water samples collected from the Norman Landfill research site in central Oklahoma were analyzed as part of the USGS Toxic Substances Hydrology Program's national reconnaissance of pharmaceuticals and other organic waste water contaminants (OWCs) in ground water. Four analytical methods were used to determine the environmental extent of 76 OWCs in these ground water samples. The analyzed compounds can be divided into groups based on their association with human, industrial, and agricultural waste waters, and include antibiotics, prescription and nonprescription drugs, steroids, personal care products, products of oil use and combustion, and other extensively used chemicals. Twenty-one antibiotic compounds were extracted and analyzed by tandem solid-phase extraction (SPE) and single quadrapole, liquid chromatography/mass spectrometry with electrospray ionization set in positive mode and selected ion monitoring (SIM). Eighteen human prescription and nonprescription drugs, and selected metabolites, were extracted by SPE and measured by high performance liquid chromatography (HPLC/MS) using a polar reverse-phase octylsilane (C8) HPLC column. Forty-three OWCs were extracted using continuous liquid-liquid extraction and measured by capillary-column gas chromatography/mass spectrometry (GC/MS). Two steroid compounds were analyzed by GC/MS. Eight compounds were analyzed by more than one method.

**Behnisch et al 2001.** The estrogenic activity (by E-screen bioassay), the concentrations of PCBs, PCDDs/PCDFs and several EDCs were analyzed from leachates of each step (before treatment, after biodegradation/sedimentation and after charcoal treatment) of a controlled landfill leachate treatment plant. In order to obtain a sufficient risk assessment of this wastewater recycling process, they established and applied a bioassay (the E-screen) to measure substances exhibiting an endocrine receptor (ER) mediated activity, regardless of their chemical structure, in combination with chemical analysis of some already evaluated ER-agonists. Other receptor agonists, like the Ah receptor mediated coplanar PCBs and dioxins were also considered and detected by chemical analysis. Performance of the E-screen assay: Cultivation of the MCF-7 cells and performance of the proliferation experiment were carried out according to the principal method described by Korner et al. (1999, 2000).

**Belfroid et al. 1999.** An analytical procedure was developed that enables routine analysis of four estrogenic hormones in concentrations below 1 ng/L in surface water and waste water. The formed complex was measured with GC-MS/MS.

**Bláha et al. 2006.** Used the human adrenocortical carcinoma cell line H295R to screen for endocrine disrupting chemicals that affect the expression of genes important in steroidogenesis. Organic contaminants from freshwater pond sediments was evaluated.

**Brain et al. 2004.** Used 15 to 12,000 L aquatic microcosms (contains rooted and floating plants) treated with eight common pharmaceuticals (atorvastatin, acetaminophen, caffeine, sulfamethoxazole, carbamazepine, levofloxacin, sertraline, and trimethoprim). The microcosm data suggest that at an ecological effect size of >20%, biologically significant risks are low for *L. gibba* and *M. sibiricum* exposed to similar mixtures of pharmaceutical compounds.

**Cordy et al 2004.** A proof-of-concept experiment was devised to determine if pharmaceuticals and other organic waste water compound (OWCs), as well as pathogens, found in treated effluent could be transported through a 2.4 m soil column and, thus, potentially reach ground water under recharge conditions similar to those in arid or semiarid climates. Samples were analyzed using a variety of experimental methods developed for the USGS Toxic Substances Hydrology Program. An overview of each of the analytical methods used in this experiment is in Koplin et al. (2002).

**Degitz et al. 2005.** Xenopus tail resorption assay for methimazole (control, 6.25, 12.5, 25, 50, 100 mg/l), 6-propylthiouracil (PTU) (control, 1.25, 2.5, 5, 10, and 20 mg/l), and thyroxine (T4) (0.25, 0.5, 1, 2, 4 μg/l) in water. Both compounds caused concentration dependent changes in thyroid gland morphology. These changes were characterized as reduced colloid, glandular hypertrophy, and cellular hyperplasia and hypertrophy. Treatment failed to negatively affect growth, even in tadpoles that experienced significant metamorphic inhibition. T4 treatment resulted in a concentration dependent increase in developmental rate, as would be expected. Similar to studies with methimazole, there were no differences in sensitivity among the two developmental stages examined. These results indicate that tadpoles in the early stages of metamorphosis are sensitive to thyroid axis disruption and that development of a short-term, diagnostic amphibian-based thyroid screening assay shows considerable promise.

**Elsalini and Rohr 2002.** Commonly used doses of 0.003% PTurea abolish T4 immunoreactivity of the thyroid follicles of zebrafish larvae. As development of the thyroid gland is not affected, these data suggest that PTurea blocks thyroid hormone production. Like other goitrogens, PTurea causes delayed hatching, retardation and malformation of embryos or larvae with increasing doses. At doses of 0.003% PTurea, however, toxic side effects seem to be at a minimum, and the maternal contribution of the hormone might compensate for compromised thyroid function during the first days of development.

**Estevez et al. 2006.** The development of an enzyme-linked immunosorbent assay (ELISA) for the dectection of technical nonylphenol (NP) is reported. A reproducible and sensitive indirect competitive ELISA has been finally obtained, reaching a limit of detection of $2.3 \pm 0.9$ µg/L. A preliminary evaluation of the analytical protocol established has been performed using real water samples.

**Fawell et al. 2000.** Evaluated surface and drinking water using trout vitellogenin and YES assays. There was little or no evidence of estrogenicity, even with high amounts of sewage effluent.

**Fukushima and Freyberger 2003**. Study protocols for the characterization of endocrine active compounds presented in Workshop 4 included the enhanced Organization for Economic Cooperation and Development (OECD) test guideline (TG) 407, the medium-term rat liver and rat multi-organ carcinogenicity assays, and an enhanced one-generation reproduction study.

**Graumann et al. 1999.** This study uses a β-gal reporter assay trasfected with a human estrogen receptor gene in yeast to determine estrogenic activity of pesticides (endosulfan, dieldrin, atrazine, and the main metabolites, desethylatrazine and desisopropylatrazine) along with phytoestrogens (betulinic acid, curcurmin, sitosterol, phloretin, diosmin, sarsasapogenin, podocarpic acid, naringenin). They found that these compounds are either weak estrogens or completely lack estrogenic potential suggesting endocrine disrupting potential in more complex organisms is due to mechanisms other than transactivation of the estrogen receptor.

**Houtman et al. 2006.** Used ER and dioxin response chemical activated luciferase reporter expression (CALUX) assay to test levels of EDCs and dioxin-like chemicals in a harbor in the Netherlands. ER-CALUX was then followed by GC-MS/MS. The main contributors to the estrogenicity were 17-β-estradiol and its metabolite estrone.

**Hu et al. 2003.** Assessed the estrogenic activity from E2 in drinking water using YES and ESI-LC-MS. The estrogenic activity after 120 and 180 minutes appears to be due to 2,4-dichloro-E2 and 2,4-dichloro-E1.

**Jos et al. 2003.** The effects of carbamazepine were investigated using a variety of systems including vegetables, bacteria, a crustacean, and cell cultures from monkey and fish origin were evaluated at 24, 48 and 72 hours. Monkey cells were the most sensitive with an EC50 of 19uM.

**Körner et al. 1999.** E-screen is used to study nine effluent samples and 1 influent sample from municipal sewage plants in South Germany. All samples strongly induced cell proliferation in a dose-dependent manner which was completely inhibited with an estrogen receptor inhibitor. The proliferative effect relative to 17-β–estradiol was between 30 and 101% and the 17-β–estradiol equivalent concentrations were between 2.5 and 25 ng/l.

**Leusch et al. 2005.** Estrogenic and androgenic activities from raw and treated sewage were determined with sheep estrogen receptor and rainbow trout androgen receptor binding assays, respectively. The raw sewage influents contained significant levels of both estrogenic

(<4–185 ng/L estradiol equivalents) and androgenic (1920–9330 ng/L testosterone equivalents) activity. Subsequent treatment of raw sewage successfully removed most of the activity so that the estrogenicity and androgenicity associated with the final effluents were very low (<1–4.2 ng/L estradiol equivalents and <6.5–736 ng/L testosterone equivalents, respectively). Secondary treatment was the most effective treatment step to remove estrogenic and androgenic activity from sewage water. Activated sludge treatment in particular removed 92% to >99% of the estrogenic activity and 82% to >99% of the androgenic activity in sewage.

**Liao et al. 2006.** ELISAs used to determine vitellogenin in rare minnow fed tubifex from wastewater treatment plant, Artemia nauplii and commercial pellet food. Lowest observed effect concentrations were 0.8 ng EE2/l for rare minnow compared to 4 ng EE2/l for zebrafish.

**Mackenzie et al. 2005.** Assesses the decline in male to female birth ration in the Aamjiwnaang First Nation during the time period 1984 to 2003. The decline was statistically significant over the most recent 10 years with the most pronounced decline over the most recent 5 years. It is acknowledged that there are several potential factors contributing to this phenomenon and may include a number of environmental and occupational chemical exposures.

**Manzon and Youson 2002.** An in vitro experimental system was devised to assess the direct effects of the goitrogen, potassium perchlorate (KClO(4)), on radioiodide uptake and organification by the larval lamprey endostyle. When KClO(4) was added to the incubation medium, both iodide uptake and organification by the endostyle were significantly reduced relative to controls as determined by gamma counting, and gel-autoradiography and densitometry, respectively.

**Martin et al. 2005.** Validates existing estrogenic in vitro assays using ROC curves to indicate efficacy. These tests were the ER binding assay, yeast estrogen screen (YES), human reporter gene assay and E-screen and were referent to the rodent uterotrophic assay. All were found to be effective diagnostic tests.

**Matsuoka et al. 2005.** Evaluates estrogenic activity in river water by E-screen Ishikawas cell-ALP and YES. The samples were taken from river water a three locations on the Muko River in Japan. The values ranged from a non-detect to 32.9 ng/l E2 equivalents and were similar between all tests except for YES. Yeast cells are less permeable to estrogenic compounds and were not able to penetrate the cell wall as effectively. As such, all YES tests were below the non-detect.

**Murk et al. 2002.** A study was performed to optimize sample preparation and application of three in vitro assays for measuring estrogenic potency in environmental extracts. The three assays applied were an estrogen receptor (ER)-binding assay and two reporter gene effect assays: a yeast estrogen screen (YES) and the ER-mediated chemically activated luciferase gene expression (ER-CALUX) assay. All assays were able to detect estrogenicity, but the amounts of material needed for the assays differed greatly between the three assays (ER-binding assay YES > ER-CALUX). In addition, in the ER-binding assay, both agonists and antagonists give an estrogenic response, resulting in higher estradiol equivalency (EEQ) levels than both the ER-CALUX and the YES assay for the same samples. The EEQs found in wastewater treatment plants (WTPs) with the ER-CALUX assay were in the range of 4 to 440 and 0.11 to 59 pmol/L for influent and effluent, respectively. Water extracts from four large rivers had levels ranging from 0.25 to 1.72 pmol/L. Extracts from suspended matter and sludge contained estrogenic potency of 0.26 to 2.49 and 1.6 to 41 pmol EEQ/g dry weight, respectively. In WTPs, the average reduction of estrogenic potency in effluent compared to influent was 90 to 95% in municipal WTPs and about 50% in industrial WTPs. In influent, 30% of the ER-CALUX activity could not be explained by the calculated potencies based on chemical analysis of a number of known (xeno)estrogens; in effluent the unexplained fraction was 80%. These first results of analyzing estrogenic potency in WTP water and surface water in The Netherlands indicate that further studies are warranted to investigate the actual risks for aquatic systems.

**Oh et al. 2006.** Evaluated river water from the Youngsan River tributaries in Korea for estrogenic and dioxin-like activity. Tests used were the ER binding assay, E-screen, aromatase assay, and EROD. Total toxic effects of downstream water were higher than that of upstream water. Estrogenic activity in upstream water was between 0.005 and 0.049 ng-EEQ/l with no CYP1A activity. In downstream water, the estrogenic activity was between 0.021 ng-EEQ/l and 1.918 ng-EEQ/l with a CYP1A activity between 0.63 and 3.55 μg/MEQ/L.

**Petterson et al. 2006.** To evaluate the effluents from two modern Swedish sewage treatment plants, trout vitellogenin was used as a bioassay and then analyzed bile fluid for estrone, BPA, and 4-nonylphenol by GC/MS. Higher concentrations of estrone, BPA and 4-nonylphenol were found in the fish with elevation of vitellogenin.

**Quanrud et al. 2004.** Estrogenic activity was tested in waste water effluent along the Santa Cruz River in Arizona using the ER competitive binding assay. Estrogenic activity decreased by ~60% over the 25 mile length of river below the effluent discharge points in Tucson. Ground water estrogenic activity was highly correlated to fractional waste water content.

**Routledge 2003.** The study aimed to identify and quantify the estrogenic substances present in domestic sewage effluent by fractionating it into simpler and simpler fractions, and assess their effects on fish at environmentally relevant concentrations. Estrogenic activity was assessed with the YES assay. Three estrogenic steroids (E2, E1, and EE2) were found to be active in the effluent fractions. They were not able to conclude whether the concentrations of E2 and E1 would be estrogenic to fish.

**Saradha and Mathur 2006.** Reviews androgen disruptor mechanisms: interactions with steroid receptors leading to interference with developmental and functional aspects of testes, epididymes, and accessory sex organs and induction of ROS and the role of oxidative stress in defective sperm function and male infertility.

**Soto et al. 2004.** Assessed whether feedlot effluent contaminates watercourses by measuring *a*) total androgenic [methyltrienolone (R1881) equivalents] and estrogenic (17ß-estradiol equivalents) activity using the A-SCREEN and E-SCREEN bioassays and *b*) concentrations of anabolic agents via gas chromatography-mass spectroscopy and enzyme-based immunoassays. Tap water was devoid of hormonal activity, but other samples had varying amounts of estrogenicity and adrogenicity. They conclude that feedlot effluents contain sufficient levels of hormonally active agents to warrant further investigation of possible effects on aquatic ecosystem health.

**Svenson et al. 2003.** The human estrogen receptor α-test, hosted in a yeast strain, was used to quantify estrogenicity in samples of untreated and treated effluents from 20 Swedish municipal sewage treatment plants. The discharge from Swedish domestic sewage treatment plants contained estrogenic compounds corresponding to <0.1–15 ng estradiol equivalents/L. Low levels of estrogenic activity were also found in a river receiving municipal effluents, 3.5–35 km downstream the outlet from a sewage treatment works. The range of estrogenicity in untreated, raw sewage effluents was found to be 1–30 ng estradiol equivalents/L.

**Ternes et al. 1999a.** Describes methodology for quantification of estrogens in sewage in river water.

**Ternes et al. 1999b.** Measured persistence of natural estrogens in a diluted slurry of activated sludge from a sewage treatment plant in Germany.

**Wang et al. 2005.** Developed and evaluated a screening assay that could be employed to detect juvenoid-related endocrine-modulating activity in an invertebrate species. Juvenoid activity, anti-juvenoid activity, and juvenoid potentiator activity of chemicals was assessed using the water flea Daphnia magna. Male sex determination is under the regulatory control of juvenoid hormone, presumably methyl farnesoate, and this endpoint was used to detect juvenoid modulating activity of chemicals. Eighteen chemicals were evaluated for juvenoid agonist activity. Positive responses were detected with the juvenoid hormones methyl farnesoate and juvenile hormone III along with the insect growth regulating insecticides pyriproxyfen, fenoxycarb, and methoprene. Weak juvenoid activity also was detected with the cyclodiene insecticide dieldrin. Assays performed repetitively with compounds that gave either strong positive, weak positive, or negative response were 100% consistent indicating that the assay is not prone to false positive or negative responses. Five candidate chemicals were evaluated for anti-juvenoid activity and none registered positive. Four chemicals (all trans-retinoic acid, methoprene, kinoprene, bisphenol A) also were evaluated for their ability to potentiate the activity of methyl farnesoate. All registered positive. Results demonstrate that an in vivo assay with a crustacean species customarily employed in toxicity testing can be used to effectively screen chemicals for juvenoid-modulating activity.

**Ying et al. 2004.** Investigated the role of waste water injected into an aquifer on the degredation of two EDCs, E2 and EE2. EE2 was resistant to biodegradation where E2 degraded in about 2 days under aerobic conditions, but slowly in anaerobic conditions.

**Zuehlke et al. 2004.** Describes a methodology for determination of E1, E2, and EE2 using solid-phase extraction and LC/MS. This methodology was used on samples from purified sewage and surface, ground, and drinking water. Samples were taken from municipal waste water treatment plants, a surface water treatment plant, a bank filtration site and a ground water enrichment pond.

# REFERENCES

ACGIH (American Conference of Governmental Industrial Hygienists). 2006. TLVs and BEIs. Threshold Limit Values for Chemical Substances and Physical Agents and Biological Exposure Indices. Cincinnati, OH.

ACS (American Cancer Society). 2006. Cancer Facts & Figures 2006. Atlanta, GA. [Online] Available at : <www.cancer.org/downloads/STT/CAFF2006PWSecured.pdf>

(UBA), U. (2003). "Bewertung der Anwesenheit teil- oder nicht bewertbarer Stoffe im Trinkwasser aus gesundheitlicher Sicht." Bundesgesundheitsblatt-Gesundheitsforsch-Gesundheitsschutz **46**: 249-251.

Acero, J. L., E. Rodriguez, et al. (2005). "Kinetics of reactions between chlorine and the cyanobacterial toxins microcystins." Water Research **39**: 1628-1638.

Adams, C., Y. Wang, et al. (2002). "Removal of antibiotics from surface and distilled water in conventional water treatment processes." Journal of Environmental Engineering **128**(3): 253-260.

Ahrer, W., E. Scherwenk, et al. (2001). "Determination of drug residues in water by the combination of liquid chromatography or capillary electrophoresis with electrospray mass spectrometry." Journal of Chromatography A **910**: 69-78.

Alum, A., Y. Yoon, et al. (2004). "Oxidation of bisphenol A, 17.beta.-estradiol, and 17.alpha.-ethynyl estradiol and byproduct estrogenicity." Environmental Toxicology **19**(3): 257-264.

Alvarez, D. A., J. D. Petty, et al. (2004). "Development of a passive, in situ, integrative sampler for hydrophilic organic contaminants in aquatic environments." Environmental Toxicology and Chemistry **23**(7): 1640-1648.

Alvarez, D. A., P. E. Stackelberg, et al. (2005). "Comparison of a novel passive sampler to standard water-column sampling for organic contaminants associated with wastewater effluents entering a New Jersey stream." Chemosphere **61**: 610-622.

Anderson, P. D. (2005). Technical Brief: Endocrine Disrupting Compounds and Implications for Wastewater Treatment. Alexandria, Virginia, Water Environment Research Foundation (WERF)**:** 36.

Andreozzi, R., R. Marotta, et al. (2002). "Carbamazepine in water: persistence in the environment, ozonation treatment and preliminary assessment on algal toxicity." Water Research **36**(11): 2869-2877.

Andreozzi, R., M. Raffaele, et al. (2003). "Pharmaceuticals in STP effluents and their solar photodegradation in aquatic environment." Chemosphere **50**(10): 1319-1330.

Arthur, C. L., L. M. Killam, et al. (1992). "Analysis of Substituted Benzene Compounds in Groundwater Using Solid-Phase Microextraction." Environmental Science & Technology **26**(5).

Arthur, C. L. and J. Pawliszyn (1990). "Solid phase microextraction with thermal desorption using fused silica optical fibers." Analytical Chemistry **62**: 2145-2148.

Ashton, D., M. Hilton, et al. (2004). "Investigating the environmental transport of human pharmaceuticals to streams in the United Kingdom." Science of the Total Environment **333**: 167-184.

Ayscough, N. J., J. Fawell, et al. (2000). Review of human pharmaceuticals in the environment. Bristol, Environment Agency.

Ayscough, N. J., J.Fawell, et al. (2000). Review of Human Pharmaceuticals in the Environment. R&D Tech. Report P390. WRc Report No. EA4761. WRc-NSF Ltd, Buckinghamshire, Bristol: Environment Agency.

Balk, F. and R. Ford (1999). "Environmental risk assessment for the polycyclic musks, AHTN and HHCB. II. Effect assessment and risk characterisation." Toxicology Letters **111**(1-2): 81-94.

Ballard, B. D. and A. A. MacKay (2005). "Estimating the Removal of Anthropogenic Organic Chemicals from Raw Drinking Water by Coagulation Flocculation." Journal of Environmental Engineering (Reston, VA, United States) **131**(1): 108-118.

Barber, L. B., E. T. Furlong, et al. (2003). Natural and contaminant organic compounds in the Boulder Creek watershed, Colorado during high-flow and low-flow conditions, 2000. Comprehensive Water Quality of the Boulder Creek Watershed, Colorado, During High-Flow and Low-Flow Conditions, 2000. Denver, Colorado, USGS.

Barnes, K. L., D. W. Kolpin, et al. (2002). Water-Quality Data for Pharmaceuticals, Hormones, and Other Organic Wastewater Contaminants in U.S. Streams, 1999-2000 Iowa City, Iowa, USGS.

Baronti, C., R. Curini, et al. (2000). "Monitoring natural and synthetic estrogens at activated sludge sewage treatment plants and in a receiving river water." Environmental Science & Technology **34**(24): 5059-5066.

Bedner, M. and W. A. MacCrehan (2006). "Reactions of the amine-containing drugs fluoxetine and metoprolol during chlorination and dechlorination processes used in wastewater treatment." Chemosphere **65**: 2130-2137.

Bellona, C. and J. E. Drewes (2005). "The role of membrane surface charge and solute physico-chemical properties in the rejection of organic acids by NF membranes." Journal of Membrane Science **249**: 227-234.

Bellona, C., J. E. Drewes, et al. (2004). "Factors affecting the rejection of organic solutes during NF/RO treatment - A literature review." Water Research **38**: 2795-2809.

Bennie, D. T., C. A. Sullivan, et al. (1997). "Occurence of alkylphenols and alkylphenol mono- and diethoxylates in natural waters of the Laurentian Great Lakes basin and the upper St. Lawrence River." The Science of the Total Environment **193**: 263-275.

Bevans, H. E., S. L. Goodbred, et al. (1996). "Synthetic organic compounds and carp endocrinology and histology in Las Vegas Wash and Las Vegas and Callville Bays of Lake Mead, Nevada, 1992 and 1995." Water-Resources Investigations Report 96-4266

Boernick, H., P. Eppinger, et al. (2001). "Simulation of biological degradation of aromatic amines in river bed sediments." Water Research **35**(3): 619-624.

Boreen, A. L., W. A. Arnold, et al. (2004). "Photochemical fate of sulfa drugs in the aquatic environment: sulfa drugs containing five-membered heterocyclic gorups." Environmental Science & Technology **38**(14): 3933-3940.

Boyd, G. R., S. Zhang, et al. (2005). "Naproxen removal from water by chlorination and biofilm processes." Water Research **39**: 668-676.

Brauch, H.-J., F. Sacher, et al. (2000). "Efficiency of bank filtration for the removal of hydrophilic organic micropollutants." GWF, Wasser/Abwasser **141**(4): 226-234.

Brookbanks, E. O., R. A. H. Welch, et al. (1969). "Estrogens in pasture and a possible relation with mastitis." New Zealand Veterinary Journal **17**: 159-160.

Bruce, D. P., P. Westerhoff, et al. (2002). "Removal of 2-nethylisoborneol and geosmin in surface water treatment plants in Arizona." Journal of Water Supply: Research and Technology - AQUA **51**(4): 183-198.

Bruchet, A., C. Hochereau, et al. (2005). "Analysis of drugs and personal care products in French source and drinking waters: the analytical challenge and examples of application." Water Science and Technology **52**: 53-61.

Bruchet, A., C. Prompsy, et al. (2002). "A broad spectrum analytical scheme for the screening of endocrine disruptors (EDs), pharmaceuticals and personal care products in wastewaters and natural waters." Water Science and Technology **46**(3): 97-104.

Buerge, I. J., T. Poiger, et al. (2003). "Caffeine, an Anthropogenic Marker for Wastewater Contamination of Surface Waters." Environmental Science and Technology **37**(4): 691-700.

Buser, H. R., T. Poiger, et al. (1998). "Occurrence and fate of the pharmaceutical drug diclofenac in surface waters: Rapid photodegradation in a lake." Environmental Science & Technology **32**(22): 3449-3456.

Buser, H. R., T. Poiger, et al. (1999). "Occurrence and environmental behavior of the chiral pharmaceutical drug ibuprofen in surface waters and in wastewater." Environmental Science & Technology **33**(15): 2529-2535.

Cai, Y. Q., G. B. Jiang, et al. (2004). "Solid-phase microextraction coupled with high performance liquid chromatography-fluorimetric detection for the determination of bisphenol A, 4-n-nonylphenol, and 4-tert-octylphenol in environmental water samples." Analytical Letters **37**(4): 739-753.

Carballa, M., F. Omil, et al. (2005). "Removal of cosmetic ingredients and pharmaceuticals in sewage primary treatment." Water Research **39**: 4790-4796.

Chang, S., T. D. Waite, et al. (2003). "Adsorption of the endocrine-active compound estrone on microfiltration hollow fiber membranes." Environmental Science & Technology **37**(14): 3158-3163.

D'Ascenzo, G., A. Di Corcia, et al. (2003). "Fate of natural estrogen conjugates in municipal sewage transport and treatment facilities." Science Of The Total Environment **302**(1-3): 199-209.

Daughton, C. (2001, June 2001). "PPCPs as Environmental Pollutants." Retrieved September 10, 2002, from http://www.epa.gov/nerlesd1/chemistry/pharma/faq.htm.

Daughton, C. G. (2001). Pharmaceuticals in the Environment: Overarching Issues and Overview. Pharmaceuticals and Personal Care Products in the Environment: Scientific and Regulatory Issues. C. G. Daughton and T. Jones-Lepp. Washington, D.C., American Chemical Society. **791:** 2-38.

Daughton, C. G. 2003. Cradle-to-Cradle Stewardship of Drugs for Minimizing Their Environmental Disposition While Promoting Human Health. I. Rationale for and Avenues toward a Green Pharmacy. *Environmental Health Perspectives*, 111(5):757-774.

Daughton, C. G. (2001, June 2001). "PPCPs as Environmental Pollutants." from http://www.epa.gov/esd/chemistry/pharma/faq.htm.

Daughton, C. G. and T. A. Ternes (1999). "Pharmaceuticals and personal care products in the environment: agents of subtle change?" Environmental Health Perspectives **107**(6): 907-938.

de Alda, M. J. L. and D. Barcelo (2001). "Determination of steroid sex hormones and related synthetic compounds considered as endocrine disrupters in water by fully automated on-line solid-phase extraction-liquid chromatography-diode array detection." Journal of Chromatography A **911**: 203-210.

Deborde, M., S. Rabouan, et al. (2005). "Kinetics of Aqueous Ozone-Induced Oxidation of Some Endocrine Disruptors." Environmental Science and Technology **39**(16): 6086-6092.

Desbrow, C., E. J. Routledge, et al. (1998). "Identification of Estrogenic Chemicals in STW Eflleunt. 1. Chemical Fractionation and *in vitro* Biological Screening." Environmental Science & Technology **32**(11): 1549-1558.

Diaz-Cruz, M. S. and D. Barcelo (2005). "LC-MS2 trace analysis of antimicrobials in water, sediment and soil." Trac-Trends in Analytical Chemistry **24**(7): 645-657.

Dodd, M., M. Buffle, et al. (2006). "Oxidation of antibacterial molecules by aqueous ozone: moiety-specific reaction kinetics and application to ozone-based wastewater treatment." Environmental Science & Technology **40**(6): 1969-77.

Dodd, M. C., A. D. Shah, et al. (2005). "Interactions of fluoroquinolone antibacterial agents with aqueous chlorine: reaction kinetics, mechanisms, and transformation pathways." Environmental Science & Technology **39**(18): 7065-7076.

Drewes, J. E., C. Bellona, et al. (2005). "Rejection of wastewater-derived micropollutants in high-pressure membrane applications leading to indirect potable reuse." Environmental Progress **24**(4): 400-409.

Drewes, J. E., P. Fox, et al. (2001). "Occurrence of iodinated X-ray contrast media in domestic effluents and their fate during indirect potable reuse." Journal of Environmental Science and Health Part a- Toxic/Hazardous Substances & Environmental Engineering **36**(9): 1633-1645.

Drewes, J. E., T. Heberer, et al. (2003). "Fate of pharmaceuticals during groundwater recharge." Ground Water Monitoring & Remediation **23**(3): 64-72.

Drewes, J. E. and M. Jekel (1996). "Simulation of groundwater recharge with advanced treated wastewater." Water Science and Technology **33**(10-11): 409-418.

Drewes, J. E. and L. S. Shore (2001). "Concerns about pharmaceuticals in water reuse, groundwater recharge, and animal waste." ACS Symposium Series **791**: 206-228..

EMEA, 2006. European Medicine Agency Guideline on the Environmental Risk Assessment of Medicinal Products for Human Use. EMEA/CHMP/SWP/4447/00.

Ericson, J. F., R. Laenge, et al. (2002). "Comment on "Pharmaceuticals, hormones, and other organic wastewater contaminants in U.S. streams, 1999-2000. A national reconnaissance"." Environmental Science & Technology **38**(18): 4005-4006.

Fatoki, O. S. and R. O. Awofolu (2003). "Methods for selective determination of persistent organochlorine pesticide residues in water and sediments by capillary gas chromatography and electron-capture detection." Journal of Chromatography A **983**(1-2): 225-236.

Ferguson, P. L., C. R. Iden, et al. (2000). "Analysis of alkylphenol ethoxylate metabolites in the aquatic environment using liquid chromatography-electrospray mass spectrometry." Analytical Chemistry **72**(18): 4322-4330.

Ferguson, P. L., C. R. Iden, et al. (2001). "Determineation of steroid estrogens in wastewater by immunoaffinity extraction coupled with HPLC-electrospray-MS." Analytical Chemistry **73**: 3890-3895.

Folmar, L. C., N. D. Denslow, et al. (1996). "Vitellogenin induction and reduced serun testosterone concentrations in feral male carp (*Cyprinus carpio*) captured near a major metropolitan sewage treatment plant." Environmental Health Perspectives **104**(10): 1096-1101.

Folmar, L. C., M. J. Hemmer, et al. (2002). "A comparison of the estrogenic potencies of estradiol, ethynylestradiol, diethylstilbestrol, nonylphenol and methoxychlor *in vivo* and *in vitro*." Aquatic Toxicology **60**(1-2): 101-110.

Fono, L. J., E. P. Kolodziej, et al. (2006). "Attenuation of wastewater-derived contaminants in an effluent-dominated river." Environmental Science & Technology **40**: 7257-7262.

Fowler, B. R., L. Haviland, et al. (1998). "Analysis of alkylphenols in Fraser river water and suspended sediments." poster

Fries, E. and W. Puettmann (2001). "Occurrence of organophosphate esters in surface water and ground water in Germany." Journal of Environmental Monitoring **3**(6): 621-626.

Fries, E. and W. Puettmann (2003). "Monitoring of the three organophosphate esters TBP, TCEP and TBEP in river water and ground water (Oder, Germany)." Journal of Environmental Monitoring **5**(2): 346-352.

Fuerhacker, M., A. Dürauer, et al. (2001). "Adsorption isotherms on 17ß-estradiol on granular activated carbon (GAC)." Chemosphere **44**: 1573-1579.

Gallard, H., A. Leclercq, et al. (2004). "Chlorination of bisphenol A: kinetics and by-products formation." Chemosphere. **56**(5): 465-73.

Gao, H. and R. G. Zepp (1998). "Factors influencing photoreactions of dissolved organic matter in a coastal river of the Southeastern United States." Environmental Science & Technology **32**(19): 3632-3638.

Garrison, A. W., J. D. Pope, et al. (1975). GC/MS analysis of organic compounds in domestic wastewaters. Chem. Congr. North Am. Cont., Ann Arbor Science.

Garrison, A. W., J. D. Pope, et al. (1976). GC/MS Analysis of Organic Compounds in Domestic Wastewaters. Identification and Analyses of Organic Pollutants in Water. L. H. Keith. Ann Arbor MI, Ann Arbor Publishers Inc: 517-556.

Gentili, A., D. Perret, et al. (2002). "Analysis of free estrogens and their conjugates in sewage and river waters by solid-phase extraction then liquid chromatography-electrospray-tandem mass spectrometry." Chromatographia **56**(1/2): 25-32.

Giesy, J. P., S. L. Pierens, et al. (2000). "Effects of 4-nonylphenol on fecundity and biomarkers of estrogenicity in fathead minnows (*Pimephales promelas*)." Environmental Toxicology and Chemistry **19**(5): 1368-1377.

Giesy, J. P., S. L. Pierens, et al. (2000). "Effects of 4-nonylphenol on fecundity and biomarkers of estrogenicity in fathead minnows (*Pimephales promelas*)." Environ Toxicol Chem **19**(5): 1368-1377.

Glassmeyer, S. T., E. T. Furlong, et al. (2005). "Transport of chemical and microbial compounds from known wastewater discharges: potential for use as indicators of human fecal contamination." Environmental Science & Technology **39**: 5157-5169.

Gold, L.S., Slone, T.H., Ames, B.N., and Manley, N.B. Pesticide Residues in Food and Cancer Risk: A Critical Analysis. In: Handbook of Pesticide Toxicology, Second Edition (R. Krieger, ed.), San Diego, CA: Academic Press, pp. 799-843 (2001).

Gray, G. M. and D. P. Ropeik. 2002. Dealing with the dangers of fear: the role of risk communication. *Health Aff.*, 6:106-116.

Grischek, T., K. M. Hiscock, et al. (1998). "Factors affecting denitrification during infiltration of river water into a sand and gravel aquifer in Saxony, Germany." Water Research **32**(2): 450-460.

Gros, M., M. Petrovic, et al. (2006). "Development of a multi-residue analytical methodology based on liquid chromatography-tandem mass spectrometry (LC-MS/MS) for screening and trace level determination of pharmaceuticals in surface and wastewaters." Talanta **70**(4): 678-690.

Gros, M., M. Petrovic, et al. (2006). "Multi-residue analytical methods using LC-tandem MS for the determination of pharmaceuticals in environmental and wastewater samples: a review." Analytical and Bioanalytical Chemistry **386**(4): 941-952.

Halden, R. U. and D. H. Paull (2005). "Co-Occurrence of Triclocarban and Triclosan in U.S. Water Resources." Environmental Science and Technology **39**(6): 1420-1426.

Harries, J. E., D. A. Sheahan, et al. (1997). "Estrogenic activity in five United Kingdom rivers detected by measurement of vitellogenesis in caged male trout." Environ Toxicol Chem **16**(3): 534-542.

Heberer, T. (2002). "Occurrence, fate, and removal of pharmaceutical residues in the aquatic environment: a review of recent research data." Toxicology Letters **131**(1-2): 5-17.

Heberer, T., S. Gramer, et al. (1999). "Occurrence and distribution of organic contaminants in the aquatic system in Berlin, Germany. Part 3. Determination of synthetic musks in Berlin surface water applying solid-phase microextraction (SPME) and gas chromatography-mass spectrometry (GC-MS)." Acta Hydrochimica et Hydrobiologica **27**(3): 150-156.

Heberer, T., K. Schmidt-Baeumler, et al. (1998). "Occurrence and distribution of organic contaminants in the aquatic system in Berlin, Germany. Part 1. Drug residues and other polar contaminants in Berlin surface and groundwater." Acta Hydrochimica et Hydrobiologica **26**(5): 272-278.

Hemming, J. M., W. T. Waller, et al. (2001). "Assessment of the estrogenicity and toxicity of a domestic wastewater effluent flowing through a constructed wetland system using biomarkers in male fathead minnows (*Pimephales promelas* Rafinesque, 1820)." Environ Toxicol Chem **20**(10): 2268-2275.

Hernando, M. D., E. Heath, et al. (2006). "Trace-level determination of pharmaceutical residues by LC-MS/MS in natural and treated waters. A pilot-survey study." Analytical and Bioanalytical Chemistry **385**: 985-991.

Hignite, C. and D. L. Azarnoff (1977). "Drugs and drug metabolites as environmental contaminants: chlorophenoxyisobutyrate and salicylic acid in sewage water effluent." Life Sciences **20**(2): 337-341.

Ho, L., G. Onstad, et al. (2006). "Differences in the chlorine reactivity of four microcystin analogues." Water Research **40**: 1200-1209.

Holm, J. V., K. Rügge, et al. (1995). "Occurence and distribution of pharmaceutical organic compounds in the groundwater downgradient of a landfill (Grindsted, Denmark)." Environmental Science & Technology **29**(5): 1415-1420.

Huber, M. M., S. Canonica, et al. (2003). "Oxidation of pharmaceuticals during ozonation and advanced oxidation processes." Environmental Science & Technology **37**(5): 1016-1024.

Huber, M. M., A. Göbel, et al. (2005). "Oxidation of pharmaceuticals during ozonation of municipal wastewater effluents: a pilot study." Environmental Science & Technology **39**(11): 4290-4299.

Huber, M. M., S. Korhonen, et al. (2005). "Oxidation of pharmaceuticals during water treatment with chlorine dioxide." Water Research Water Research **39**: 3607-3617.

Ikeda, M. 1988. Multiple exposure to chemicals. *Regul Toxicol Pharmacol.*, 8(4):414-421.

Isobe, T., H. Shiraishi, et al. (2003). "Determination of estrogens and their conjugates in water using solid-phase extraction followed by liquid chromatography-tandem mass spectrometry." Journal of Chromatography A **984**(2): 195-202.

Jobling, S., M. Nolan, et al. (1998). "Widespread sexual disruption in wild fish." Environ Sci Technol **32**(17): 2498-2506.

Jobling, S., M. Noylan, et al. (1998). "Widespread sexual disruption in wild fish." Environmental Science & Technology **32**: 2498-2506.

Jobling, S. and C. R. Tyler (2003). "Endocrine disruption in wild freshwater fish." Pure and Applied Chemistry **75**(11-12): 2219-2234.

Johnson, A. C., A. C. Belfroid, et al. (2000). "Estimating steroid oestrogen inputs into activated sludge treatment works and observations on their removal from the effluent." The Science of the Total Environment **256**(2-3): 163-173.

Jones, O. A. H., N. Voulvoulis, et al. (2002). "Aquatic environmental assessment of the top 25 English prescription pharmaceuticals." Water Research **36**: 5013-5022.

Khim, J. S., K. Kannan, et al. (1999). "Characterization and distribution of trace organic contaminants in sediment from Masan Bay, Korea. 1. Instrumental analyses." Environmental Science & Technology **33**(23): 4199-4205.

Khim, J. S., D. L. Villeneuve, et al. (1999). "Characterization and distribution of trace organic contaminants in sediment from Masan Bay, Korea. 2. *In vitro* gene expression assays." Environmental Science & Technology **33**(23): 4206-4211.

Khim, J. S., D. L. Villeneuve, et al. (1999). "Alkylphenols, polycyclic aromatic hydrocarbons, and organochlorines in sediment from Lake Shihwa, Korea: Instrumental and bioanalytical characterization." Environmental Toxicology and Chemistry **18**(11): 2424-2432.

Kidd, K., P. J. Blanchfield, et al. (2007). Collapse of a fish population after exposure to a synthetic estrogen. National Academy of Sciences of the United States of America, National Academy of Sciences of the United States of America.

Kim, S.-C. and K. Carlson (2006). "Occurrence of ionophore antibiotics in water and sediments of a mixed-landscape watershed." Water Research **40**: 2549-2560.

Kim, S. D., J. Cho, et al. (2007). "Occurrence and removal of pharmaceuticals and endocrine disruptors in South Korean surface, drinking, and waste waters." Water Research **41**(5): 1013-1021.

Kiso, Y., T. Kon, et al. (2001). "Rejection properties of alkyl phthalates with nanofiltration membranes." Journal of Membrane Science **182**(1-2): 205-214.

Kiso, Y., Y. Nishimura, et al. (2000). "Rejection properties of non-phenylic pesticides with nanofiltration membranes." Journal of Membrane Science **171**(2): 229-237.

Kiso, Y., Y. Sugiura, et al. (2001). "Effects of hydrophobicity and molecular size on rejection of aromatic pesticides with nanofiltration membranes." Journal of Membrane Science **192**(1-2): 1-10.

Kloepfer, A., J. Quintana, et al. (2005). "Operational options to reduce matrix effects in liquid chromatography-electrospray ionization-mass spectrometry analysis of aqueous environmental samples." Journal of Chromatography A **1067**(1-2): 153-60.

Knappe, D. R. U., Y. Matsui, et al. (1998). "Predicting the capacity of powder activated carbon for trace organic compounds in natural waters." Environmental Science & Technology **32**(11): 1694-1698.

Koester, C. J. and A. Moulik (2005). "Trends in environmental analysis." Analytical Chemistry **77**: 3737-3754.

Kolpin, D. W., E. T. Furlong, et al. (2002). "Pharmaceuticals, hormones, and other organic waste contaminants in U.S. streams, 1999-2000: a national reconnaissance." Environmental Science & Technology **36**(6): 1202-1211.

Kolpin, D. W., E. T. Furlong, et al. (2002). "Response to comment on "Pharmaceuticals, hormones, and other organic wastewater contaminants in U.S. streams, 1999-2000: A national reconnaissance"." Environmental Science & Technology **36**(18): 4007-4008.

Kolpin, D. W., M. Skopec, et al. (2004). "Urban contribution of pharmaceuticals and other organic wastewater contaminants to streams during differing flow conditions." Science Of The Total Environment **328**(1-3): 119-130.

Kosky, P. G., J. M. Silva, et al. (1991). "The aqueous phase in the interfacial synthesis of polycarbonates. Part 1. Ionic equilibria and experimental solubilities in the BPA-sodium hydroxide-water system." Industrial & Engineering Chemistry Research **30**(3): 462-7.

Krause, V. 2006. Risk communication for salmon aquaculture: A submission to the special committee on sustainable aquaculture of the legislative assembly of British Columbia.

Kreuzinger, N., M. Clara, et al. (2004). "Investigation on the behaviour of selected pharmaceuticals in the groundwater after infiltration of treated wastewater." Water Science and Technology **50**(2): 221-228.

Kubeck, E. and C. G. Naylor (1990). "Trace analysis of alkylphenol ethoxylates." JAOCS **67**(6): 400-405.

Kuehn, W. and U. Mueller (2000). "Riverbank filtration. An overview." Journal - American Water Works Association **92**(12): 60-69.

Laganá, A., A. Bacaloni, et al. (2000). "Trace analysis of estrogenic chemicals in sewage effluent using liquid chromatography combined with tandem mass spectrometry." Rapid Commun. Mass Spectrom. **14**: 401-407.

Lamas, J. P., C. Salgado-Petinal, et al. (2004). "Solid-phase microexraction-gas chromatography-mass spectrometry for the analysis of selective serotonin reuptake inhibitors in environmental water." Journal of Chromatography A **1046**: 241-247.

Lazorchak, J. M., J. M. Hemming, et al. (2001). Evaluation of a wastewater discharge using vitellogenin gene expression and plasma protein levels in fathead minnows. SETAC, Baltimore, MD.

Lee, B.-C., M. Kamata, et al. (2004). "Effects of chlorine on the decrease of estrogenic chemicals." Water Research **38**(3): 733-739.

Lee, K. E., L. B. Barber, et al. (2004). Presence and Distribution of Organic Wastewater Compounds in Wastewater, Surface, Ground, and Drinking Waters, Minnesota, 2000-02. Reston, Virginia, United States Geological Survey: 47.

Liebig, M., J. F. Moltmann, et al. (2006). "Evaluation of measured and predicted environmental concentrations of selected human pharmaceuticals and personal care products." Environmental Science and Pollution Research **13**: 110-119.

Lindsey, M. E., M. Meyer, et al. (2001). "Analysis of trace levels of sulfonamide and tetracycline antimicrobials in groundwater and surface water using solid-phase extraction and liquid chromatography/mass spectrometry." Analytical Chemistry **73**(19): 4640-4646.

Lindstrom, A., I. J. Buerge, et al. (2002). "Occurrence and environmental behavior of the bactericide triclosan and its methyl derivative in surface waters and in wastewater." Environmental Science & Technology **36**(11): 2322-2329.

Loffler, D., J. Rombke, et al. (2005). "Environmental fate of pharmaceuticals in water/sediment systems." Environmental Science & Technology **39**(14): 5209-5218.

Lopez de Alda, M. J. and D. Barcelo (2001). "Review of analytical methods for the determination of estrogens and progestogens in waste waters." Fresenius Journal of Analytical Chemistry **371**: 437-447.

Mansell, J. and J. Drewes (2004). "Fate of steroidal hormones during soil-aquifer treatment." Ground Water Monitoring & Remediation **24**(2): 94-101.

Mansell, J., J. E. Drewes, et al. (2004). "Removal mechanisms of endocrine disrupting compounds (steroids) during soil aquifer treatment." Water Science and Technology **50**(2): 229-237.

McDowell, D. C., M. M. Huber, et al. (2005). "Ozonation of carbamazepine in drinking water: identification and kinetic study of major oxidation products." Environmental Science & Technology **39**(20): 8014-8022.

Miao, X.-S. and C. D. Metcalfe (2003). "Determination of carbamazepine and its metabolites in aqueous samples using liquid chromatography-electrospray tandem mass spectrometry." Analytical Chemistry **75**: 3731-3738.

Moldovan, Z. (2006). "Occurrences of pharmaceutical and personal care products as micropollutants in rivers from Romania." Chemosphere **2006**: 1808-1817.

Montgomery-Brown, J., J. Drewes, et al. (2003). "Behavior of alkylphenol polyethoxylate metabolites during soil aquifer treatment." Water Research **37**: 3672-3681.

Nevskaia, D. M., A. Sepulveda-Escribano, et al. (2001). "Surface properties of activated carbons in relation to their ability to adsorb nonylphenol aqueous contaminant." Physical Chemistry Chemical Physics **3**: 463-468.

Ng, H. Y. and M. Elimelech (2004). "Influence of colloidal fouling on rejection of trace organic contaminants by reverse osmosis." Journal of Membrane Science **244**: 215-226.

Ngheim, L. D., A. I. Schaefer, et al. (2004). "Removal of natural hormones by nanofiltration membranes: measurement, modeling, and mechanisms " Environmental Science & Technology **38**: 1888-1896.

Nghiem, L., A. Schaefer, et al. (2005). "Nanofiltration of hormone mimicking trace organic contaminants." Separation Science and Technology **40**(13): 2633-2649.

Nghiem, L. D., A. I. Schaefer, et al. (2005). "Pharmaceutical retention mechanisms by nanofiltration membranes." Environmental Science & Technology **39**: 7698-7705.

Nikolai, L. N., E. L. McClure, et al. (2006). "Stereoisomer quantification of the .beta.-blocker drugs atenolol, metoprolol, and propranolol in wastewaters by chiral high-performance liquid chromatography-tandem mass spectrometry." Journal of Chromatography A **1131**(1-2): 103-109.

*Oxford English Dictionary*. 1989. 2nd ed. New York City, New York: Oxford Unveristy Press.

Ozaki, H. and H. Li (2002). "Rejection of organic compounds by ultra-low pressure reverse osmosis membrane." Water Research **36**: 123-130.

Paul, S., H. Boernick, et al. (2001). "The relevance of aliphatic amines for drinking water production from river Elbe bank filtrate." Vom Wasser **97**: 125-133.

Penalver, A., E. Pocurull, et al. (2002). "Method based on solid-phase microextraction-high-performance liquid chromatography with UV and electrochemical detection to determine estrogenic compounds in water samples." Journal of Chromatography A **964**(1-2): 153-160.

Pepich, B. V., M. M. Domino, et al. (2004). "Validating sample preservation techniques and holding times for the approved compliance monitoring methods for haloacetic acids under the US EPA's stage 1 D/DBP rule." Water Research **38**(4): 895-902.

Petrovic, M. and D. Barcelo (2007). Analysis, fate and removal of pharmaceuticals in the water cycle. Amsterdam, The Netherlands, Elsevier Science.

Petrovic, M., M. D. Hernando, et al. (2005). "Liquid chromatography-tandem mass spectrometry for the analysis of pharmaceutical residues in environmental samples: a review." Journal of Chromatography A **1067**(1-2): 1-14.

Pinkston, K. E. and D. L. Sedlak (2004). "Transformation of aromatic ether- and amine-containing pharmaceuticals during chlorine disinfection." Environmental Science & Technology **38**(4019-4025).

Pryor, S. W., A. G. Hay, et al. (2002). "Nonylphenol in anaerobically digested sewage sludge from New York state." Environmental Science & Technology **36**(17): 3678-3682.

Purdom, C. E., P. A. Hardiman, et al. (1994). "Estrogenic effects of effluents from sewage treament works." Chemistry and Ecology **8**: 275-285.

Quintana, J. B., R. Rodil, et al. (2004). "Suitability of hollow fibre liquid-phase microextraction for the determination of acidic pharmaceuticals in wastewater by liquid chromatography-electrospray tandem mass spectrometry without matrix effects." Journal of Chromatography A **1061**: 19-26.

Rauch, T. and J. E. Drewes (2004). "Assessing the removal potential of soil-aquifer treatment systems for bulk organic matter." Water Science and Technology **50**(2): 245-253.

Rauch, T., J. Munoz, et al. (2004). Metabolic removal of organic micropollutants in artificial groundwater recharge systems using water of impaired quality. WEFTEC. New Orleans, Louisiana.

Ray, C., D. T. W. Soong, et al. (1998). "Agricultural chemical: Impacts on riparian wells during floods." Journal of the American Water Works Association **90**(7): 90-100.

Reemtsma, T. (2003). "Liquid-chromatography-mass spectrometry and strategies for trace-level analysis of polar organic pollutants." Journal of Chromatography A **1000**: 477-501.

Reemtsma, T. and M. Jekel (2006). Organic Pollutants in the Water Cycle: Properties, Occurrence, Analysis and Environmental Relevance of Polar Compounds. Weinheim, Wiley-VCH.

Richardson, S. D. (2000). "Environmental Mass Spectrometry." Analytical Chemistry **72**: 4477-4496.

Richardson, S. D. (2002). "Environmental Mass Spectrometry: Emerging Contaminants and Current Issues." Analytical Chemistry **74**(12): 2719-2742.

Richardson, S. D. (2004). "Environmental mass spectrometry: Emerging contaminants and current issues." Analytical Chemistry **76**(12): 3337-3364.

Richardson, S. D. (2006). "Environmental mass spectrometry: Emerging contaminants and current issues." Analytical Chemistry **78**: 4021-4045.

Richardson, S. D. and T. A. Ternes (2005). "Water analysis: emerging contaminants and current issues." Analytical Chemistry **77**: 3807-3838.

Rimkus, G. G. (1999). "Polycyclic musk fragrances in the aquatic environment." Toxicology Letters **111**(1-2): 37-56.

Rodgers-Gray, T. P., S. Jobling, et al. (2001). "Exposure of juvenile roach (*Rutilus rutilus*) to treated sewage effluent induces dose-dependent and persistent disruption in gonadal duct development." Environ Sci Technol **35**(3): 462-470.

Rodríguez, I., J. Carpinteiro, et al. (2004). "Solid-phase microextraction with on-fiber derivatization for the analysis of anti-inflammatory drugs in water samples." Journal of Chromatography A **1024**(1-2): 1-8.

Romero, J., F. Ventura, et al. (2002). "Characterization of paint samples used in drinking water reservoirs: Identification of endocrine disruptor compounds." Journal of Chromatographic Science **40**(4): 191-197.

Rosenfeldt, E. and K. G. Linden (2004). "Degradation of endocrine disrupting chemicals bisphenol A, ethinyl estradiol, and estradiol during UV photolysis and advanced oxidation processes." Environmental Science & Technology **38**(20): 5476-5483.

Routledge, E. J., D. Sheahan, et al. (1998). "Identification of estrogenic chemicals in STW effluent. 2. *In vivo* responses in trout and roach." Environ Sci Technol **32**(11): 1559-1565.

Routledge, E. J., D. Sheahan, et al. (1998). "Identification of estrogenic chemicals in STW effluent. 2. *In vivo* responses in trout and roach." Environmental Science & Technology **32**(11): 1559-1565.

Santos, J. L., I. Aparicio, et al. (2005). "Simultaneous determination of pharmaceutically active compounds in wastewater samples by solid phase extraction and high-performance liquid chromatography with diode array and fluorescence detectors." Analytica Chimica Acta **550**: 116-122.

Sandman, Peter M. 1993. Responding to Community Outrage: Strategies for effective risk communication. Fairfax, VA: American Industrial Hygiene Association.

Schmidt, C., F. Lange, et al. (2004). Assessing the impact of different redox conditions and residence times on the fate of organic micropollutants during riverbank filtration. 4th International Symposium on Pharmaceutically Active Compounds and Endocrine Disrupting Chemicals, Minneapolis, Minnesota.

Schoenfuss, H. L., J. T. Levitt, et al. (2002). "Ten-week exposure to treated sewage discharge has relatively minor, variable effects on reproductive behavior and sperm production in goldfish." Environ Toxicol Chem **21**(10): 2185-2190.

Segner, H., J. M. Navas, et al. (2003). "Potencies of estrogenic compounds in *in vitro* screening assays and in life cycle tests with zebrafish *in vivo*." Ecotoxicology and Environmental Safety **54**(3): 315-322.

Singer, H., S. Mueller, et al. (2002). "Triclosan: occurrence and fate of a widely used biocide in the aquatic environment: field measurements in wastewater treatment plants, surface waters, and lake sediments." Environmental Science & Technology **36**(23): 4998-5004.

Smyth, H. F., C. S. Weil, J. S. West and C. P. Carpenter. An exploration of joint toxic action II. Equitoxic versus equivolume mixtures. *Toxicology and Applied Pharmacology*, 17(2):498-503.

Snyder, E. M., S. A. Snyder, et al. (2004). "Reproductive responses of common carp (*Cyprinus carpio*) exposed in cages to influent of the Las Vegas Wash in Lake Mead, Nevada from late winter to early spring." Environmental Science & Technology **38**(23): 6385-6395.

Snyder, S. A., T. L. Keith, et al. (2001). "Identification and quantification method for nonylphenol and lower oligomer nonylphenol ethoxylates in fish tissues." Environmental Toxicology and Chemistry **20**(9): 1870-1873.

Snyder, S. A., T. L. Keith, et al. (2001). "Bioconcentration of nonylphenol in fathead minnows (*Pimephalas promelas*)." Chemosphere **44**(8): 1697-1702.

Snyder, S. A., T. L. Keith, et al. (1999). "Analytical methods for detection of selected estrogenic compounds in aqueous mixtures." Environmental Science & Technology **33**(16): 2814-2820.

Snyder, S. A., K. L. Kelly, et al. (2001). Pharmaceuticals and personal care products in the waters of Lake Mead, Nevada. Pharmaceuticals and Personal Care Products in the Environment: Scientific and Regulatory Issues. C. G. Daughton and T. L. Jones-Lepp. Washington,D.C., American Chemical Society. **Symposium Series 791:** 116-140.

Snyder, S. A., B. J. Vanderford, et al. (2003). "Analytical methods used to measure endocrine disrupting compounds in water." Practice Periodical of Hazardous,Toxic, and Radioactive Waste Management **7**(4): 224-234.

Snyder, S. A., D. L. Villeneuve, et al. (2001). "Identification and quantification of estrogen receptor agonists in wastewater effluents." Environmental Science & Technology **35**(18): 3620-3625.

Snyder, S. A., P. Westerhoff, et al. (2003). "Pharmaceuticals, personal care products, and endocrine disruptors in water: implications for the water industry." Environmental Engineering Science **20**(5): 449-469.

Sontheimer, H. (1991). Trinkwasser aus dem Rhein? Sankt Augustin, Germany, Academia Verlag.

Stackelberg, P. E., E. T. Furlong, et al. (2004). "Persistence of pharmaceutical compounds and other organic wastewater contaminants in a conventional drinking-water-treatment plant." Science of the Total Environment **329**(1-3): 99-113.

Stumm-Zollinger, E. and G. M. Fair (1965). "Biodegradation of steroid hormones." Journal of the Water Pollution Control Federation **37**: 1506-1510.

Swayne, M. D., G. H. Boone, et al. (1980). Wastewater in Receiving Waters at Water Supply Abstraction Points. Final rept. Aug 77-Jan 79. Cincinnati, Sponsor: Municipal Environmental Research Lab**:** 200.

Tabak, H. H. and R. L. Bunch (1970). "Steroid hormones as water pollutants. I. Metabolism of natural and synthetic ovulation-inhibiting hormones by microorganisms of activated sludge and primary settled sewage." Dev. Ind. Microbiol. **11**: 367-376.

Tanghe, T. and W. Verstraete (2001). "Adsorption of nonylphenol onto granular activated carbon." Water Air and Soil Pollution **131**: 61-72.

Ternes, T., M. Bonerz, et al. (2001). "Determination of neutral pharmaceuticals in wastewater and rivers by liquid chromatography-electrospray tandem mass spectrometry." Journal of Chromatography a **938**(1-2): 175-185.

Ternes, T. and A. Joss (2006). Human pharmaceuticals, hormones and fragrances - the challenge of micropollutants in urban water management. London, IWA Publishing.
Ternes, T. A. (1998). "Occurrence of drugs in German sewage treatment plants and rivers." Water Research **32**(11): 3245-3260.
Ternes, T. A. (2001). "Analytical methods for the determination of pharmaceuticals in aqueous environmental samples." Trends in Analytical Chemistry **20**(8): 419-434.
Ternes, T. A., R. Hirsch, et al. (1998). "Methods for the determination of neutral drugs as well as betablockers and $\beta_2$-sympathomimetics in aqueous matrices using GC/MS and LC/MS/MS." Fresenius' Journal of Analytical Chemistry **362**: 329-340.
Ternes, T. A., M. Meisenheimer, et al. (2002). "Removal of pharmaceuticals during drinking water treatment." Environmental Science & Technology **36**(17): 3855-3863.
Tixier, C., H. P. Singer, et al. (2002). "Phototransformation of triclosan in surface waters: a relevant elimination process for this widely used biocide-laboratory studies, field measurements, and modeling." Environmental Science & Technology **36**(16): 3482-3489.
Tixier, C., H. P. Singer, et al. (2003). "Occurrence and fate of carbamazepine, clofibric acid, diclofenac, ibuprofen, ketoprofen, and naproxen in surface waters." Environmental Science & Technology **37**(6): 1061-1068.
Trenholm, R. A., B. J. Vanderford, et al. (2006). "Broad range analysis of endocrine disruptors and pharmaceuticals using gas chromatography and liquid chromatography tandem mass spectroscopy." Chemosphere **65**: 1990-1998.
Tsuda, T., A. Takino, et al. (2001). "Evaluation of 4-nonylphenols and 4-tert-octylphenol contamination of fish in rivers by laboratory accumulation and excretion experiments." Water Research **35**(7): 1786-1792.
Tswett, M. (1906). "Physical chemical studies on chlorophyll adsoprtions." Berichte der Deutschen Botanischen Gesellschaft **24**: 316-323.
U.S. EPA (2004). Guidelines for Water Reuse. Washington, DC, USA, U.S. Environmental Protection Agency.
U.S. EPA (2005). Ambient Water Quality Criteria - Nonylphenol. Final. Washington, DC, U.S. Environmental Protection Agency (U.S. EPA), Office of Water: 88.
U.S. EPA. (2005). "Glossary of IRIS terms. Revised December 2005." Retrieved April 2, 2006, from http://www.epa.gov/iris/gloss8.htm.
van Aerle, R., N. Pounds, et al. (2002). "Window of sensitivity for the estrogenic effects of ethinylestradiol in early life-stages of fathead minnow, *Pimephales promelas*." Ecotoxicology **11**(6): 423-434.
Vanderford, B. J., R. A. Pearson, et al. (2003). "Analysis of endocrine disruptors, pharmaceuticals, and personal care products in water using liquid chromatography/tandem mass spectrometry." Analytical Chemistry **75**(22): 6265-6274.
Vanderford, B. J. and S. A. Snyder (2006). "Analysis of pharmaceuticals in water by isotope dilution liquid chromatography/tandem mass spectrometry." Environmental Science & Technology **40**: 7312-7320.
Velagaleti, R., P. K. Burns, et al. (2002). "Impact of current good manufacturing practices and emission regulations and guidances on the discharge of pharmaceutical chemicals into the environment from manufacturing, use, and disposal." Environmental Health Perspectives **110**(3): 213-220.

Vermeirssen, E. L. M., O. Koerner, et al. (2005). "Characterization of environmental estrogens in river water using a three pronged approach: active and passive water sampling and the analysis of accumulated estrogens in the bile of caged fish." Environmental Science and Technology **39**: 8191-8198.

Verstraeten, I. I., J. D. Carr, et al. (1999). "Surface-water/groundwater interaction: Herbicide transport into municipal collector wells." Journal of Environmental Quality **28**(5): 1396-1405.

Verstraeten, I. I., T. Heberer, et al. (2003). "Occurrence of endocrine-disrupting and other wastewater compounds during water treatment with case studies from Lincoln, Nebraska and Berlin, Germany." Practice Periodical of Hazardous,Toxic, and Radioactive Waste Management **7**(4): 253-263.

Vervoort, R. J. M., E. Ruyter, et al. (2001). "Characterisation of reversed-phase liquid chromatography stationary phases for the analysis of basic pharmaceuticals: eluent properties and comparison of empirical test methods." Journal of Chromatography A **931**: 67-79.

Vieno, N., T. Tuhkanen, et al. (2006). "Analysis of neutral and basic pharmaceuticals in sewage treatment plants and in recipient waters using solid phase extraction and liquid chromatography-tandem mass spectrometry detection." Journal of Chromatography A **1134**: 101-111.

Vieno, N., T. Tuhkanen, et al. (2006). "Removal of pharmaceuticals in drinking water treatment: effect of chemical coagulation." Enivronmental Technology **27**: 183-192.

Vieno, N. M., T. Tuhkanen, et al. (2005). "Seasonal variation in the occurrence of pharmaceuticals in effluents from a sewage treatment plant and in the recipient water." Environmental Science & Technology **39**(21): 8220-8226.

Westerhoff, P., Y. Yoon, et al. (2005). "Fate of endocrine-disruptor, pharmaceutical, and personal care product chemicals during simulated drinking water treatment processes " Environmental Science & Technology **39**(17): 6649-6663.

WHO (2002). Global Assessment of the State-of-the-Science of Endocrine Disruptors. T. B. Damstra, S.; Bergman, A.; Kavlock, R.; Van Der Kraak, G., World Health Organization (WHO).

Wiegel, S., A. Aulinger, et al. (2004). "Pharmaceuticals in the river Elbe and its tributaries." Chemosphere **57**(2): 107-26.

Winkler, M., J. R. Lawrence, et al. (2001). "Selective degradation of ibuprofen and clofibric acid in two model river biofilm systems." Water Research **35**(13): 3197-3205.

Xu, P., J. E. Drewes, et al. (2005). "Rejection of emerging organic micropollutants in nanofiltration-reverse osmosis membrane applications." Water Environment Research **77**(1): 40-48.

Xu, P., J. E. Drewes, et al. (2006). "Effect of membrane fouling on transport of organic contaminants in NF/RO membrane applications." Journal of Membrane Science **279**: 165-175.

Ye, Z., H. S. Weinberg, et al. (2007). "Trace analysis of trimethoprim and sulfonamide, macrolide, quinolone, and tetracycline antibiotics in chlorinated drinking water using liquid chromatography electrospray tandem mass spectrometry." Analytical Chemistry **79**(3): 1135-1144.

Ying, G.-G., R. S. Kookana, et al. (2002). "On-line solid-phase extraction and fluorescence detection of selected endocrine disrupting chemicals in water by high-performance liquid chromatography." J. Environ. Sci. Heal. B. **37**(3): 225-234.

Yook, K.-S., S.-M. Hong, et al. (1994). "Comparison of liquid-liquid extraction and solid-phase extraction coupled with GC/MS for determination of priority pollutants in water." Analytical Science Technology **7**(4): 441-453.

Yoon, B. O., S. Koyanagi, et al. (2003). "Removal of endocrine disruptors by selective sorption method using polydimethylsiloxane membranes." Journal of Membrane Science **213**(1-2): 137-144.

Yoon, Y., P. Westerhoff, et al. (2003). "HPLC-fluorescence detection and adsorption of bisphenol A, 17β-estradiol, and 17α-ethynyl estradiol on powdered activated carbon." Water Research **37**(14): 3530-3537.

Yoon, Y., P. Westerhoff, et al. (2006). "Nanofiltration and ultrafiltration of endocrine disrupting compounds,pharmaceuticals and personal care products." Journal of Membrane Science **270**: 88-100.

Yoon, Y., P. Westerhoff, et al. (2007). "Removal of endocrine disrupting compounds and pharmaceuticals by nanofiltration and ultrafiltration membranes." Desalination **202**(1-3): 16-23.

Zhao, Y.-Y., J. Boyd, et al. (2006). "Characterization of new nitrosamines in drinking water using liquid chromatography tandem mass spectrometry." Environmental Science & Technology **40**: 7636-7641.

Zwiener, C. and F. H. Frimmel (2000). "Oxidative treatment of pharmaceuticals in water." Water Research **34**(6): 1881-1885.

Zwiener, C. and F. H. Frimmel (2004). "LC-MS analysis in the aquatic environment and in water treatment - a critical review. Part I: Instrumentation and general aspects of analysis and detection." Analytical and Bioanalytical Chemistry **378**(4): 851-861.

Zwiener, C. and F. H. Frimmel (2004). "LC-MS analysis in the aquatic environment and in water treatment technology - a critical review. Part II: Applications for emerging contaminants and related pollutants, microorganisms and humic acids." Analytical and Bioanalytical Chemistry **378**(4): 862-874.

# ABBREVIATIONS AND ACRONYMS

| | |
|---|---|
| ADI | Acceptable Daily Intake |
| Aeq | androgen equivalents |
| AERS | Adverse Event Reporting System |
| ALP | alkaline phosphatase |
| AR | androgen receptor |
| ATSDR | Agency for Toxic Substances and Disease Registry |
| AwwaRF | American Water Works Association Research Foundation |
| Ca DHS | California Department of Health Services |
| CALUX | chemically-activated luciferase gene expression |
| CAT | chloramphenicol acetyltransferase |
| cDNA | complementary deoxyribonucleic acid |
| CHO | Chinese hamster ovary |
| DDT | dichlorodiphenyl trichloroethane |
| DES | diethylstilbestrol |
| DNA | deoxyribonucleic acid |
| DWEL | Drinking water equivalent level |
| E1 | estrone |
| E2 | 17ß-estradiol |
| E3 | estriol |
| EAC | endocrine active chemical (contaminant) |
| EAT | estrogenic, androgenic, and thyroidal |
| EC50 | effective concentration 50% |
| EDC | endocrine disrupting chemical |
| EDMVAC | Endocrine Disruptor Methods Validation Advisory Committee |
| EDMVS | Endocrine Disruptor Methods Validation Subcommittee |
| EDSP | Endocrine Disruptor Screening Program |
| EDSTAC | Endocrine Disruptor Screening and Testing Advisory Committee |
| EE2 | 17α-ethynylestradiol |
| EEQ | estradiol equivalents |
| ELISA | enzyme-linked immunosorbent assay |
| EMEA | European Medicines Agency |
| ER | estrogen receptor |
| ER-CALUX | estrogen receptor-chemically activated luciferase gene expression |
| ERE | estrogen response element |
| ER-ß | estrogen receptor ß isoform |
| ER-α | estrogen receptor α isoform |
| EU | European Union |

| | |
|---|---|
| FAO | Food and Agriculture Organization of the United Nations |
| FDA | U.S. Food and Drug Administration |
| FFDCA | Federal Food, Drug, and Cosmetic Act |
| FSH | follicle-stimulating hormone |
| GFP | green fluorescent protein |
| GnRH | gonadotropin-releasing hormone |
| GWRC | Global Water Resource Coalition |
| HAA | hormonally active agent |
| HPG-axis | hypothalamic-pituitary-gonad axis |
| HPT-axis | hypothalamic-pituitary-thyroid axis |
| hsp90 | heat shock protein 90 |
| IEH | Institute for Environment and Health |
| IPCS | International Programme on Chemical Safety |
| JECFA | Joint FAO/WHO Expert Committee on Food Additives |
| LH | leutenizing hormone |
| LOAEL | lowest observed adverse effect level |
| LOEL | lowest observed effect level |
| MOA | mechanism (or mode) of action |
| mRNA | messenger ribonucleic acid |
| MTD | maximum tolerated dose |
| NOAEL | no observed adverse effect level |
| NOEL | no observed effect level |
| NRC | National Research Council |
| NTP | National Toxicology Program |
| OECD | European Organisation for Economic Co-operation and Development |

| | |
|---|---|
| PAH | polycyclic aromatic hydrocarbon |
| PCB | polychlorinated biphenyl |
| PCDD | polychlorinated dibenzodioxin |
| PCDF | polychlorinated dibenzofuran |
| PhAC | pharmaceutically-active chemical (contaminant) |
| POD | point of departure |
| PPCP | pharmaceuticals and personal care products |
| QSAR | quantitative structure-activity relationship |
| REACH | Registration, Evaluation & Authorization of Chemicals |
| RIA | radioimmunoassay |
| RT-PCR | reverse-transcription polymerase chain reaction |
| rTTR | recombinant transthyretin |
| SDWA | Safe Drinking Water Act |
| SHBG | sex hormone binding globulin |
| T3 | triiodothyrone |
| T4 | tetraiodothyronine, or thyroxine |
| TBG | thyroid binding globulin |
| TRH | thyrotropin-releasing hormone |
| TSCA | Toxic Substances Control Act |
| TSH | thyroid-stimulating hormone |
| TTC | threshold of toxicologic concern |
| TTR | transthyretin |
| U.S. EPA | United States Environmental Protection Agency |
| WHO | World Health Organization |
| WWTP | wastewater treatment plant |

IWA Publishing's authorised EU representative for General Product Safety Regulations is Diane D'Arras, 15 rue Duret, 75116 Paris, France, e-mail: safety@iwap.co.uk.

Printed and bound by CPI Group (UK) Ltd, Croydon, CR0 4YY
17/09/2026
02218802-0004